THE ROTATING
DISC ELECTRODE

STUDIES IN SOVIET SCIENCE

PHYSICAL SCIENCES

1973

DENSIFICATION OF METAL POWDERS DURING SINTERING
V. A. Ivensen
THE TRANSURANIUM ELEMENTS
V. I. Goldanskii and S. M. Polikanov
GAS-CHROMATOGRAPHIC ANALYSIS OF TRACE IMPURITIES
V. G. Berezkin and V. S. Tatarinskii
A CONFIGURATIONAL MODEL OF MATTER
G. V. Samsonov, I. F. Pryadko, and L. F. Pryadko
COMPLEX THERMODYNAMIC SYSTEMS
V. V. Sychev
CRYSTALLIZATION PROCESSES UNDER HYDROTHERMAL CONDITIONS
A. N. Lobachev
MIGRATION OF MACROSCOPIC INCLUSIONS IN SOLIDS
Ya. E. Geguzin and M. A. Krivoglaz

1974

THEORY OF PLASMA INSTABILITIES
Volume 1: Instabilities of a Homogenous Plasma
A. B. Mikhailovskii
THEORY OF PLASMA INSTABILITIES
Volume 2: Instabilities of an Inhomogenous Plasma
A. B. Mikhailovskii
NONEQUILIBRIUM STATISTICAL THERMODYNAMICS
D. N. Zubarev
REFRACTORY CARBIDES
G. V. Samsonov
WAVES AND SATELLITES IN THE NEAR-EARTH PLASMA
Ya. L. Al'pert

1975

ENVIRONMENTAL HAZARDS OF METALS:
Toxicity of Powdered Metals and Metal Compounds
I. T. Brakhnova
DOMAIN ELECTRICAL INSTABILITIES IN SEMICONDUCTORS
V. L. Bonch-Bruevich, I. P. Zvyagin, and A. G. Mironov

1976

THE ROTATING DISC ELECTRODE
Yu. V. Pleskov and V. Yu. Filinovskii

STUDIES IN SOVIET SCIENCE

THE ROTATING DISC ELECTRODE

Yu. V. Pleskov and V. Yu. Filinovskii

Academy of Sciences of the USSR
Moscow, USSR

Translated from Russian by
Halina S. Wroblowa
University of Pennsylvania

Translation edited by
Halina S. Wroblowa and
B. E. Conway
University of Ottawa

CONSULTANTS BUREAU • NEW YORK AND LONDON

Library of Congress Cataloging in Publication Data

Pleskov, I͡Urii Viktorovich.
  The rotating disc electrode.

  (Studies in Soviet science)
  Translation of Vrashchai͡ushchiĭsi͡a diskovyĭ elektrod.
  Includes bibliographical references and index.
    1. Electrodes. 2. Electrochemistry. I. Filinovskiĭ, Vladislav I͡Ul'evich, joint
author. II. Title. III. Series.
  QD571.P5413   541'.3724   76-22774
  ISBN 978-1-4615-8563-3   ISBN 978-1-4615-8561-9 (eBook)
  DOI 10.1007/978-1-4615-8561-9

Yurii Viktorovich Pleskov was born in 1933 and graduated from the Chemical
  Department of Moscow University. His entire scientific activities have been
  connected with the Institute of Electrochemistry of the Academy of Sciences
  of the USSR, where he is employed at present as a Senior Scientist.

Vladislav Yul'evich Filinovskii graduated from the Department of Physics of
  Moscow University. He has been working for more than 15 years in the
  Institute of Electrochemistry of the Academy of Sciences of the USSR. He
  recently became a Senior Scientist in the Institute.

The original Russian text, published by Nauka Press in Moscow in 1972, has been
corrected by the authors for the present edition. This translation is published
under an agreement with the Copyright Agency of the USSR (VAAP).

ВРАЩАЮЩИЙСЯ ДИСКОВЫЙ ЭЛЕКТРОД
Ю. В. ПЛЕСКОВ, В. Ю. ФИЛИНОВСКИЙ
VRASHCHAYUSHCHIISYA DISKOVYI ELECTROD
Yu. V. Pleskov, V. Yu. Filinovskii

© 1976 Consultants Bureau, New York
Softcover reprint of the hardcover 1st edition 1976

A Division of Plenum Publishing Corporation
227 West 17th Street, New York, N.Y. 10011

Foreword

Important advances in a subject are as often promoted by a new technique as by new concepts and theories. In the study of electrode reactions which involve diffusion in a primary or a secondary step, the development and use of techniques involving rotating disc electrodes and derived instrumentation based on ring–disc and split-ring systems has enabled advances of great importance to be made in the quantitative examination of diffusion processes at electrodes and their role in electrode processes generally. The technique allows precisely defined mass–transport conditions to be set up which can be subjected to exact mathematical analysis so that quantitative treatment of hydrodynamic and diffusion behavior can be made. Of special interest for electrochemists is the opportunity which the rotating ring–disc system offers for studying solution-soluble intermediates in sequential electrode processes and the kinetics of their reactions in solution.

In this book by Pleskov and Filinovskii, both the experimental techniques and the mathematical analysis for the treatments of results for various conditions and types of reaction are described in detail. We believe that presentation of work that has been carried out by means of rotating electrode techniques, to a large extent by Russian workers, in the form of a concise book will be of great value both to electrochemists and kineticists, and those interested in the physics of fluid motion. In the present form of an English translation, this work will be made more accessible and conveniently available to workers less familiar with the Russian language and scientific literature.

H. Wroblowa
B. E. Conway

Introduction

It was shown in 1942 by V. G. Levich that a disc-shaped electrode, rotating in a liquid, has an important property: it is uniformly accessible to diffusion. This property makes a rotating disc a unique tool for investigating kinetics of electrochemical reactions at solid electrodes. It is superior to the dropping mercury electrode in that it allows measurements to be carried out at much faster reaction rates than at the mercury electrode.

Within three decades after the appearance of Levich's paper, the rotating disc electrode became one of the principal methods in experimental electrochemistry. It is used for mechanistic determinations and kinetic studies of electrode reactions, for studies of chemical bulk reactions accompanied by electrode processes, and for measurements of diffusion coefficients of dissolved species. The application of the method became considerably more extensive after the development of the rotating ring—disc assembly. Studies of critical phenomena in liquids should be mentioned as an example of a nonelectrochemical application of the rotating disc.

The rotating-disc method is one of the results of development of physicochemical hydrodynamics — a field bordering on mechanics of liquids and physical chemistry — due to V. G. Levich. A considerable contribution to the development and the practical applications of this method was made by a large number of Soviet electrochemists. Theoretical and experimental work concerning applications of the rotating disc has also been extensively carried out abroad (in England, USA, East Germany, and in other countries).

In recent years work on the rotating disc electrode and the rotating ring−disc assembly has been reviewed several times†; this book contains, however, the first full systematic presentation of the theory and methodology of rotating-disc experiments, as well as of the important experimental results obtained using this method.

Academician A. N. Frumkin

---

† A. C. Riddiford, in: Advances in Electrochemistry and Electrochemical Engineering, Vol. 4, P. Delahay, ed., Interscience Publishers (1966), p. 47; J. Newman, in: Advances in Electrochemistry and Electrochemical Engineering, Vol. 5, C. W. Tobias, ed., Interscience Publishers (1967), p. 87; R. N. Adams, Electrochemistry at Solid Electrodes, Marcel Dekker, New York (1969); M. Bierowski, M. Pawelkowa, and Z. Zembura, Wiadom. Chem., 16:497 (1962); 18:215 (1964).

Preface to the American Edition

The present monograph covers the theory, experimental techniques, and practical applications of the rotating disc electrode, which has been widely used in kinetic studies of electrochemical reactions.

The number of publications concerning the rotating disc continuously increases — it exceeded 1000 in 1973. In fact, these publications started a new chapter of experimental electrochemistry in which a considerable part has been played by Soviet electrochemistry.

The authors express their gratitude to Plenum Press for enabling them to present the book to the English-speaking reader. The present edition is fully equivalent to the Russian one, which was published in 1972 by "Nauka." The only differences in the two editions involve some corrections of printing errors and extension of the additional references which now include material published up to October 1973.

Yu. V. Pleskov
V. Yu Filinovskii

Preface to the ninth edition

Preface

Until the middle forties theoretical electrochemistry was developed mainly in terms of the electrochemistry of the mercury electrode. The development of the rotating-disc method provided the opportunity for extensive quantitative studies of solid electrodes which have been important both from the theoretical and practical points of view.

The rotating disc electrode method appeared as a result of the application of concepts of theoretical thermodynamics to the requirements of electrochemical experiments. By considering mass-transfer phenomena in flowing liquids, V. G. Levich demonstrated that the rotating disc has a unique set of important properties: high accuracy in the measurement of diffusional fluxes combined with the uniform accessibility of the surface and stationary conditions. These properties in conjunction with the experimental simplicity of the system resulted in wide application of the method. This is illustrated by the fact that the number of publications in this area increased from a few per year in 1951–1957 to about 100 in 1970.

We are deeply grateful to Academician A. N. Frumkin, whose benevolent insistence resulted finally in the appearance of this book, and to Corresponding Member of the Academy of Sciences of the USSR V. G. Levich for discussions of the book's outline which were important for us. We express also our thanks to B. M. Grafov, L. N. Nekrasov, N. V. Nikolaeva-Fedorovich, M. R. Tarasevich, and Yu. A. Chimadzhev who read the manuscript and made helpful suggestions.

Preface

Contents

Chapter 3
Mixed Kinetics of Heterogeneous
      Reactions

Chapter 4
Electric Current in a Cell with a
      Rotating Disc Electrode

Chapter 7
Application of the Rotating Disc Electrode
    to Various Electrochemical Problems

Chapter 8
Rotating Ring–Disc Electrode

Chapter 9
Design of the Rotating Disc Assembly

Notation

c — concentration of dissolved species in dimensional units

c^* — bulk concentration

c_S — concentration in the vicinity of the disc surface

c_O, c_R — concentration of the oxidized or reduced form, respectively

c_M — moment of drag

C — dimensionless concentration

D — diffusion coefficient

E — vector of the intensity of the electric field (Chap. 4)

E_z — component of E normal to the surface of the disc (Chap. 4)

$\mathscr{E} = nF\delta_d E_z/RT$ — dimensionless intensity of electric field (Chap. 4)

$F = u/r\omega$ — dimensionless function

$G = v/r\omega$ — dimensionless function

$H = w/\sqrt{\nu\omega}$ — dimensionless function

i — vector of the current density

$i_z = i$ — component of i normal to the surface

i_d — diffusion-limiting current density

i_{dO}, i_{dR} — diffusion-limiting current density of oxidized or reduced form, respectively

i_k — limiting kinetic current density

i_0 — exchange current density

$\bar{\imath}$ — total current

$\bar{\imath}_{\mathscr{D}}$ — total current to the disc electrode

$\bar{\imath}_{\mathscr{R}}$ — total current of the ring electrode

$\hat{\imath} = i/i_d$ — dimensionless current density (Chap. 4)

1

I — flux density of the dissolved species to the rotating disc

j — total flux of the dissolved species

k — rate constant of the heterogeneous process

k_f, k_b — rate constants of the forward and backward reactions, respectively

k_f^*, k_b^* — potential-independent rate constants

$K = k\delta_d c^{*(\mu-1)}/D$ — dimensionless rate constant

l_0 — characteristic geometric dimension of the system

m — number of disc revolutions per minute

M — moment of viscous friction forces

n — number of electrons transferred in the electrochemical reaction

$N = -\bar{i}_\mathscr{R}/\bar{i}_\mathscr{D}$ — collection efficiency of the ring electrode

N_c — collection efficiency of the ring electrode in the presence of chemical reactions of intermediates in the bulk solution

p — hydrodynamic pressure

$P = p/\rho\nu\omega$ — dimensionless function

q — rate of heterogeneous reaction

$\mathbf{r}$ — radius vector

r — radial axis

r_0, r_{10} — radius of the rotating disc electrode

r_{20}, r_{30} — internal and external radii of the rotating disc, respectively

$\mathbf{R} = \mathbf{r}/l_0$ — dimensionless radius vector (§ § 1.3 and 2.2)

s — surface area of the disc

t — time

T — temperature

u — radial component of $\mathbf{v}$

$\mathbf{v}$ — vector of the liquid velocity

v — azimuthal component of $\mathbf{v}$

v_0 — characteristic flow velocity

$\mathbf{V} = \mathbf{v}/v_0$ — dimensionless vector of the velocity of the liquid (§ § 1.3 and 2.2)

$V_M = nF\varphi_M/RT$ — dimensionless electrode potential (Chap. 4)

$V_e = nF\varphi_e/RT$ — dimensionless equilibrium potential (Chap. 4)

w — normal component of $\mathbf{v}$

z — axis normal to the disc surface

α — transfer coefficient

$\gamma = c_2^*/c_1^*$ — relative composition of the inert electrolyte (Chap. 4)

$\delta_0 = 3.6\sqrt{\nu/\omega}$ — thickness of the hydrodynamic boundary layer

$\delta_d = 1.61\,(D/\nu)^{1/3}(\nu/\omega)^{1/2}$ — thickness of the diffusion layer according to Levich

δ_N — thickness of the diffusion layer according to Nernst

ε_0 — dielectric permeability

$\varepsilon_a = (RT\varepsilon_0/4\pi n^2 F c_1^* \delta_d^2)^{1/2}$ — dimensionless small parameter (Chap. 4)

$\varepsilon_c = \mu_c/\delta_d$ — dimensionless small parameter (Chap. 6)

$\eta = Dt/\delta_d^2$ — dimensionless time (Chap. 5)

$\varkappa$ — electrical conductance

μ — order of the heterogeneous process

μ_c — thickness of the kinetic reaction layer (§ 6.17)

$\mu_1, \ldots, \mu_N$ — stoichiometric coefficient

ν — kinematic viscosity

ν_{turb} — turbulent viscosity coefficient

ρ — electric charge density (Chap. 4)

ρ_c — rate constant of the chemical reaction in the bulk

σ — chemical equilibrium constant

τ — transition time

φ — electric potential in dimensionless units

φ_e — equilibrium potential

$\Delta\varphi_{ohm}$ — ohmic potential drop in the solution

$\Phi = nF\varphi/RT$ — dimensionless electric potential

ψ_1 — potential at the Helmholtz plane

$\Psi_1 = nF\psi_1/RT$ — (Chap. 4)

ω — angular velocity of the disc

Ω — frequency of variations of the polarizing voltage

Liquid Flow at a Rotating Disc

§ 1.1. Basic Equations of Flow of a Viscous Incompressible Liquid

The quantitative theory of mass-transfer processes in stirred media is based on modern developments in hydrodynamics. Therefore, it is useful first to present the basic concepts involved in hydrodynamics of a viscous liquid which are necessary for the theoretical analysis of mass transport to the surface of a rotating disc.

A detailed and systematic treatment of hydrodynamic problems may be found in other special monographs [1, 2].

Here, special attention will be focused on the flow of a non-compressible liquid, i.e., a liquid having a constant density ρ. A full description of the flow of such a liquid at each point requires knowledge only of the velocity **v** of the liquid and the pressure p. Generally, these quantities depend on the three spatial coordinates and on time. The four unknown functions (the three components of velocity and the pressure) can be determined from a system of four equations describing the flow of an arbitrary element of the liquid. Those relations are the continuity equation and the three equations of motion.

The continuity equation expresses the law of mass conservation for the moving liquid. It can be described by a vectorial expression

$$\operatorname{div} \mathbf{v} = 0. \qquad (1.1)$$

The equations of motion of a unit volume of liquid follow the well known laws of mechanics, namely, the product of the mass of a unit liquid volume and velocity should equal the sum of all forces acting on the unit volume. These are external forces (gravitational, magnetic, etc.), the force due to pressure, and the internal friction forces.

The equation of motion for a unit volume of a viscous incompressible liquid has the form

$$\rho \left[\frac{\partial v}{\partial t} + (v \, \mathrm{grad}) \, v \right] = f - \mathrm{grad} \, p + \mu \Delta v. \tag{1.2}$$

The vector f designates external forces; grad p is the gradient of pressure acting in the liquid; $\mu \Delta v$ (where μ is the dynamic viscosity coefficient) characterizes internal friction forces caused by motion of the liquid.

It must be stressed that viscosity considerably affects the flow characteristics. Forces of internal friction (arising from the motion of separate liquid regions to each other) result in a momentum transfer between those regions. The magnitude of the friction forces is determined by the liquid properties and by the velocity gradients due to motion of the liquid.

Since the viscous force acting on a unit volume of the liquid is expressed by $\mu \Delta v$, the analysis is restricted to the case of incompressible, Newtonian (or normal) liquids. In addition to exhibiting isotropic properties, Newtonian fluids are characterized by a linear dependence of viscous stresses on the velocity gradient (Newton's law of friction). The viscosity μ of Newtonian fluids depends only on temperature T and pressure p and is independent of the velocity and its derivatives.

Obviously, Newton's law of friction must be considered as one of the simplest friction laws. It is valid only for a certain range of flow rates. In practice, it is obeyed by all gases, as well as by liquids and solutions of low molecular weight.

There is, however, a broad class of liquids the rheology of which deviates strongly from Newtonian behavior. Their viscosity depends not only on temperature and pressure, but also on the previous history of the fluid and on the way the apparatus is constructed. Their viscosity is a nonlinear function of the velocity gradients, etc. The theoretical description of the motion of non-Newtonian liquids is thus extremely complex.

In the following, basically Newtonian liquids† will be treated. Such fluids exhibit viscosity which depends little on pressure but varies considerably with temperature; for example, for water, a temperature increase from 0 to 100°C brings about a decrease of viscosity by approximately one order of magnitude.

Dynamic viscosity is often replaced in the literature by the kinematic viscosity

$$\nu = \mu/\rho. \tag{1.3}$$

Equations (1.1) and (1.2) represent the Navier–Stokes equations which fully describe the behavior of a viscous incompressible fluid.

Depending on the system considered, the Navier–Stokes equations must be supplemented by suitable initial and boundary conditions. The initial conditions (which must be necessarily stated in the case of nonsteady-state liquid flow) are usually formulated as follows: at a fixed time (e.g., t = 0) the velocity and pressure distribution is considered to be known, i.e., at t = 0

$$v = v^{(0)}(\mathbf{r}), \qquad p = p^{(0)}(\mathbf{r}). \tag{1.4}$$

Boundary conditions characterize the behavior of the liquid at the boundaries of the liquid-occupied space. These conditions depend considerably on the properties of the liquid and of its surroundings and are formulated in different ways, e.g., for a liquid–gas, liquid–liquid, or liquid–solid interface, and for permeable or impermeable solid walls, etc.

The subsequent material is concerned primarily with liquid flow in the vicinity of an impermeable solid surface. Particles of a viscous liquid adhere to this surface, and the velocity of layers immediately adjacent to the wall is equal to the velocity of the solid surface. Thus, the normal (v_n) and tangential (v_t) velocity

---

†A few papers pertaining to the rotating disc in non-Newtonian fluids are cited in Appendix 1.

components disappear at the interface of the immobile solid wall, i.e.,

$$v_n = 0, \quad v_t = 0.$$

$$(1.5)$$

A liquid in contact with a mobile solid wall (as, e.g., in the case of a rotating disc) has at each point close to the interface a velocity vector **v** equal, in respect to its value and direction, to the velocity of the corresponding point on the solid surface, $\mathbf{v}_S$, i.e.,

$$\mathbf{v} = \mathbf{v}_S.$$

$$(1.6)$$

In the case of liquid flow at permeable walls, an additional hydrodynamic flux has to be considered. Sometimes, however, boundary conditions can be reduced to the same type as those of Eq. (1.6).

Solving the equations of motion for an incompressible viscous liquid with suitable initial and boundary conditions is usually very difficult. An exact solution can be obtained only for a limited number of cases, while approximate methods have been developed for other cases; the most productive of these is the method of the boundary layer [3].

§1.2. Solution of Equation of Motion for a Viscous, Incompressible Liquid at the Surface of a Rotating Disc

The flow of a viscous, incompressible liquid at the surface of a disc under conditions of uniform axial rotation represents one of the cases for which an exact solution of the Navier−Stokes equation is possible. The solution was first obtained by von Kármán [4] and later by Cochrane [5].

Consider a disc of a rather large radius† rotating in a viscous incompressible liquid at a constant angular velocity ω. The liquid layer immediately adjacent to the disc surface adheres to it and takes part in the rotational motion. The layers not imme-

---

†So that edge effects become negligible.—Editor.

diately adjacent to the disc must also rotate owing to the viscous forces. However, with increasing distance from the disc, the rotational motion becomes progressively attenuated. The centrifugal forces tend to throw elements of the rotating liquid toward the periphery of the disc. They are replaced by new elements of liquid arriving from the bulk which in turn are entrained by the disc and transported by acquisition of momentum toward its edges. Continuous supply of liquid to the disc surface requires that a vertical flow be maintaned at large distances from the disc.

The problem will be solved using cylindrical coordinates (r, φ, z) (cf. Fig. 1.1). Velocity components in radial, azimuthal, and axial directions are designated by $v_r = u$, $v_\varphi = v$, and $v_z = w$, respectively. The Navier–Stokes equations have the form

$$\frac{\partial u}{\partial r} + \frac{u}{r} + \frac{\partial w}{\partial z} = 0,$$

$$u\frac{\partial u}{\partial r} - \frac{v^2}{r} + w\frac{\partial u}{\partial z} = -\frac{1}{\rho}\frac{\partial p}{\partial r} + v\left[\frac{\partial^2 u}{\partial r^2} + \frac{\partial}{\partial r}\left(\frac{u}{r}\right) + \frac{\partial^2 u}{\partial z^2}\right],$$

$$u\frac{\partial v}{\partial r} + \frac{uv}{r} + w\frac{\partial v}{\partial z} = v\left[\frac{\partial^2 v}{\partial r^2} + \frac{\partial}{\partial r}\left(\frac{v}{r}\right) + \frac{\partial^2 v}{\partial z^2}\right], \qquad (1.7)$$

$$u\frac{\partial w}{\partial r} + w\frac{\partial w}{\partial z} = -\frac{1}{\rho}\frac{\partial p}{\partial z} + v\left[\frac{\partial^2 w}{\partial r^2} + \frac{1}{r}\frac{\partial w}{\partial r} + \frac{\partial^2 w}{\partial z^2}\right].$$

[Owing to the axial symmetry of the flow, the azimuthal coordinate φ does not appear in Eqs. (1.7).]

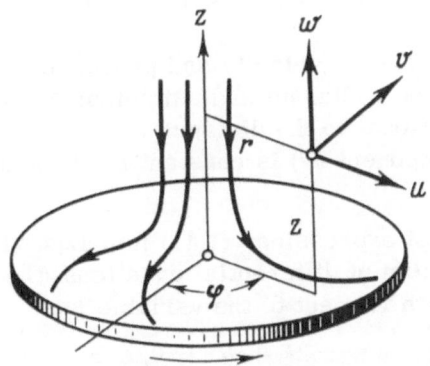

Fig. 1.1. Coordinate system used in calculations of liquid flow near a rotating disc. Solid curves are streamlines of liquid flow.

The above form of the Navier–Stokes equations implies that no external forces other than gravity act on the system. The external pressure is included, as usually, with the gravitational forces. Thus, p in Eqs. (1.7) represents the excess pressure over that exerted in the stationary liquid.

Let the origin of the coordinate system be placed at the center of the disc, and the z axis be taken normal to its surface. The liquid is placed above the plane of the disc (i.e., where z > 0).

The following condition should obtain [cf. Eq. (1.6)] at the surface of a uniformly rotating disc:

$$u = 0, \quad v = r\omega, \quad w = 0 \quad \text{for } z = 0. \tag{1.8}$$

The flux far removed from and normal to the disc is given by

$$u \to 0, \quad v \to 0 \quad \text{as } z \to \infty. \tag{1.9}$$

Integration is facilitated by introduction of a dimensionless reduced distance from the disc:

$$\zeta = z \sqrt{\frac{\omega}{\nu}}. \tag{1.10}$$

Further, let the velocity and pressure components be given by

$$u = r\omega F(\zeta), \quad v = r\omega G(\zeta), \quad w = \sqrt{\nu\omega} H(\zeta), \quad p = \rho\nu\omega P(\zeta). \tag{1.11}$$

The above choice of velocity and pressure is based on the assumption that the radial (u) and azimuthal (v) velocity components are proportional to the distance from the axis of rotation, and the axial component (w) is constant in all planes parallel to that of disc.

Substitution of expressions (1.11) into Eqs. (1.7) results in the following system of differential equations (the prime represents differentiation with respect to the variable ζ):

$$2F + H' = 0,$$
$$F^2 + F'H - G^2 - F'' = 0,$$
$$2FG + HG' - G'' = 0,$$
$$P' + HH' - H'' = 0. \tag{1.12}$$

Boundary conditions are expressed as

$$F = 0, \quad G = 1, \quad H = 0 \quad \text{for } \zeta = 0,$$
$$F \to 0, \quad G \to 0 \quad \text{as } \zeta \to \infty. \tag{1.13}$$

The first three equations of (1.12) describe the velocity distribution, i.e., the functions F, G, and H, and the last equation describes the pressure P.

Von Kármán [4] integrated the set of equations (1.12) using his approximate integral method. Cochrane [5] used a numerical integration method for this purpose.

Using Eqs. (1.12) and boundary conditions at $\zeta \to \infty$, an asymptotic expansion of functions F, G, and H can be obtained for regions far from the surface of the disc.

Since the radial and azimuthal velocity components die out with increasing distance from the disc, the series expansions are given in powers of $\exp(-c\zeta)$, where c is a constant. The first terms of these expansions are

$$F = A \exp(-c\zeta) - \frac{A^2 + B^2}{2c^2} \exp(-2c\zeta) + \frac{A(A^2 + B^2)}{4c^4} \exp(-3c\zeta) + \ldots$$

$$G = B \exp(-c\zeta) - \frac{B(A^2 + B^2)}{12c^4} \exp(-3c\zeta) + \ldots$$

$$H = -c + \frac{2A}{c} \exp(-c\zeta) - \frac{A^2 + B^2}{2c^3} \exp(-2c\zeta) + \tag{1.14}$$
$$+ \frac{A(A^2 + B^2)}{6c^5} \exp(-3c\zeta) + \ldots$$

Close to the disc surface ($\zeta = 0$) the unknown functions can be represented in the form of power series. Series which satisfy the original differential equations and the boundary conditions at $\zeta = 0$ are as follows:

$$F = a_0\zeta - \frac{\zeta^2}{2} - \frac{b_0\zeta^3}{3} - \frac{b_0^2\zeta^4}{12} - \ldots$$

$$G = 1 + b_0\zeta + \frac{a_0\zeta^3}{3} + \frac{(a_0b_0 - 1)\zeta^4}{12} - \ldots \tag{1.15}$$

$$H = -a_0\zeta^2 + \frac{\zeta^3}{3} + \frac{b_0\zeta^4}{6} + \ldots$$

The constants A, B, c, a_0, and b_0 must be chosen so that the solutions continuously transform one to another. It is sufficient for this purpose that the functions F, G, H, F', and G' satisfy continuity conditions. Cochrane found that these conditions are satisfied for the following values of the constants:

$$A = 0.934, \quad B = 1.208, \quad c = 0.886, \\ a_0 = 0.510, \quad b_0 = -0.616. \hspace{1cm} \Big\} \hspace{1cm} (1.16)$$

Values of the functions F, G, H, H', G', and P calculated by Cochrane [5] are given in Table 1.1 and plotted in Fig. 1.2.

It can be seen from the graphs that the azimuthal component rapidly decreases with distance from the disc. Thus, at $\zeta = 3.6$, the value of function $G(\zeta)$ is only 0.05 of that at the surface of the disc. It is useful to refer to the liquid layer of thickness

$$\delta_0 = 3.6 \sqrt{\frac{\nu}{\omega}} \hspace{2cm} (1.17)$$

as the "hydrodynamic boundary layer" at the rotating disc.†

It is easy to show by analysis of the velocity distribution shown in Fig. 1.2 that the physical picture of flow resulting from rotation of a disc is the same as that due to a centrifugal fan. Indeed, outside the hydrodynamic boundary layer ($\zeta \to \infty$) liquid flows towards the disc surface with a constant velocity w_∞, equal to $0.886(\nu_w)^{1/2}$ and the velocity of this flux is independent of the distance from the rotation axis. Close to the surface of the disc (within the hydrodynamic boundary layer) liquid is slowed down and thrown towards the edges of the disc. The streamlines in the boundary layer form a family of logarithmic spirals.

It follows from Eq. (1.17) that the thickness of the hydrodynamic boundary layer, δ_0, is independent of the distance from the rotation axis and increases with decreasing angular velocity of the disc. In water, values of δ_0 are 0.36 cm for 1 rev sec^{-1}; 0.11 cm for 10 rev sec^{-1}; 0.036 cm for 100 rev sec^{-1} and 0.011 cm for 10^3 rev sec^{-1}.

All the preceding results were obtained for a disc of infinite radius. Nevertheless, they can be used also for a disc having

---

†In a number of papers [3, 6], the thickness of the hydrodynamic boundary layer is defined in a different way. Therefore, the numerical factor of Eq. (1.17) has a different value there.

TABLE 1.1. Values of Functions Determining the
Velocity and Pressure Distribution in the Vicinity of
a Disc Rotating in a Stationary Liquid

$\zeta = z\sqrt{\dfrac{\omega}{\nu}}$	F	G	$-H$	P	F'	$-G'$
0	0	1.0	0	0	0.510	0.616
0.1	0.046	0.939	0.005	0.092	0.416	0.611
0.2	0.084	0.878	0.018	0.167	0.334	0.599
0.3	0.114	0.819	0.038	0.228	0.262	0.580
0.4	0.136	0.762	0.063	0.275	0.200	0.558
0.5	0.154	0.708	0.092	0.312	0.147	0.532
0.6	0.166	0.656	0.124	0.340	0.102	0.505
0.7	0.174	0.607	0.158	0.361	0.063	0.476
0.8	0.179	0.561	0.193	0.377	0.032	0.448
0.9	0.181	0.517	0.230	0.388	0.006	0.419
1.0	0.180	0.468	0.266	0.395	−0.016	0.391
1.1	0.177	0.439	0.301	0.400	−0.033	0.364
1.2	0.173	0.404	0.336	0.403	−0.046	0.338
1.3	0.168	0.371	0.371	0.405	−0.057	0.313
1.4	0.162	0.341	0.404	0.406	−0.064	0.290
1.5	0.156	0.313	0.435	0.406	−0.070	0.268
1.6	0.148	0.288	0.466	0.405	−0.073	0.247
1.7	0.141	0.264	0.495	0.404	−0.075	0.228
1.8	0.133	0.242	0.522	0.403	−0.076	0.210
1.9	0.126	0.222	0.548	0.402	−0.075	0.193
2.0	0.118	0.203	0.572	0.401	−0.074	0.177
2.1	0.111	0.186	0.596	0.399	−0.072	0.163
2.2	0.104	0.171	0.617	0.398	−0.070	0.150
2.3	0.097	0.156	0.637	0.397	−0.067	0.137
2.4	0.091	0.143	0.656	0.396	−0.065	0.126
2.5	0.084	0.131	0.674	0.395	−0.061	0.116
2.6	0.078	0.120	0.690	0.395	−0.058	0.106
2.8	0.068	0.101	0.721	0.395	−0.052	0.089
3.0	0.058	0.083	0.746	0.395	−0.046	0.075
3.2	0.050	0.071	0.768	0.395	−0.040	0.063
3.4	0.042	0.059	0.786	0.394	−0.035	0.053
3.6	0.036	0.050	0.802	0.394	−0.030	0.044
3.8	0.031	0.042	0.815	0.393	−0.025	0.037
4.0	0.026	0.035	0.826	0.393	−0.022	0.031
4.2	0.022	0.029	0.836	0.393	−0.019	0.026
4.4	0.018	0.024	0.844	0.393	−0.016	0.022
∞	0	0	0.886	0.393	0	0

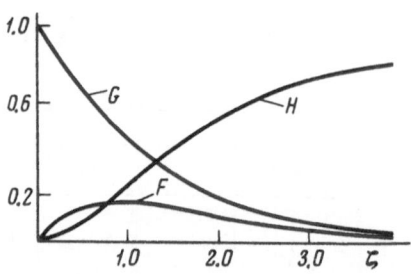

Fig. 1.2. Velocity distribution around the disc rotating in a stationary liquid [8]. F is the radial, G the azimuthal, and H the axial component [Eqs. (1.11)].

a finite radius, r_0, if the edge effect is negligible. This effect disappears when $r_0 \gg \delta_0$.

The moment of the viscous friction forces acting on the rotating disc will now be calculated.

The azimuthal component of the tangential stress arising at the surface of the disc is given by

$$\tau_{z\varphi} = \mu \left(\frac{\partial v}{\partial z}\right)_0 = r\rho v^{1/2}\omega^{3/2}G'(0). \qquad (1.18)$$

The moment of the forces acting on an annular ring having a width dr situated at a distance r from the rotation axis is given by

$$dM = r\tau_{z\varphi} 2\pi r \, dr.$$

The total moment of forces acting on a disc of radius r_0 is obtained by integration of the above expression over the disc radius:

$$M = \int_0^{r_0} r\tau_{z\varphi} 2\pi r \, dr = \frac{\pi\rho v^{1/2}\omega^{3/2}G'(0) r_0^4}{2}.$$

Substituting the value of $G'(0)$ from Table 1.1, we find

$$M = -0.308 \, \pi\rho v^{1/2}\omega^{3/2}r_0^4. \qquad (1.19)$$

Instead of Eq. (1.19), hydrodynamic problems usually involve a dimensionless coefficient of the moment of drag, defined as follows:

$$c_M = -\frac{2M}{\frac{\rho}{2}\omega^2 r_0^5} = \frac{3.87}{\mathrm{Re}^{1/2}}, \qquad (1.20)$$

where $Re = r_0^2 \omega / \nu$ is the dimensionless Reynolds number, the physical meaning of which is discussed in detail in the next section.

Experimental data pertaining to measurements of the drag coefficient at a rotating disc electrode are shown in Fig. 1.3 (reprinted from [6]). Line 1 corresponds to the theoretical relation (1.20). It can be seen from Fig. 1.3 that in the region of laminar flow the agreement of theory with experiment is fully satisfactory.

The flow of a viscous incompressible liquid in the vicinity of an infinite rotating disc (the problem treated by von Kármán) is one of the classical problems in hydrodynamics. The great interest it evokes in the mechanics of liquids is due not only to the important practical applications of such a case, but also to the fact that the analysis of the von Kármán case facilitates the approach to the description of other hydrodynamic problems. Therefore, the work of von Kármán [4] and of Cochrane [5] has stimulated numerous authors to study liquid flow in the vicinity of a rotating disc using a number of different methods. Their results are partly described in Dorfman's monograph [6] and in reviews by Moore [7] and Shlichting [8]. Among more recent studies, those

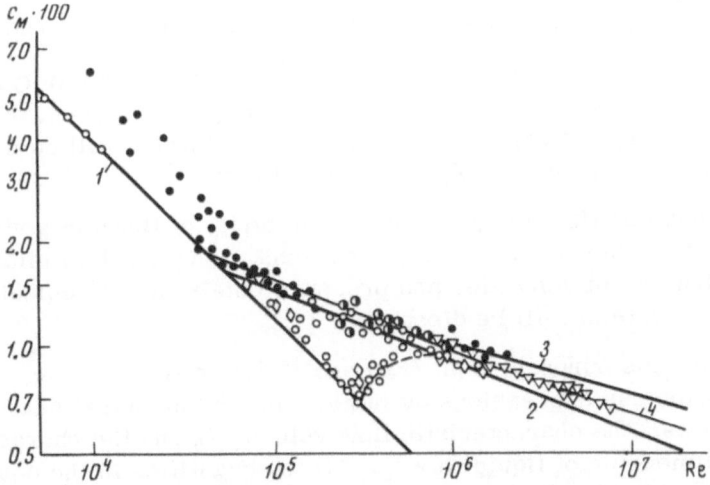

Fig. 1.3. Drag coefficient of a disc rotating in a stationary liquid [6]. Solid lines are calculated: 1) laminar flow [Eq. (1.20)]; 2) turbulent flow [Eq. (1.31)]; 3) from Eq. (1.32); 4) Dorfman calculations. The points are the experimental data of various authors.

of Schwiderski et al. [9], Kaminski et al. [10], and Benton [11] should be mentioned. Calculations in [10] and [11] were computer assisted. In particular, Benton [11] showed that the numerical values of (1.16) of constants calculated by Cochrane should be somewhat revised:

$$A = 0.91772, \quad B = 1.20211, \quad c = 0.88446,$$
$$a_0 = 0.510233, \quad b_0 = -0.61592. \tag{1.21}$$

§ 1.3. "Similarity" of Hydrodynamic Phenomena

Studies of hydrodynamic problems, as well as of heat and mass-transfer processes, commonly utilize methods of dimensional analysis and similarity theory. They allow a relatively easy formulation to be made of general relations governing the given process and make possible the definition of parameters controlling it. In several cases, particularly in studies of complex physical phenomena (e.g., turbulent flow), dimensional analysis coupled with certain model considerations or experimental facts enables important practical conclusions to be reached. The methods of dimensional analysis and similarity theory are especially important in the development of models for various phenomena and processes.

A detailed description of the dimensional and similarity methods can be found in special monographs (cf.,e.g., [12]). Here only the simplest concepts will be discussed which will be needed later for interpretation of experimental data.

Consider first the problem of similarity of flows of a viscous incompressible liquid. First, conditions under which similar distributions of velocities and pressure obtain in two geometrically similar systems will be discussed.

For this purpose, Eqs. (1.1) and (1.2) are transformed into dimensionless expressions by introducing the principal scaling quantities: the characteristic flow velocity v_0 and the characteristic dimension of liquid flux l_0. All the quantities in the equations of motion (1.1) and (1.2) can be expressed in units of v_0 and l_0 as follows:

$$\mathbf{v} = \mathbf{V}v_0, \quad \mathbf{r} = \mathbf{R}l_0, \quad p = P\rho v_0^2, \tag{1.22}$$

where $\mathbf{V}$, $\mathbf{R}$, and P are dimensionless quantities.

Introducing additionally a dimensionless parameter, the Reynolds number,

$$\text{Re} = \frac{v_0 l_0}{\nu}, \tag{1.23}$$

the equations of motion (1.1) and (1.2) can be rewritten for steady-state flow of an incompressible liquid in the absence of external forces as

$$\text{div } \mathbf{V} = 0,$$
$$(\mathbf{V} \text{ grad}) \mathbf{V} = - \text{grad } P + \frac{1}{\text{Re}} \Delta \mathbf{V}. \tag{1.24}$$

Differentiation of Eqs. (1.24) is carried out with respect to the dimensionless variable $\mathbf{R}$.

It is obvious that Eqs. (1.24) must be supplemented by suitable boundary conditions.

Since the Reynolds number is the only dimensionless parameter in Eq. (1.24), the velocity distribution $\mathbf{V}$ and the pressure distribution P are functions of $\mathbf{R}$ and Re:

$$\mathbf{V} = \mathbf{V}(\mathbf{R}, \text{Re}),$$
$$P = P(\mathbf{R}, \text{Re}). \tag{1.25}$$

It can be concluded on the basis of Eq. (1.25) that distributions of velocity and pressure are the same for the same type of flow having equal Reynolds numbers ("similarity" principle).

In order to obtain an insight into the similarity principle, consider viscous flow in the vicinity of two rotating discs. One disc has a radius r_{10} and rotates with an angular velocity ω_1 in a liquid with a kinematic viscosity ν_1. The second disc of radius r_{20} rotates with a velocity ω_2 in a liquid with a kinematic viscosity ν_2. If the characteristic dimension l_0 is chosen as the disc radius r_0, and the characteristic velocity v_0 is taken as the quantity $r_0\omega$, the following Reynolds numbers for the two systems are found: $\text{Re}_1 = r_{10}^2 \omega_1 / \nu_1$ and $\text{Re}_2 = r_{20}^2 \omega_2 / \nu_2$. If $R_1 = R_2$, the distribution of velocities and of pressure are similar in both cases, i.e., if the distribution at one of the discs is known, it is possible to describe the flow at the second disc by suitable adjustment of the distance scale and velocity.

The condition of "hydrodynamic similarity" for two rotating discs is thus expressed by $Re_1 = Re_2$, or $r_{10}^2 \omega_1 / \nu_1 = r_{20}^2 \omega_2 / \nu_2$.

"Similarity" of hydrodynamic flows is a very important concept. Two simple consequences of the similarity principle will be mentioned. First, the similarity principle removes the necessity for seeking anew velocity and pressure distributions when the flux parameters are changed. Second, the similarity principle makes modeling of hydrodynamic systems possible; i.e., study of the hydrodynamic flow characteristics under convenient experimental conditions can be made.

The motion of solids in liquids is often described by mechanical characteristics (drag forces, their associated momentum, etc.) of solid−liquid interactions. Similarity of hydrodynamic phenomena also makes possible comparison of mechanical flow characteristics. It has been shown in the previous section that a suitably chosen drag coefficient of the disc, c_M, is a function of only one parameter, namely, Re [Eq. (1.20)]. This means that flow at two rotating discs results in the same drag coefficients when $Re_1 = Re_2$.

When the viscous flow is significantly affected by external forces, other dimensionless parameters may be added to the Reynolds number (Froude number in the case of considerable gravitational effects, magnetic Reynolds number in the case of magnetic forces, etc.).

§ 1.4. Turbulent Flow

The results discussed above pertain to laminar flow of liquid at a rotating disc surface. As can be seen from Fig. 1.3, experimental values of the drag coefficient of a rotating disc, c_M, start deviating from relation (1.20) when Re becomes higher than 10^5. The sharpest transition occurs at $Re = 3 \cdot 10^5$. In this region the relation $c_M \sim Re^{-0.5}$ is replaced by $c_M \sim Re^{-0.2}$, and the laminar-flow regime becomes turbulent.

Turbulent flow is characterized by chaotic movements of liquid particles superposed on a background of smooth motion of the medium. Turbulent flow can be represented by an infinite amount of eddies of various dimensions, forms, and angular velocities, which are transported together with the flux of the fluid. The instantaneous values of velocities and pressure in a turbulent flux undergo sharp irregular changes in time and space. The established direction of flow is observed only for time-averaged velocities and pressures. The appearance of turbulent eddies has a considerable effect on the basic flow characteristics. Relations describing laminar flow are invalid in the turbulent regime.

Study of the laws of turbulent flow is a large and important part of hydrodynamics. Attempts at description of turbulent flow utilize several statistical and semiempirical methods. Nevertheless, a satisfactory quantitative theory of turbulence does not exist at present. A review of contemporary concepts concerning turbulent flow can be found in special monographs [13-15].

Studies of turbulent flow are directed toward two basic problems: (1) understanding of the process of the appearance of turbulent flow and of the transition from laminar to turbulent conditions, and (2) establishment of laws governing the turbulent flow.

The most interesting question in the study of the laminar − turbulent transition is the value of the Reynolds number (or range of Reynolds numbers) at which this transition occurs. This value is called the critical Reynolds number, Re_{cr}. It determines the limits of stability of laminar flow. When $Re < Re_{cr}$, small disturbances, which can always appear in a laminar flow, are rapidly damped out upon further advance of the liquid. When $Re > Re_{cr}$, the disturbances not only are not damped, but their amplitude increases. Disturbances reinforce each other by superposition and result in a chaotic movement of the liquid.

The value of Re_{cr} depends to a considerable extent on a number of factors (geometry of the system, roughness of solid walls, presence of external forces, supply or removal of the liquid through walls, etc.) and varies greatly between various hydrodynamic systems. In spite of the existence of recently developed special methods, a theoretical calculation of Re_{cr} is extremely difficult and does not give reliable results. An experimental determination of Re_{cr} is also rather difficult. Therefore, the literature data concerning Re_{cr} have only an approximate character.

It may be concluded by analysis of Fig. 1.3 that in the case of a rotating disc $Re \sim 3 \cdot 10^5$. Several authors have studied directly the appearance of turbulence near a rotating disc using optical, acoustic, and thermal methods. Values of Re_{cr} thus obtained are in the range $3 \cdot 10^5 \le \mathrm{Re} \le 1.8 \cdot 10^6$.† It has been found, however, that the Re_{cr} value depends considerably on the roughness of the disc surface. Increase of the latter results in a decrease of Re_{cr}. This is obviously connected with the fact that roughness serves as an additional source of disturbances. A decrease of Re_{cr} occurs also when a thin stream is directed at the disc along the rotation axis. Sucking off the liquid from the surface usually stabilizes the flow.

The ensuing discussion of elementary concepts in the semi-empirical theory of turbulence is based on the following physical fact. Turbulent eddies (apart from the chaotic movement of liquid particles) also bring about an additional momentum transfer between various regions of the fluid flux. This causes additional stresses to arise in the liquid, the so-called Reynolds stresses. Introduction of these stresses into Navier—Stokes equations (1.2) makes the mathematical problem of obtaining distributions of velocities and pressures under turbulent flow conditions insolvable, since the number of unknown functions exceeds the number of existing equations. In order to overcome this difficulty, several additional concepts can be utilized, which are based usually on experimental data only.

If the Reynolds stresses are taken into account, Eq. (1.18) must be replaced by

$$\overline{\tau_{z\varphi}} = \mu \left(\frac{\partial \bar{v}}{\partial z}\right) - \rho \left(\overline{v'w'}\right), \tag{1.26}$$

where v' and w' are the azimuthal and axial components of the velocities of turbulent eddies and the prime signifies a quantity averaged over a sufficiently long period of time.

According to the most fruitful and well-known semiempirical turbulence theory of Prandtl and von Kármán [14, 16] Reynolds stresses are related to the characteristics of the averaged motion in the following way:

$$-\rho \left(\overline{v'w'}\right) = \rho \nu_{turb} \left(\frac{\partial \bar{v}}{\partial z}\right), \tag{1.27}$$

---

†See Dorfman's monograph [6].

where ν_{turb} is the coefficient of the effective eddy viscosity. Substitution of (1.27) into (1.26) results in

$$\overline{\tau_{z\varphi}} = \rho\left(\nu + \nu_{turb}\right)\frac{\partial \overline{v}}{\partial z}. \tag{1.28}$$

As opposed to the usually encountered (molecular) viscosity ν, the eddy viscosity ν_{turb} is determined not by physical liquid properties but by the character of the periodic eddy motion. Therefore, in general, the value of ν_{turb} can vary in space.

It is obvious that in the region of a developed turbulent flow far from the solid—liquid interface $\nu_{turb} \gg \nu$. Viscous forces play a negligible role here. A solid wall has a retarding effect on turbulent eddies, and thus they are progressively damped with decreasing distance from the wall. In the region of the turbulent boundary layer, ν_{turb} decreases linearly with distance ($\nu_{turb} \sim z$). Still closer to the wall the rate of damping increases and $\nu_{turb} \sim z^m$, where $m \geq 3$ (viscous sublayer). Finally, there exists a thin liquid layer (laminar sublayer) immediately adjacent to the surface into which turbulent eddies do not penetrate. Here $\nu_{turb} \ll \nu$.

The above four-layer model of turbulent flow was first proposed by Landau and Levich [17]. The model is constructed on the basis of certain general concepts concerning the character of damping of turbulent eddies. The authors conclude that the most probable value of the power m is 4 and the distribution of velocities they obtained for turbulent flow at a solid wall is in good agreement with experiment.

It should be mentioned that besides the ideas presented above, several other theories exist concerning liquid motion in turbulent boundary layers [18]. Considerations leading to these theories have, however, a somewhat more intuitive character.

In 1921 von Kármán [4] calculated the basic characteristic parameters of the boundary layer at a rotating disc using his own integral method. Assuming, on the basis of experimental data, that the azimuthal velocity component in a turbulent boundary layer is a function of distance to the power 1/7, von Kármán postulated the following velocity profile:

$$u = \alpha r\omega \left(z/\delta_0\right)^{1/7}\left(1 - z/\delta_0\right),$$
$$v = r\omega \left[1 - \left(z/\delta_0\right)^{1/7}\right], \tag{1.29}$$

where α is a coefficient determined in the course of the calculations. The following result was obtained for the thickness of the hydrodynamic boundary layer:

$$\delta_0 = 0.525 \, r \, (\nu/r_0^2 \omega)^{1/4}. \tag{1.30}$$

Thus, as opposed to Eq. (1.17) which is valid for laminar flow, under turbulent conditions δ_0 varies with distance from the rotation axis. The drag coefficient was shown to be

$$c_M = \frac{0.146}{\mathrm{Re}^{0.2}}. \tag{1.31}$$

Subsequently Goldstein [19] replaced relation (1.29) by a logarithmic profile and obtained the following expression for the drag coefficient:

$$\frac{1}{\sqrt{c_M}} = 1.97 \, \log\left(\frac{\mathrm{Re}}{\sqrt{c_M}}\right) + 0.03. \tag{1.32}$$

Relations (1.31) and (1.32) are shown in Fig. 1.3 (curves 2 and 3). Curve 4 describes a dependence derived by Dorfman [6] on the basis of a more complex functional form of the velocity distribution in a turbulent boundary layer.

References

1.   N. E. Kochin, I. A. Kibel', and N. V. Roze, Theoretical Hydromechanics, Vols. 1 and 2, Fizmatgiz, Moscow (1963).
2.   S. Gol'dstein, ed., Modern Hydrodynamics of Viscous Fluids, Vols. 1 and 2, IL, Moscow (1948).
3.   L. G. Loitsyanskii, The Laminar Boundary Layer, Fizmatgiz, Moscow (1962).
4.   T. von Kármán, Z. Angew. Math. Mech., 1:244 (1921).
5.   W. G. Cochran, Proc. Cambr. Phil. Soc., 30:365 (1934).
6.   L. A. Dorfman, Hydrodynamic Drag and Heat Transfer of Rotating Bodies, Fizmatgiz, Moscow (1960).
7.   F. Moore, in: Problems of Mechanics, Vol. 2, H. Dryden and T. von Kármán, eds., IL, Moscow (1959).
8.   H. Shlichting, Mekhanika, No. 3(85), p. 151 (1964).
9.   E. W. Schwiderski and H. J. Lugt, Phys. Fluids, 7:867 (1964).
10.   T. L. Kaminski and P. Sverkar, Trans. ASAE, 9:875 (1966).
11.   E. R. Benton, J. Fluid. Mech., 24:781 (1966).
12.   L. I. Sedov, Similarity and Dimensional Methods in Mechanics, Nauka, Moscow (1965).

13.   A. S. Monin and A. M. Yaglom, Statistical Hydromechanics, Vols. 1 and 2, Nauka, Moscow (1965, 1967).

14.   Pai Shih-I, Turbulent Flow of Liquids and Gases, IL, Moscow (1962).

15.   J. O. Hinze, Turbulence, McGraw-Hill, New York (1959).

16.   T. von Kármán, in: Problems of Turbulence, ONTI, Moscow—Leningrad (1936), p. 271.

17.   V. G. Levich, Physicochemical Hydrodynamics, Prentice Hall, Inc. (1962).

18.   L. G. Loitsyanskii, in: Transactions of the All-Union Conference on Theoretical and Applied Mechanics, Izd. AN SSSR, Moscow—Leningrad (1962), p. 145.

19.   S. Goldstein, Proc. Cambr. Phil. Soc., 31:232 (1935).

Convective Diffusion

§ 2.1. Basic Equations of Convective Diffusion in a Moving Liquid

Heterogeneous transformations at phase boundaries (including electrochemical reactions as a special case) consist of several consecutive stages: transport of reactants to the reaction site and the heterogeneous reaction itself followed by removal of reaction products from the reaction surface. Sometimes heterogeneous processes are accompanied by bulk reactions between species taking part in the transformation and by heat-transfer processes.

The net rate of a heterogeneous reaction depends on the rates of one (or more) of the steps listed above. If the supply of reactants or removal of products is the slowest step, the heterogeneous reaction obeys diffusion-controlled kinetics. If, conversely, the heterogeneous transformation itself is the slowest step in the sequence, the process is controlled by the chemical kinetics of the surface process.

Heterogeneous transformation can often proceed under mixed kinetic control, particulary in the case of mass transport in a liquid phase. (The term "mixed kinetics" refers to conditions where the rate constants for the mass transport process and for the heterogeneous reaction itself are commensurate.) Therefore, studies of the true kinetics of heterogeneous reactions themselves require elimination of limitations introduced by mass transport of the reacting species. A number of successful experimental methods have been developed in order to examine effects of mass transport on the net rate of heterogeneous processes. Evaluation of mass transport contributions is especially simple in the case where the rate of reactant mass transport can be theoretically calculated.

Modern methods of calculation of mass-transport rates will be exemplified below by the case of reactions proceeding at a rotating-disc surface.

The discussion will be restricted to the system containing as the diffusing species a small admixture of solute in the liquid phase. Its concentration, i.e., the number of moles of the solute per unit liquid volume, is designated by c.

At equilibrium, the system should satisfy the conditions of thermodynamic equilibrium; also, the following conditions are assumed: the liquid is at rest; the temperature T, the pressure p, and the reactant concentration c are uniform throughout the liquid volume; finally, the external forces f are fully balanced.

Mass transport in the liquid can originate in various ways, e.g., from mechanical liquid motion, from an existing concentration gradient of the transported species (grad c), from a temperature (grad T) or pressure (grad p) gradient, or from external forces (f), etc.

For small deviations from equilibrium, the flux of the solute can be represented by a linear dependence on the above factors:

$$\mathbf{j} = c\mathbf{v} - \alpha_c \operatorname{grad} c - \alpha_T \operatorname{grad} T - \alpha_p \operatorname{grad} p + \alpha_f c\mathbf{f}, \qquad (2.1)$$

where $\mathbf{j}$ is the number of moles of the dissolved substance crossing an area of 1 cm^2 in 1 sec and α_c, α_T, α_p, and α_f are proportionality coefficients. The minus sign on the left side of Eq. (2.1) indicates that reactant particles are moving in a direction opposite to that of the respective gradients.

Except for § 7.6, the material following is concerned with iso-thermal systems, i.e., it is assumed that everywhere T = const. It is also assumed that the flux due to the pressure gradient (grad p) is small in comparison with the last term in Eq. (2.1). Actually, in stirred solutions a certain pressure drop always exists owing to the liquid motion. Usually, however, it is quite small.

TABLE 2.1. Diffusion Coefficients of Some Substances

Diffusing substance	$D \cdot 10^5$ (cm$^2 \cdot$ sec^{-1})
Acetic acid in water (12.5°C; 0.01 M)	0.91
Glycerin in water (10°C; 0.125 M)	0.63
Ethanol in water (15°C)	1.00
Silver in amalgam (16°C)	1.11
Gold in amalgam (11°C)	0.83
AgBr in molten salt (780°C)	4.92
Interdiffusion in the water vapor—air system (20°C)	$0.25 \cdot 10^5$

Taking into account the above limitations, Eq. (2.1) can be rewritten as follows:

$$\mathbf{j} = c\mathbf{v} - D\,\mathrm{grad}\,c + \alpha_f c\mathbf{f}. \qquad (2.2)$$

The value of the diffusion coefficient D introduced in Eq. (2.2) depends, in general, on the solution composition, temperature T, and pressure p. In sufficiently dilute solutions, D can be considered at constant T and p to be constant and independent of concentration c.† Values of D for some substances are given in Table 2.1.

The temperature dependence of the diffusion coefficient is usually expressed in the form

$$D = D_0 \exp\left(-\Delta U_d/RT\right), \qquad (2.3)$$

where D_0 is a constant factor, ΔU_d is the activation energy for diffusion, and R is the gas constant. The value of D varies little within a narrow temperature range (10°C). Variations of pressure p also have an insignificant effect on the diffusion coefficient. Therefore, it will be further assumed that D can be taken as constant.

An equation will now be derived which allows the distribution of concentration of the diffusing substance to be calculated. The

---

†Several cases where D varies with concentration are discussed below (§ 7.5).

mass balance equation for diffusing particles in a unit volume will first be considered. It is obvious that the change of the number of particles in this volume must be equal to the number of particles brought in by flux **j** through the surface of the volume,† i.e.,

$$\frac{\partial c}{\partial t} = -\operatorname{div} \mathbf{j}. \qquad (2.4)$$

Equations (2.2) and (2.4) combined with the continuity equation (1.1) give

$$\frac{\partial c}{\partial t} + (\mathbf{v}\,\operatorname{grad})\,c = \operatorname{div}\,(D\,\operatorname{grad} c - \alpha_f c\mathbf{f}). \qquad (2.5)$$

In the absence of external forces (**f** = 0), Eq. (2.5) has a particularly simple form:

$$\frac{\partial c}{\partial t} + (\mathbf{v}\,\operatorname{grad})\,c = D\Delta c. \qquad (2.6)$$

Relation (2.6) is the equation for convective diffusion of a substance dissolved in a moving liquid. It should be remembered that within the frame of previously made approximations, the flow rate of the liquid **v** is independent of mass-transport processes and is considered to be known from the solution of the hydrodynamic problem.

Equation (2.6) must be supplemented by suitable initial and boundary conditions. The form of the latter is determined by the kinetics of any processes occurring at the solid−solution interface.

§ 2.2. Theory of "Similarity" of

Mass-Transport Processes

It is useful to consider first some general aspects of the equation of convective diffusion (2.6). For simplicity, the stationary case ($\partial c/\partial t = 0$) will be discussed.

---

†If the substance takes part in a homogeneous bulk reaction, or if the volume contains a source of the particles in question, the removal or appearance of new particles on account of these processes must, of course, be taken into account.

Similarly to the procedure adopted in § 1.3, Eq. (2.6) is transformed using dimensionless parameters. In addition to the characteristic parameters of the hydrodynamic flux, l_0 and v_0, a characteristic concentration c_0 (which can be chosen as, e.g., bulk concentration c*) is introduced. If concentration is measured in c_0 units, i.e., by defining $c = c_0 C$, where C is a dimensionless concentration, Eq. (2.6) leads to

$$(\mathbf{V} \, \text{grad}) C = \frac{1}{\text{Pe}} \Delta C. \tag{2.7}$$

Differentiation of Eq. (2.7) is carried out with respect to the dimensionless distance $\mathbf{R}$ as in Eq. (1.24). The quantity Pe = $v_0 l_0 / D$ is the Peclet number.

First, the parameters which determine the value of concentration obtained from solution of Eq. (2.7) are considered. It has been shown in § 1.3 that the distribution of velocity $\mathbf{V}$ is a function of distance $\mathbf{R}$ and of the Reynolds number Re. Equation (2.7) contains an additional parameter, the Peclet number Pe. Therefore, the concentration distribution should depend on $\mathbf{R}$, Re, and Pe, i.e.,

$$C = C(\mathbf{R}, \text{Re}, \text{Pe}). \tag{2.8}$$

However, the Peclet number can be represented by the product of two dimensionless parameters

$$\text{Pe} = \frac{v_0 l_0}{D} = \frac{v_0 l_0}{\nu} \frac{\nu}{D} = \text{Re} \cdot \text{Sc},$$

where Sc = ν / D is called the Schmidt number.† Therefore, a more correct version of Eq. (2.8) is

$$C = C(\mathbf{R}, \text{Re}, \text{Sc}), \tag{2.9}$$

since it explicitly separates the hydrodynamic factor (Re) from the molecular characteristics of the medium (Sc).

---

† The Schmidt number Sc has been widely used in recent years in the mass-transport literature. Using this criterion, authors try to separate mass- and heat-transfer processes. Since, however, the laws of both processes are similar in most cases, the distinction between Sc and the heat number Pr = ν / χ (where χ is the thermal conductivity of the medium) is largely only a formal matter.

It can be seen from Eq. (2.9) that "similarity" conditions are here somewhat more complex than those for the purely hydrodynamic case. However, similar concentration distributions arise in two systems if the requirements of geometric similarity and equality of Reynolds numbers sufficient for hydrodynamic similarity of fluxes are met and complemented by the additional condition of equality of Schmidt numbers ($Sc_1 = Sc_2$).

When Sc = 1, any dependence of concentration on Sc disappears, and similarity of momentum and mass transport processes is observed (Reynolds analogy). Such systems include gases for which numerical values of D and ν are of the same order.

In liquids the diffusion and viscosity coefficient usually differ greatly. In the majority of cases, Schmidt numbers are much greater than unity. Therefore, even at relatively low flow rates (Re $\approx$ 1), the Peclet number (Pe = Sc $\cdot$ Re) can be a large quantity. The 1/Pe coefficient on the right side of Eq. (2.7) is hence very small. The Pe value determines the relative contributions of the convective and diffusional mechanism of transport processes.

When Pe < 1 (e.g., for mass transport processes in a slowly moving gas) the concentration distribution depends almost exclusively on the rate of molecular diffusion. This can easily be established in a formal way by inspecting Eq. (2.7) the left side of which becomes much smaller than the right when Pe < 1.

In liquids for which Pe > 1, the concentration distribution is qualitatively indicated by the following factors: in the bulk the distribution is fully determined by the convective stirring, and diffusion effects are negligible. Stirring results in a uniform concentration distribution in the bulk since, owing to the slow change of C, the right side of Eq. (2.7) is much smaller than the left one.

Close to the interface at which the heterogeneous reaction proceeds, the concentration varies in a very sharp manner. In the diffusion boundary layer δ_d, rates of molecular diffusion and of convective transport become commensurate. Moreover, with decreasing distance from the solid wall, the rate of convective stirring decreases, and the relative contribution of diffusion increases [both sides of Eq. (2.7) are then of the same order of magnitude].

The above picture served as a basis for the Nernst theory of heterogeneous processes [1].

According to the Nernst theory, the liquid can be divided in two regions: 1) a thin layer of thickness δ_N immediately adjacent to the solid surface and 2) the remaining part of the liquid. Inside the layer δ_N, the liquid is motionless and the mass transfer to the surface is by diffusion only. In the second region, stirring is so intensive that throughout the whole region the concentration is constant and equal to that in the bulk solution.

The limitations and the approximate character of the Nernst theory were clearly demonstrated by Levich [2, 2a] who also formulated a consistent quantitative theory of mass transfer in moving media. The basic shortcomings of the Nernst theory consist in that: 1) it totally ignores liquid motion (to and along the solid surface) inside the layer δ_N; 2) the theory assumes a sharp change of concentration profile at the boundary between the δ_N layer and the bulk of the solution, which cannot happen in reality; and 3) numerical values of δ_N in the Nernst theory cannot be calculated.

In spite of the obvious shortcomings, the Nernst theory has proved to have useful descriptive value and can often be utilized for qualitative interpretations of mass-transfer processes.

For comparison of mass-transfer data obtained in various systems, the rate of removal of solute from the surface can often be represented by a dimensionless mass-transfer coefficient, i.e., the diffusional Nusselt number Nu_d (or Sherwood number† Sh) or the diffusional Stanton number St. The coefficients depend on Re and Sc only. The Sherwood and Stanton numbers will be defined in § 2.4 where mass transport to the rotating disc surface is discussed.

The analogy between momentum and mass transport may also be extended to heat-transfer processes. When the heat released in viscous energy dissipation and the possible variations of density and of other characteristic properties of the medium accompanying the heat transfer are neglected, the equations describing mass transport fully coincide (within the accuracy of replacing concentration by temperature) with the equations of convective diffusion.

---

†The Sherwood number Sh has been introduced to stress the differences between mass- and heat-transfer processes. The formerly used Nusselt number Nu is commonly used now only in descriptions of heat-transfer processes.

The monograph by Frank–Kamenetskii [3] is recommended for more detailed information concerning similarity of mass-transport processes and the analogy between mass and heat transfer.

§ 2.3. Solution of the Equation
of Convective Diffusion to the
Surface of a Rotating Disc

One of the simplest examples of convective diffusion is mass transfer in systems for which heterogeneous reaction occurs at the surface of an infinite rotating disc.

It was shown in the solution of the hydrodynamic problem that, far from the surface of a rotating disc, liquid flows in the direction perpendicular to the surface of the disc. Reactant entrained by this flux compensates the amount that disappears at the disc surface in the heterogeneous process.

Consider steady-state mass transfer to the disc surface.

Under stationary conditions, the convective diffusion equation (2.6), written for cylindrical coordinates, has the form

$$u \frac{\partial c}{\partial r} + \frac{v}{r} \frac{\partial c}{\partial \varphi} + w \frac{\partial c}{\partial z} = D \left(\frac{\partial^2 c}{\partial z^2} + \frac{\partial^2 c}{\partial r^2} + \frac{1}{r} \frac{\partial c}{\partial r} + \frac{1}{r^2} \frac{\partial^2 c}{\partial \varphi^2} \right). \qquad (2.10)$$

Far from the disc, the concentration of the solute is equal to the bulk value

$$c \to c^* \quad \text{as} \quad z \to \infty. \qquad (2.11)$$

Then, assuming that the heterogeneous reaction results in the concentration of reactant c_S at the surface of the rotating disc, we have

$$c = c_S \quad \text{for} \quad z = 0. \qquad (2.12)$$

If the properties of the disc surface are uniform with respect to the heterogeneous reaction, c_S can be considered constant over the whole surface of the disc. Its value is determined by the mechanism of the heterogeneous reaction. The effect of various electrochemical mechanisms on the functional dependences of c_S will be discussed later.

   The axial symmetry causes concentration distribution to be independent of the azimuthal coordinate φ, i.e., $\partial c / \partial \varphi = 0$. For an infinite rotating disc, this distribution can be assumed independent of distance r from the rotation axis, i.e., $\partial c / \partial r = 0$. Since the boundary concentrations (2.11) and (2.12) as well as the normal component of velocity w [cf. Eq. (2.10)] do not depend on the coordinate r, c = c(z) is the solution of Eq. (2.10) satisfying both boundary conditions (2.11) and (2.12).†

   Introduction of a dimensionless concentration C,

$$C = \frac{c^{*} - c}{c^{*} - c_S},$$

(2.13)

reduces the problem of concentration distribution to one of solving the ordinary differential equation

$$w(z) \frac{dC}{dz} = D \frac{d^2 C}{dz^2}$$

(2.14)

with the boundary conditions

$$C = 1 \quad \text{for} \quad z = 0 \quad \text{and} \quad C \to 0 \quad \text{as} \quad z \to \infty.$$

(2.15)

It should be remembered that the form of the function w(z) is known from the solution of the hydrodynamic problem (§ 1.2).

   Integration of Eq. (2.14) gives

$$C = A_1 \int_0^z \exp\left\{\frac{1}{D} \int_0^\tau w(\eta)\, d\eta\right\} d\tau + A_2.$$

(2.16)

   Constants A_1 and A_2 can be determined from the boundary condition (2.15). After substitution, we obtain

$$C(z) = 1 - \frac{\int_0^z \exp\left\{\frac{1}{D} \int_0^\tau w(\eta)\, d\eta\right\} d\tau}{\int_0^\infty \exp\left\{\frac{1}{D} \int_0^\tau w(\eta)\, d\eta\right\} d\tau}.$$

(2.17)

---

†It must be mentioned that the relative simplicity of the expression obtained here for concentration distribution of the transferred substance results from the simple boundary condition (2.12) postulated for the disc surface. If the surface were nonuniform, a more complex relation would result (cf. Chap. 8).

Calculation of the integrals in Eq. (2.17) generally requires utilization of values of the axial velocity components listed in Table 1.1. However, in the case of convective diffusion in liquids (when Sc ≫ 1), expansions (1.14) and (1.15) can be used. Indeed, as has been shown in § 2.2 for high Schmidt numbers, the main variations of concentration of the reactant occur in a thin layer immediately adjacent to the disc surface. Values of the axial velocity component can be found for this region from Eq. (1.15).

It can therefore be assumed that the axial velocity component can be represented by

$$w(z) = \begin{cases} \sqrt{v\omega}\left[-\dfrac{0.51\omega z^2}{v} + \dfrac{\omega^{3/2}z^3}{3v^{3/2}} - \dfrac{0.616\omega^2 z^4}{6v^2} + \cdots\right] & \text{for } z < \delta_0 \ (2.18a) \\ -0.886\sqrt{v\omega} & \text{for } z > \delta_0 \ (2.18b) \end{cases}$$

where δ_0 is the thickness of the hydrodynamic boundary layer defined by (1.17).

Following Levich [2], the integral in the denominator of Eq. (2.17)

$$J = \int_0^\infty \exp\left\{\frac{1}{D}\int_0^\tau w(\eta)\,d\eta\right\} d\tau, \qquad (2.19)$$

can be rewritten in the following form:

$$J = J_1 + J_2 = \int_0^{\delta_0} \exp\left\{\frac{1}{D}\int_0^\tau w_1(\eta)\,d\eta\right\} d\tau +$$

$$+ \int_{\delta_0}^\infty \exp\left\{\frac{1}{D}\int_0^{\delta_0} w_1(\eta)\,d\eta + \frac{1}{D}\int_{\delta_0}^\tau w_2(\eta)\,d\eta\right\} d\tau,$$

where functions $w_1(\eta)$ and $w_2(\eta)$ are determined by expressions (2.18a) and (2.18b), respectively.

Considering

$$J_1 = \int_0^{\delta_0} \exp\left\{\frac{1}{D}\int_0^\tau w_1(\eta)\,d\eta\right\} d\tau$$

and taking into account the first term only of the series (2.18a), the integral J_1 can be transformed into

$$J_1 = \int_0^{\delta_0} \exp\left(-\frac{0.51\omega^{3/2}}{3Dv^{1/2}}\,\tau^3\right) d\tau =$$

$$= \frac{1}{3}\left(\frac{3Dv^{1/2}}{0.51\omega^{3/2}}\right)^{1/3} \int_0^{\frac{0.51\omega^{3/2}\delta_0^3}{3Dv^{1/2}}} \exp(-\eta)\,\eta^{-2/3}\,d\eta.$$

Then, substituting the value of δ_0 from Eq. (1.17), we obtain

$$J_1 = 0.602\,\frac{D^{1/3}v^{1/6}}{\omega^{1/2}} \int_0^{7,92\,\frac{v}{D}} \exp(-\eta)\,\eta^{-2/3}\,d\eta.$$

Since the integrand decreases rapidly with increasing η, the upper integration limit can be replaced by infinity; i.e.,

$$J_1 \approx 0.602\,\frac{D^{1/3}v^{1/6}}{\omega^{1/2}} \int_0^{\infty} \exp(-\eta)\,\eta^{-2/3}\,d\eta.$$

The integral in the above expression is the gamma-function [4]

$$J_1 = 0.602\,\frac{D^{1/3}v^{1/6}}{\omega^{1/2}}\,\Gamma(1/3) = 1.61\,\frac{D^{1/3}v^{1/6}}{\omega^{1/2}}.$$

Levich's calculations [2] demonstrated that for high Schmidt numbers the value of integral

$$J_2 = \int_{\delta_0}^{\infty} \exp\left[\frac{1}{D}\int_0^{\delta_0} w_1(\eta)\,d\eta + \frac{1}{D}\int_{\delta_0}^{\tau} w_2(\eta)\,d\eta\right] d\tau$$

is negligible in comparison with J_1, so that $J \approx J_1$.

The integral J (which has the dimensions of length) is called the thickness of the diffusion boundary layer δ_d at a rotating disc. Thus,

$$\delta_d = 1.61\,D^{1/3}v^{1/6}\omega^{-1/2}. \tag{2.20}$$

The thickness of the diffusion layer δ_d is an important quantity in the description of convective diffusion processes at a rotating disc and will be further discussed in detail.

The integral in the denominator of Eq. (2.17) can be calculated using the same approximation for $w(z)$. It can easily be shown that

$$\int_0^z \exp\left\{ \frac{1}{D} \int_0^\tau w(\eta)\, d\eta \right\} d\tau \approx 0.602 \frac{D^{1/3} \nu^{1/6}}{\omega^{1/2}} \int_0^{\frac{0.51\omega^{3/2}z^3}{3D\nu^{1/2}}} \exp(-\eta)\, \eta^{-2/3} d\eta.$$

Function $C(z)$ describing the concentration distribution of the diffusing species close to the disc surface is obtained in the form

$$C(z) = 1 - \frac{1}{\Gamma(1/3)} \int_0^{\frac{0.51\omega^{3/2}z^3}{3D\nu^{1/2}}} \exp(-\eta)\, \eta^{-2/3} d\eta = 1 - \frac{\gamma\left(1/3; \dfrac{0.51\omega^{3/2}z^3}{3D\nu^{1/2}}\right)}{\Gamma(1/3)}, \quad (2.21)$$

where $\gamma(1/3;\ 0.51\omega^{3/2}z^3/3D\nu^{1/2})$ is an incomplete gamma–function of the arguments shown. The definition of this function can be found in [4].

The function $C(z)$ is graphically presented in Fig. 2.1 (curve 1), where the dimensionless concentration C is plotted for convenience as a function of a dimensionless parameter $\xi = z/\delta_d$.

For comparison, the concentration profile obtained on the basis of the Nernst theory is plotted (curve 2) on the same graph.

Figure 2.1 shows that the main variations of concentration occur within the range of the diffusion boundary layer δ_d (at $z = \delta_d\, C(\delta_d) \sim 0.1$). It can easily be verified by comparing Eq. (2.20) with Eq. (1.17), which describes the thickness of the hydrody-

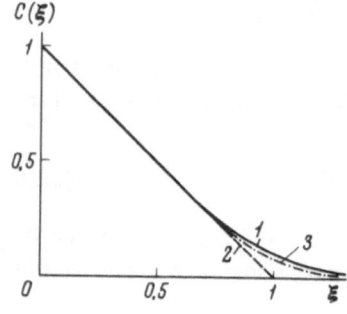

Fig. 2.1. Distribution of dimensionless concentration at the rotating–disc surface: 1) Eq. (2.21); 2) Eq. (2.42); 3) Eq. (2.53).

namic boundary layer, that

$$\delta_d/\delta_0 \approx 0.45\,(D/\nu)^{1/3}. \tag{2.22}$$

When convective diffusion takes place in water for which Sc = $\nu/D \approx 10^3$, the diffusion-layer thickness δ_d is only 5% of the δ_0 value. This shows that the approximations made in the choice of expressions (2.18a) and (2.18b) for the normal component of the liquid velocity, w(z), were valid.

When Sc = 10^3, the diffusion-layer thickness is equal to 0.015 cm at 1 rev·sec^{-1}, 0.005 cm at 10 rev·sec^{-1}, 0.0015 cm at 100 rev·sec^{-1}, and $5 \cdot 10^{-4}$ at 10^3 rev·sec^{-1}.

The concentration profile C(z) corresponding to Eq. (2.21) is independent of the distance from the rotation axis, and the diffusion-layer thickness δ_d is constant over the whole disc surface. This property of the rotating disc, usually referred to as uniform accessibility of its surface, results from two causes: first, from the characteristics of the hydrodynamic flow discussed in § 1.2; second, from the assumption of homogeneity of disc surface [i.e., of the condition (2.12) assumed above], which was that the rate of the heterogeneous reaction is constant over the whole surface of the disc.

In the case of deviations from the latter condition exemplified by a "coated" disc (§ 2.8) or by a rotating disc electrode with a ring (cf. Chap.8), the condition of uniform diffusional accessibility becomes invalid.†

Uniform accessibility is a property rarely encountered in other systems with forced convection. It therefore distinguishes conditions at the rotating disc from those in other hydrodynamic systems and simplifies considerably the analysis of mass transfer to the disc surface.

§ 2.4. Diffusional Flux

The flux of the substance to the rotating disc will now be calculated:

$$j = D\left(\frac{dc}{dz}\right)_S = -\,D\,(c^* - c_S)\left(\frac{dC}{dz}\right)_S. \tag{2.23}$$

---

† The above calculations are also invalid at the edges of the rotating disc where the surface homogeneity condition is naturally inapplicable.

Differentiation of Eq. (2.21) yields

$$j = \frac{D(c^* - c_S)}{\delta_d},$$

(2.24)

where δ_d is defined by Eq. (2.20).

A term which will often be encountered in the subsequent text is the "limiting diffusion flux" to the disc surface, j_d. It follows from (2.24) that the maximum flux corresponds to zero surface concentration ($c_S = 0$). Limiting current conditions arise in the case of fast heterogeneous reactions where all of the substance transported can be assumed to be immediately transformed at the disc surface by the fast electrode reaction. The limiting diffusion flux is

$$j_d = \frac{Dc^*}{\delta_d}.$$

(2.25)

When the reaction of each particle at the disc surface is accompanied by transfer of n electrons, the current density i at a rotating disc electrode is given by

$$i = nFj = \frac{nFD(c^* - c_S)}{\delta_d}.$$

(2.26)

Correspondingly, the limiting diffusion current density i_d is

$$i_d = \frac{nFDc^*}{\delta_d}.$$

(2.27)

If the heterogeneous reaction at the disc surface results in an increase of concentration of a species over that in the bulk (i.e., $c_S > c^*$), convective diffusion proceeds in the opposite direction. The species is transferred to the bulk solution. Equation (2.23) still obtains, with the opposite sign on the right side, indicating the reverse direction of the diffusional flux. It is obvious that the value of the limiting diffusion flux (for $c^* = 0$) is given in this case by

$$j_d = -\frac{Dc_S}{\delta_d}.$$

(2.28)

The total diffusion flux to a disc of radius r_0 is obtained simply by multiplication of the flux density j [Eq. (2.24)] by the area of the disc s, i.e.,

$$\bar{j} = js = \pi r_0^2 \frac{D(c^* - c_S)}{\delta_d} . \qquad (2.29)$$

The total current $\bar{i}$ passing at a rotating disc electrode is found in the same way. In particular, the limiting diffusion current is

$$\bar{i}_d = \frac{nFsDc^*}{\delta_d} , \qquad (2.30)$$

where $s(=\pi r_0^2)$ is the area of the rotating disc electrode.

Expression (2.29) for the total diffusional flux to a disc can be easily presented in terms of the diffusional Nusselt number Nu_d (or Sherwood number Sh). By definition [3], in the case considered,

$$Nu_d = \frac{\bar{j} r_0}{D \Delta c} , \qquad (2.31)$$

where $\Delta c = c^* - c_S$ is the concentration drop. Substituting in Eq. (2.31) the expressions for j (2.29) and δ_d (2.20), we find

$$Nu_d = 0.62\, Sc^{1/3}\, Re^{1/2}. \qquad (2.32)$$

It should be remembered that for a rotating disc $Re = r_0^2 \omega / \nu$, and the Stanton criterion St is related to the diffusional Nusselt number Nu_d by the following relation [3]:

$$St = \frac{Nu_d}{Sc\ Re} .$$

In the case of a rotating disc,

$$St = 0.62\, Sc^{-2/3}\, Re^{-1/2}. \qquad (2.33)$$

A comparison of the predictions of the relations obtained with experimental data should be preceded by the following remarks.

Calculations of the distribution of the transferred species were made assuming all terms except the first (proportional to z^2)

in expansion (2.18a) to be negligible. It has been mentioned that this approximation is valid for very large Schmidt numbers (Sc > 10^3) where concentration varies only within a very narrow distance from the disc surface and $\delta_d \ll \delta_0$. However, at lower Schmidt numbers, the region of varying concentration widens (δ_d increases). Therefore, calculations of concentration distribution then require a more accurate description of the hydrodynamic flow.

Gregory and Riddiford [5] calculated the second term (proportional to z^3) of the expansion (2.18a) and determined the integral J by numerical graphical integration for $2.5 \cdot 10^2 \leq Sc \leq 10^3$. These authors described the obtained J = δ_{dR} values by the following interpolation formula:

$$\delta_{dR} = 1.611 \, (D/\nu)^{1/3} (\nu/\omega)^{1/2} [1 + 0.3539 \, (D/\nu)^{0.36}]. \tag{2.34}$$

Taking into account further terms in the w(z) expansion results in changes of δ_{dR} value which do not exceed 1%.

Newman [6] calculated J values for a wider range of Schmidt numbers by expanding the integrand in Eq. (2.19) in a series and integrating the latter term by term. He obtained the following dependence of J = δ_{dN} on Sc:

$$\delta_{dN} = 1.611 \, (D/\nu)^{1/3} (\nu/\omega)^{1/2} [1 + 0.2980 \, (D/\nu)^{1/3} + 0.14514 \, (D/\nu)^{2/3}]. \tag{2.35}$$

It has been pointed out by Newman that Eq. (2.35) gives J values for Sc > 100 with an accuracy equal or better than 0.1%.

Recently Kassner [7] calculated the third term of the expansion $w_1(z)$, proportional to z^4. The results of his computer calculations coincide with those of Gregory and Riddiford [5] and Newman [6] (for Sc > 100). According to Kassner for Sc numbers in the range 4 to ∞, δ_{dK} can be represented by

$$\delta_{dK} = 1.611 \, (D/\nu)^{1/3} (\nu/\omega)^{1/2} \{1.1203 \, I + 0.6977 \, (D/\nu)^{1/3} \exp \, [-3.11 \, (\nu/D)]\}, \tag{2.36}$$

where the numerical values of I for various Sc values are given in Table 2.2. The terms in powers higher than 2 in the expansion of w(z) in Eq. (2.18a) result in a negligible correction [8].

Calculation of the convective flux to a rotating disc at still lower Schmidt numbers ($0 \leq Sc \leq 100$) was carried out in order

TABLE 2.2. Values of I Corresponding to Kassner [7]

$\frac{D}{\nu} = Sc^{-1}$	I	$\frac{D}{\nu} = Sc^{-1}$	I	$\frac{D}{\nu} = Sc^{-1}$	I
0	0.8934	0.010	0.9564	0.100	1.0368
0.001	0.9209	0.020	0.9747	0.110	1.0412
0.002	0.9286	0.030	0.9877	0.120	1.0451
0.003	0.9341	0.040	0.9981	0.130	0.0488
0.004	0.9385	0.050	1.0068	0.140	1.0521
0.005	0.9424	0.060	1.0143	0.150	1.0552
0.006	0.9457	0.070	1.0209	0.160	1.0580
0.007	0.9487	0.080	1.0268	0.180	1.0631
0.008	0.9515	0'.090	1.0321	0.200	1.0675
0.009	0.9541	—	—	0.250	1.0762

to interpret data concerning heat transfer between the fluid and the rotating disc.† The heat flux values calculated by various authors for given Prandtl numbers are collected in Spalding's monograph [9].

It can easily be seen by comparison of Eq. (2.36) with that obtained by Levich (2.20) that the difference in the value of the diffusion boundary layer calculated by Levich and the more recent values does not exceed 3% for Sc $\sim 10^3$. However, in the range $10 < Sc < 100$, this difference increases, reaching about 17% for low Sc values.

Solution of the convective diffusion equation at the surface of a rotating disc and calculation of the limiting current can be made numerically. Feldberg's review [10] includes a detailed description of a Fortran IV program, convenient for numerical modeling of diffusion-kinetic problems.

§ 2.5. Quantitative Verification of the Theory; Laminar Flow

The theoretical expression derived in § 2.4 for the diffusional flux to a rotating disc is in good agreement with the general laws of mass transfer established empirically under conditions of laminar flow. In particular, the dependences of Nu_d on Re and Sc (2.32) coincide with those experimentally obtained ([3]; cf. also bibliography in Appendix 1).

---

†For heat transfer in liquids, the Prandtl number Pr (analogous to Sc) is less than 100.

Studies of mass transfer in electrochemical systems allow an exact quantitative verification to be made of the expression for diffusional flux to the rotating disc electrode. Measurements of limiting diffusion currents can be made with high accuracy (the relative error can be as low as to 0.25% [11]); thus, not only can the dependence of the limiting current on Re and Sc be determined, but also the accuracy of calculations of the numerical coefficients in Eqs. (2.32) and (2.34) can be checked.

For the purpose of comparison with experiment, the expressions for limiting current density are conveniently rewritten using Eqs. (2.32) and (2.34) in the forms

$$i_d = 0.62 \, nFc^* D^{2/3} \nu^{-1/6} \omega^{1/2} \tag{2.37}$$

and

$$i_d = 0.621 \, nFc^* D^{2/3} \nu^{-1/2} \omega^{1/2} [1 + 0.3539 \, (D/\nu)^{0.36}]^{-1}. \tag{2.38}$$

It can be seen that limiting current is proportional to the square root of the angular velocity of the rotating electrode, to the bulk concentration of the reactant, and (within a certain accuracy) to the 2/3 power of the diffusion coefficient of the reactant [cf. Eq. (2.37)].

The theory was first verified experimentally by Siver and Kabanov [12] who measured the limiting diffusion current of hydrogen ions and oxygen using a rotating disc cathode; silver or copper discs, 2.5 cm in diameter, were used. For rotation speeds 0.5–50 rpm the Reynolds numbers varied within the range $5 \cdot 10^2$–$5 \cdot 10^4$. The limiting current was found to be proportional to the square root of the angular velocity (Fig. 2.2) and to the 2/3 power of the diffusion coefficient.

It should be mentioned that any verification of the theory of convective diffusion necessarily requires that the process occur under conditions of strictly laminar flow. This was overlooked in some older papers which therefore contain incorrect quantitative results (a critical review of this literature can be found in Levich's monograph [2]).

Furthermore, equations for convective diffusion were derived assuming negligible contribution from ionic migration. The ex-

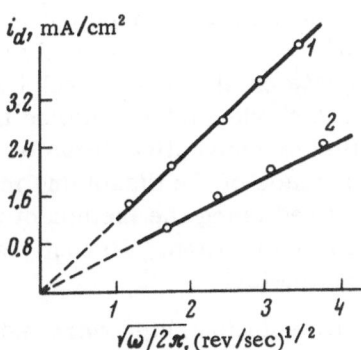

Fig. 2.2. Limiting current density of hydrogen evolution as a function of the square root of the angular velocity of the electrode [12]. 1) 0.1 N KCl; 2) 0.5 N H_2SO_4.

perimental data discussed in the present chapter were obtained under exactly these conditions (unless otherwise stated). Migration was suppressed by addition of a large excess of an inert salt the ions of which were the main current carriers.

Finally, it should be remembered that the above theory was developed for a disc of infinite radius. In the case of disc with finite dimensions, an edge diffusion effect must be expected (the diffusion flux density is higher at the edges). This results in current densities higher than calculated. However, since the width of the peripheral region is commensurate with the diffusion layer thickness δ_d, the effect is negligible for sufficiently large electrodes, for which $\delta_d(\omega, D) \ll r_0$.

A special study of the dependence of limiting current density on the radius of the rotating disc showed that for the usual range of rotation speeds deviations from theory due to edge effects are observed only in the case of very small electrodes whose radii do not exceed 0.5 mm (rotation speed 100–1000 rpm) [13].

Quantitative verification of the theory can be made in two ways. The first consists in comparison of the theoretically calculated values of diffusion flux with the experimental ones. The required values of the diffusion coefficient D and kinematic viscosity ν can be obtained for the system studied with sufficient accuracy by independent methods. Such a verification was made by Gregory and Riddiford [5, 14] who investigated dissolution of zinc and copper discs in iodine (potassium iodide and sulfuric acid) and potassium dichromate solutions, respectively. Discs 5.3 cm in diameter cut out from thin zinc or copper foil were mounted on a rotating axis. Rotation speed was varied between 70–300 rpm.

Dissolution was carried out in a cell 12.5 cm in diameter (solution volume 500 cm^3). Separately studied dependences of the dissolution rate on stirring, on solution concentration, and on the state of the electrode surface showed that the rate of the process is controlled by convective diffusion of triiodide or dichromate ions to the surface of the dissolving metal. Diffusion coefficients were measured using the method of nonstationary diffusion through a porous membrane. Viscosity was measured using an Ostwald viscosimeter.

Dissolution rates were measured by titration. The results can be presented in terms of the dependence of the quantity i_d/nFc^* on experimental conditions. It can be seen from Eq. (2.37) that this quantity represents the diffusion flux density of reacting particles when their bulk concentration is unity (referred to a transfer of 1 electron).

The dependence of i_d/nFc^* on the angular velocity of the electrode is shown in Fig. 2.3; circles correspond to experimental results. Experimental i_d values were reproducible to within 1%. The solid line corresponds to Eq. (2.38), and the dashed line to Eq. (2.37). It can be seen from Fig. 2.3 that the measured diffusional fluxes coincide, within the (very small) experimental error, with the values calculated according to Eq. (2.38). Similar results were obtained in the case of copper dissolution in sulfuric acid/ potassium dichromate solutions. One of the most accurate studies aimed at verification of the theory of convective diffusion to a rotating disc is that of triiodide reduction on platinum carried out by Newson and Riddiford [11]. These workers obtained well-defined

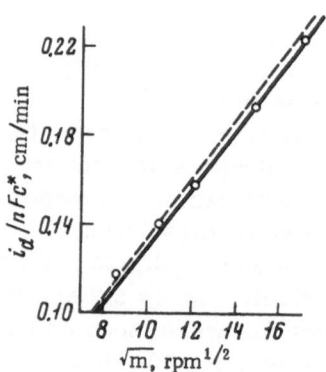

Fig. 2.3. Rate of zinc dissolution in 0.08 N KI + 0.02 N I$_2$ as a function of the square root of angular velocity of the disc [5].

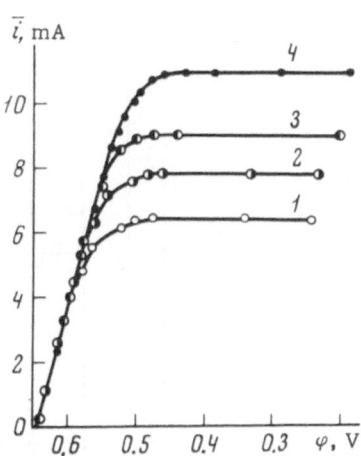

Fig. 2.4. Polarization curves of triiodide reduction at a platinum electrode for various rotation speeds of the disc [11]. Solution $2.47 \cdot 10^{-3}$ M KI_3 + 0.1 M KI (25°C); Rotation speed (rpm): 1) 66.7 ; 2) 100; 3) 133.3; 4) 200.

limiting currents for I_3^- ion reduction (Fig. 2.4) and compared them with values calculated using Eq. (2.38). The diffusion coefficient of I_3^- ions and the viscosity of the solutions were varied by changing the solution temperature as well as by addition of sucrose. Measurements were made using the same methods as in the earlier work of Gregory and Riddiford [5, 14]. The results are summarized in Tables 2.3-2.5 and exemplified in Fig. 2.5. The deviations between the theoretical and experimental data usually do not exceed 1%.

The experiments described in [11] were repeated by Blurton and Riddiford [15] using disc electrodes with various shapes of insulating coatings (cf. § 9.1 for further details), and later by other authors [16] for a wide range of Reynolds numbers (up to $3 \cdot 10^5$). Deviations of the measured limiting currents from values calculated according to Eq. (2.38) did not exceed 2%. Jahn [17] compared the measured limiting diffusion current of Ag^+ ions in KNO_3 with calculated values. The diffusion coefficient of Ag^+ ions was measured using the method of Cottrell with ±1% accuracy. Equation (2.38) was found to be in good agreement with experiment; currents calculated according to Eq. (2.37) were, however, found to be somewhat lower than the experimental ones. A similar result was obtained in studies of dissolution of benzoic acid (pressed in a disc form) in water [18].

The determination of diffusion coefficient in a rotating disc system can be used as a second method for verification of the

TABLE 2.3. Limiting Currents (mA) for Triiodide Reduction in 0.1 M KI at 25°C [11]

m, rpm	Triiodide concentration (mmoles/liter)					
	0.124	0.241	0.617	1.24	2.47	3.70
44.4	—	—	—	2.62 (2.62)	—	7.84 (7.85)
50	0.280 (0.278)	0.537 (0.539)	1.39 (1.39)	2.75 (2.78)	5.50 (5.54)	—
66.7	0.322 (0.321)	0.621 (0.624)	1.58 (1.60)	3.20 (3.21)	6.40 (6.40)	9.42 (9.61)
100	0.390 (0.393)	0.760 (0.764)	1.94 (1.96)	3.90 (3.93)	7.80 (7.84)	11.62 (11.77)
133.3	0.451 (0.454)	0.881 (0.882)	2.23 (2.27)	4.50 (4.54)	9.02 (9.05)	13.41 (13.59)
200	0.550 (0.556)	1.06 (1.08)	2.73 (2.77)	5.61 (5.56)	10.98 (11.08)	16.38 (16.65)
400	0.779 (0.785)	—	3.83 (3.92)	7.83 (7.85)	—	—

Note: Numbers in brackets correspond to calculated values.

TABLE 2.4. Limiting Currents (mA) for Triiodide Reduction in 0.1 M KI + 2.42 · 10^{-3} M KI_3 + Sucrose at 25°C [11]

m, rpm	Saccharose concentration (mmoles/liter)					
	0*	0.25	0.50	0.75	1.00	1.50
50	5.50 (5.54)	4.50 (4.50)	3.65 (3.65)	2.90 (2.89)	2.19 (2.24)	1.18 (1.17)
66.7	6.40 (6.40)	5.20 (5.20)	4.22 (4.22)	3.39 (3.34)	2.59 (2.58)	1.37 (1.35)
100	7.80 (7.84)	6.34 (6.38)	5.13 (5.16)	4.10 (4.09)	3.16 (3.17)	1.67 (1.65)
133.3	9.02 (9.05)	7.28 (7.26)	5.98 (5.96)	4.75 (4.72)	3.69 (3.66)	1.97 (1.90)
200	10.98 (11.08)	8.90 (9.01)	7.28 (7.31)	5.80 (5.78)	4.51 (4.48)	2.37 (2.33)

Note: Numbers in brackets correspond to calculated values.
*Triiodide concentration 2.47 mmole/liter.

TABLE 2.5. Limiting Currents (mA) for Triiodide
Reduction in 0.1 M KI at Various Temperatures [11]

m, rpm	Triiodide concentration, mmoles/liter				
	2.47	2.46	2.46	2.45	2.45
	25.0°C	30.0°C	34.9°C	39.8°C	44.5°C
50	5.50 (5.54)	5.98 (6.06)	6.52 (6.57)	7.10 (7.19)	7.74 (7.67)
66.7	6.40 (6.40)	6.96 (6.99)	7.54 (7.99)	8.20 (8.30)	8.86 (8.85)
100	7.80 (7.84)	8.50 (8.56)	9.30 (9.30)	10.00 (10.17)	10.86 (10.84)
133.3	9.00 (9.05)	9.80 (9.88)	10.71 (10.74)	11.56 (11.73)	12.41 (12.50)
200	10.98 (11.08)	12.00 (12.11)	13.10 (13.15)	14.16 (14.38)	15.26 (15.34)

Note: Numbers in brackets correspond to calculated values.

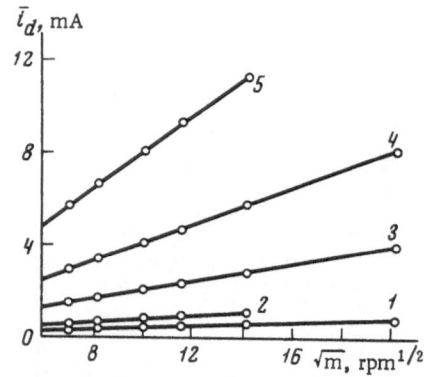

Fig. 2.5. Limiting currents for triiodide reduction
in 0.1 M KI as a function of the square root of
angular velocity at various concentrations of KI_3
[11]. KI_3 concentration (moles/liter): 1) 1.24 ·
10^{-4}; 2) 2.41 · 10^{-4}; 3) 6.14 · 10^{-4}; 4) 1.24 · 10^{-3};
5) 2.47 · 10^{-3}. Solid lines are calculated accord-
ing to Eq. (2.38).

theory. Data are obtained for various concentrations of an indifferent electrolyte and extrapolated to infinite dilution. The value
of the diffusion coefficient thus obtained for an infinitely dilute
solution, D_∞,† can be compared with the value calculated from
ionic mobility using the Einstein equation $D = (RT/F)\gamma_i$. Since

---

†The quantity D_∞ differs considerably from the value of the diffusion coefficient D
measured in the presence of excess inert electrolyte (often called the "polarographic
diffusion coefficient").

ionic mobilities have been measured with sufficiently high accuracy, D_∞ values thus calculated are very reliable. The measurement and evaluation of limiting currents carried out at low concentrations of inert electrolyte are quite difficult both experimentally (owing to the considerable ohmic drop in a poorly conducting solution) and theoretically, owing to the undefined contribution of ionic migration. Nevertheless, such measurements were carried out by Landsberg et al. [19, 20] and Jahn [17] for several ions.

Landsberg's experiments [19, 20] were carried out at a disc electrode 0.9 cm in diameter consisting of a pressed mixture of azobenzene with graphite (cf. § 9.1) rotating with a speed 1–16 rev·sec^{-1}. The anodic polarogram obtained in KI solutions containing an excess of KNO_3 at pH > 10 shows two waves corresponding to oxidation of iodide ions to mono- and pentavalent iodine. Since oxidation involves hydroxyl ions, the limiting current is determined in a certain pH range by diffusion of OH$^-$ ions to the electrode. The limiting current values served to determine the diffusion coefficient D of hydroxyl ions at various concentrations of the inert electrolyte (KNO_3). Results are presented in Fig. 2.6 as a plot of D vs square root of the ionic strength of the solution J_i. A linear dependence between D and $\sqrt{J_i}$ is observed; the extrapolated value of D_∞ (for $J_i = 0$) is practically identical with that calculated from the mobility of hydroxyl ions ($5.34 \cdot 10^{-5}$ cm^2·sec^{-1}). In Fig. 2.7 results are shown for measurements of the hydrogen ion diffusion coefficient [20] at various concentrations of the inert electrolyte (potassium chloride) (circles). Crosses correspond to D values obtained from i vs t curves [21, 22]. Extrapolation to infinite dilution results in a D_∞ value close to that calculated from the hydrogen-ion mobility ($9.3 \cdot 10^{-5}$ cm^2·sec^{-1}). A similar result was obtained by Jahn [17]. In the papers discussed above [17, 19, 20], diffusion coefficients were calculated from measured limiting currents according to Eq. (2.38).

Fig. 2.6. Diffusion coefficient of hydroxyl ions as a function of the square root of the ionic strength of the solution [19].

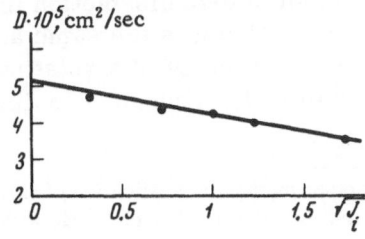

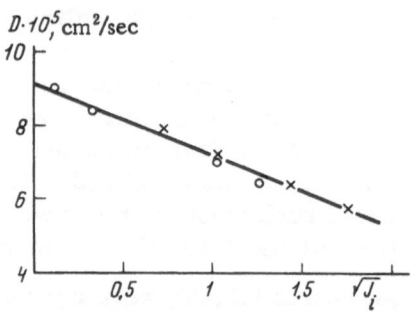

Fig. 2.7. Diffusion coefficient of hydrogen
ions as a function of the square root of the
ionic strength of the solution [20].

Summarizing the above results, it can be stated that the current is described with a high accuracy (±1%) by Eq. (2.38) if the
experimental system fulfills the conditions on which the calculations are based. It has already been mentioned that the differences between Eqs. (2.37) and (2.38) are small in the investigated
range and usually do not exceed the experimental errors. Therefore, a majority of authors still utilize the simpler expression
(2.37) for interpretation of experimental results. Precision measurements, however, require the use of Eq. (2.38). In the further
description of experimental results, the choice of the expression
for calculating limiting currents corresponds to that of the stated
author.

Special studies were made to verify the uniform accessibility
of a rotating disc. Beacom and Hollier [23] investigated copper
deposition from copper sulfate solutions at a copper disc cathode.
They showed (using interference microscopy) that the thickness
of the copper deposit and, consequently, the density of diffusion
flux are uniform over the whole electrode surface (accuracy of
measurements was ±5%). Similar results were obtained in a study
of tantalum disc dissolution in liquid tin at 1000–1200°C. The rate
of dissolution was the same at all points on the disc surface [7].
Thus, the surface of a rotating disc was unambiguously shown to
be uniformly accessible to diffusion.†

_____

†Another method of verification of uniform accessibility to the disc is discussed in
§8.8. It utilizes a disc electrode with a ring.

Direct measurements of the thickness of the diffusion layer can be made using laser interferometry [24].

Finally, some more subtle aspects of convective diffusion studies at a rotating disc will be discussed, but principally the characteristic form of the dependence of diffusion limiting current on the rotation speed of the electrode will be examined.

Simple proportionality between the limiting current density i_d and the square root of the angular velocity ω has been observed in a great majority of investigations. However, a few papers have appeared [25-27] reporting a linear relation of the type $i_d = a + b\sqrt{\omega}$. Apparently, in these cases, forced convection was considerably assisted by natural convection, which is particularly noticeable at low ω. The initial i_d vs $\sqrt{\omega}$ curve can then be easily approximated by a straight line which does not pass through the origin.[†]

The appearance of this type of relation may also be due to parallel occurrence at the electrode of two processes[‡] one of which is kinetically controlled. The rate of the diffusion-controlled process is described by $i_1 = b\sqrt{\omega}$. The rate of the second process, however, is independent of stirring: $i_2 = a = $ const. This is a very probable explanation since in a majority of cases, when upon extrapolation to $\omega \rightarrow 0$ a constant component of the limiting current is observed, the limiting current is poorly defined. Obviously, the limiting current is distorted by a "background current" increasing with overpotential, as opposed to the limiting diffusion current.

Still another explanation of the $i_d = a + b\sqrt{\omega}$ relation, based on the concept of macroscopic nonuniformity of the disc's surface, will be discussed in § 3.7.

An interesting property of diffusion-controlled kinetics at a rotating disc is that after the process of dissolution or deposition under laminar conditions is terminated, contours can often be seen on the surface, corresponding to the streamlines of liquid flow adjacent to the disc [29, 30]. As an example of this phenomenon, Fig. 2.8 shows a photograph of electrolytic copper deposits

---

[†]Additional stirring which causes deviations from linearity can be also due to intensive gas evolution at the electrode (e.g., in cathodic hydrogen evolution from acids at high current densities [28]).

[‡]See § 7.1 for further details.

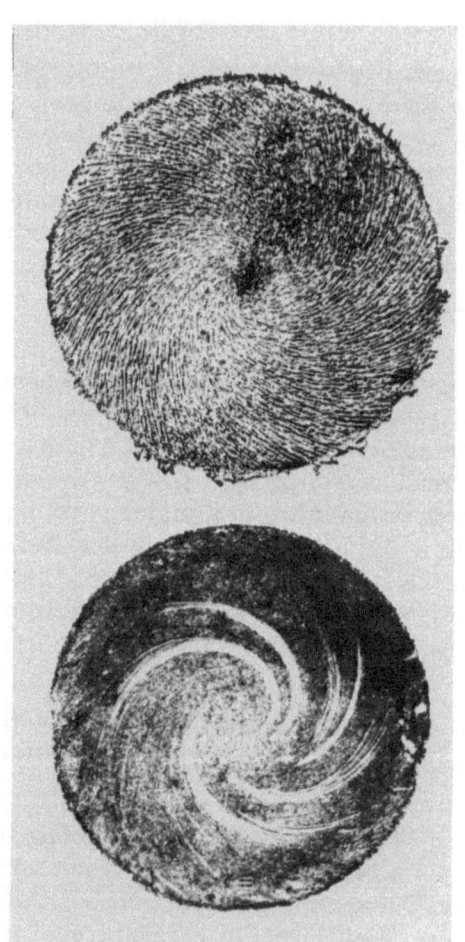

Fig. 2.8. Forms of copper deposits at the disc surface [29].

on a rotating disc cathode [29]. Similar figures — logarithmic spirals — were observed on discs pressed from an organic acid, after dissolution in aqueous solutions [31].

The appearance of these contours is rather puzzling in view of the uniform accessibility of the disc and of its axial symmetry. It has been shown [32] that the spiral contours result from the appearance of tiny protuberances on the disc's surface. These may be, for example, hydrogen bubbles evolving at the electrode surface during electrodeposition. The rate of metal deposition under the bubbles is somewhat lower than at the remaining surface and causes the appearance of spirals. This hypothesis has been confirmed [33] by a study of self-dissolution of a lead disc in nitric acid.

Appearance of spirals is sometimes connected with the transition from laminar to turbulent flow. At a certain critical angular velocity, the central part of the disc still remains under laminar conditions, whereas the peripheral part experiences turbulence. In this case, spirals are observed only within the ring separating the two regions [34, 35].

§ 2.6.   Measurements of Diffusion

Coefficients

The diffusion coefficient is an important physicochemical property of substances and is involved in a variety of calculations connected with diffusion processes. The diffusion coefficients of ions and molecules in solutions containing inert electrolytes are called the "polarographic" values since they are used in calculations of limiting currents in polarography (cf. footnote on p. 48). Diffusion coefficients can be determined using various electrochemical and nonelectrochemical methods. The former include, for example, polarography itself (with a dropping mercury electrode), evaluation of diffusion in capillaries (Cottrell's method [36]), and analysis of potentiostatic i vs t curves. The second group includes, among others, the diffraction micromethod [37] and the isotope tracer method. Although optical and physical methods give accurate results, they often require relatively sophisticated equipment.

Among electrochemical methods for determining diffusion coefficients, that using the rotating disc electrode stands out as

TABLE 2.6. Diffusion Coefficients ($D \cdot 10^6$, cm^2/sec)
Measured by Various Methods in 1 N KCl

Ion	Method			
	Rotating disc electrode [38]	Nonstationary capillary diffusion [41]	Polarography [41]	Diffraction [37]
Cd^{2+}	7.72	7.89	—	7.16
Tl^+	15.28	—	15.67	—
Pb^{2+}	8.95	9.17	—	8.06
Zn^{2+}	7.30	—	7.23	7.54

a relatively simple and highly accurate one. The exact solution of
the diffusion problem makes it superior to, for example, the usual
polarographic method utilizing an approximated Ilkovic equation
and requiring a number of empirical corrections. Finally, the
possibility of investigating anodic processes at solid electrodes
widely extends the number of electrochemical reactions which can
be used for determination of diffusion coefficients.

The results of the most precise measurements of diffusion
coefficients were discussed and compared with values calculated
from ionic mobilities in § 2.5. Below, further examples of the
application of a rotating disc to determinations of diffusion coeffi-
ents will be discussed.† The D value thus obtained will be compared with those found by classical methods.

Electroreduction of metals is usually studied at amalgamated
electrodes, the deposited metal being dissolved in mercury.
Otherwise, a loose deposit forms at the electrode surface and
causes additional stirring in the vicinity of the disc so that the
accuracy of limiting current measurements decreases.

The diffusion coefficient of Zn^{2+}, Tl^+, Cd^{2+}, and Pb^{2+} ions
were calculated from the limiting reduction currents of these ions
at an amalgamated copper electrode in 1 N KCl [38-40]. They are
presented in Table 2.6 together with D values measured by von
Stackelberg et al. [41] using Cottrell's method (nonstationary
capillary diffusion), or a polarographic method, as well as by
Hochstein [37] using the diffraction micromethod. Results obtained using the rotating disc and the capillary diffusion methods

---

†In the majority of papers reviewed below, the interpretation of experimental results
relies on Eq. (2.37).

agree within ±2%; somewhat larger differences arise with results obtained by the diffraction micromethod.

The diffusion coefficient of Ag^+ ions was measured by Kraichman and Hogge [42] in 0.2 N KNO_3 at 25°C: The value $D = 1.48 \cdot 10^5$ cm^2/sec is in very good agreement with the results of calculations made using the Onsager limiting law $(1.485 \cdot 10^{-5} \ cm^2 \cdot sec^{-1})$ [5]. Other authors [43, 44], using the same method, obtained $D = 1.53 \cdot 10^{-5} \ cm^2 \cdot sec^{-1}$. Polarographic data resulted in $D = 1.54 \cdot 10^{-5}$ $cm^2 \cdot sec^{-1}$ [41].

Nekrasov measured diffusion coefficients of mono- and divalent copper ions [45], obtaining $2.7 \cdot 10^{-6}$ (1 Na_2SO_4 and $6.6 \cdot 10^{-6} \ cm^2 \cdot sec^{-1}$ (1 N KCl), respectively, at 21°C, which may be compared with the value $D_{Cu^{2+}} = 6.3 \cdot 10^{-6} \ cm^2 \cdot sec^{-1}$ (0.1 N KCl) obtained using the capillary diffusion method [46].

A series of diffusion coefficients of simple and complex metal ions were determined using a rotating platinum or gold disc. The ions in question were silver in strongly alkaline solution [43], $AuCl_4^-$ and $PdCl_4^{2-}$ in 0.1 M $NaNO_3$ [47], Np ions [48], Ce^{4+} [49], and $HCrO_4^-$ anions [50]. The diffusion coefficient of ZnO_2^{2-} ions in NaOH solutions, measured at a rotating iron electrode, was in close agreement with that measured by diffusion through a porous membrane [51].

The diffusion coefficients of hydrogen ions were measured in a number of studies other than those mentioned in § 2.5. Stackelberg et al. [52, 53] obtained in 1 N KCl a value $7.5 \cdot 10^{-5} \ cm^2 \cdot sec^{-1}$ (23°C); under the same conditions Cottrell's method gives $7.3 \cdot 10^{-5} \ cm^2 \cdot sec^{-1}$ (25°C). Values of D_{H^+} were measured over a wide concentration range of background electrolyte in solutions of LiCl, NaCl, and $NaClO_4$ and are in good agreement with values obtained from isotopic determinations [54]. The diffusion coefficients of hydrogen ions are abnormally low [55-57] in aqueous sulfate solutions as well as in some nonaqueous solvents.

Gregory and Riddiford [5] used the experimental data of Hogge and Kraichman [58], obtained in 0.1 N KI solutions, to determine the diffusion coefficient of I_3^- ions. Their value, $1.15 \cdot 10^{-5} \ cm^2 \cdot sec^{-1}$ (26°C), is in good agreement with results obtained from measurements of diffusion through a porous membrane. Diffusion coefficients have also been measured in nonaqueous solutions (methanol,

ethanol, acetonitrile, dimethylformamide) [59–62]. The diffusion coefficients of I^+, I_2, and I^- were determined in the iodine–iodide system [19, 63], and those of Br^-, Br_3^-, and Br_2 in the bromine–bromide system [64].

Several values of diffusion coefficients have been obtained using electrooxidation reactions. Tedoradze [65], oxidizing Cl^- ions at a Pt microelectrode in 3.8 N H_2SO_4, obtained the value $D_{Cl^-} = 1.77 \cdot 10^{-5}$ $cm^2 \cdot sec^{-1}$ ($D_{Cl^-_\infty}$ calculated from mobility data = $1.9 \cdot 10^{-5}$ $cm^2 \cdot sec^{-1}$).

Manganate ion (MnO_4^{2-}) oxidizes to permanganate at a Pt anode with a sharply defined wave, which was used by Landsberg et al. [66] to calculate the diffusion coefficient of manganate ions in alkaline solutions. The diffusion coefficients of $Fe(CN)_6^{4-}$ ions in sulfate and chloride solutions [67] and in alkaline solutions [68–70] were measured by electrooxidation. The diffusion coefficients of positive and negative anthracene ions were measured at a rotating anthracene electrode [71].

The diffusion coefficients of neutral molecules can also be determined using a rotating disc electrode. Thus, Breiter and Hoffmann [72] measured the diffusion coefficients of molecular hydrogen in H_2SO_4 solutions over a wide temperature range (−10 to +60°C), and Lewis and Ruetschi [73] measured the diffusion coefficient of deuterium D_2. The diffusion coefficient of molecular oxygen was measured in KOH solutions [74, 75], while Stonehart has measured the diffusion coefficient of CO [76], and Arvia et al. that of SO_3 [77].

The diffusion coefficients of hydroquinone and quinone in Na_2SO_4 and KCl solutions were measured using a Pt electrode. Their values in 2 N KCl at 21°C are $0.91 \cdot 10^{-5}$ and $1.10 \cdot 10^{-5}$ $cm^2 \cdot sec^{-1}$, respectively, in good agreement with data obtained by the diffraction micromethod [78].

The diffusion coefficients of a series of organic compounds have been determined by their oxidation at platinum and carbon electrodes. They are in good agreement with results obtained using the iostopic and i vs curve methods [79–82].

The above results pertain to polarographic diffusion coefficients. Addition of inert electrolyte was necessary here since current measurements require high conductivity of the solution.

The diffusion coefficients of neutral species can be measured by the rotating disc method in the absence of inert electrolyte since no electrolytic migration transport is involved. Krichevskii and Tsekhanskaya [83, 84] measured the rate of dissolution of terephthalic acid in dilute solutions of ammonia and hexamethylenimine. The acid was pressed in pellets 1 cm in diameter mounted in a rotating steel holder. The dissolution rate was measured by the decrease in the weight of the pellet.† The rate is controlled by diffusion of ammonia or hexamethylenimine to the disc. The diffusion coefficients of dissolved species were calculated from the dissolution rate.

Similarly, rates of dissolution of benzoic acid were used to determine the diffusion coefficient of the latter in aqueous solutions of sucrose and glycerol (the acid concentration in solution was measured by spectrophotometry) [85]. The rates of dissolution of iron, nickel, cobalt, and other metals in molten copper served to determine their diffusion coefficients [86].

The rotating disc method can be applied also in molten electrolytes. For example, the diffusion coefficient of Ag^+ ions was determined in a molten eutectic mixture of sodium and potassium nitrates at 300°C. Its value $(0.945 \cdot 10^{-5} \ cm^2 \cdot sec^{-1})$ is in good agreement with that obtained from i vs t curves at a stationary electrode [87]. A series of papers by Delimarskii et al. [88, 89] is concerned with measurements of the diffusion coefficients of a number of metal ions (thallium, lead, cadmium, cobalt, zinc, bismuth, copper, and others) in molten eutectic mixtures ($LiNO_3 - KNO_3 - NaNO_3$, $LiCl - KCl$, etc.) using micro- and macro-Pt electrodes. The diffusion coefficients of some metals were measured in molten slugs of nonferrous metals ($CaO - SiO_2 - Al_2O_3$) at 1300-1470°C using iron and platinum electrodes with corundum insulation [90].

The above literature review demonstrates the very high accuracy which can be obtained in measurements of diffusion coefficients carried out by means of the rotating disc electrode. It should be remembered that the temperature coefficient of D (controlled by changes of solution viscosity) is of the order of 2% per degree centigrade at temperatures close to room temperature.

---

†The dissolution rate can also be measured optically by projecting the image of the dissolving pellet on a screen and measuring the decrease of its width [9, p. 162].

Accurate measurements of diffusion coefficients require thermo-
stated cells and an accuracy of 0.1°C in temperature control and
measurement.

§ 2.7. Concentration Measurements.
Analytical Applications of the
Rotating Disc Electrode

Equation (2.37) or (2.38) can be used for determination of the
concentration of the reacting substance if the diffusion coefficient
of the latter is known. No preliminary calibration involving a solution of known concentration is needed. Therefore, the rotating
disc electrode is a very convenient tool for absolute quantitative
analysis, particularly in the case of electrooxidation.

In addition to the reactions discussed in § 2.6 which may serve
both for determination of diffusion coefficients and of concentration of the reacting species, some other examples will be described. The oxidation of MnO_4^{2-} ion at a platinum electrode can be
used in the quantitative analysis of the ion within $5 \cdot 10^{-4}$-10^{-2} M
concentration range with a 1-2% accuracy. The presence in solution of MnO_4^- ion (in ninefold excess) has no effect on the accuracy
of measurements. This analytical method was used in determinations of manganate ion solubility and in studies of the mechanism
of permanganate decomposition in alkaline solutions [20, 59, 66,
91].

Landsberg et al. [92] proposed application of an anode constructed of a mixture of azobenzene, or diphenyl with graphite
(cf. § 9.1), for analytical determinations using electrooxidation.
The electrode does not passivate even at high anodic potentials.
Anodic oxidation waves of I^-, Br^-, NO_2^-, and ClO_2^- with well-
defined limiting currents can thus be obtained. It should be mentioned that oxidation of these ions (e.g., at a platinum electrode)
is accompanied by electrode passivation. Therefore, the current
measured is of a kinetic nature, so that polarograms obtained
using platinum anodes are not applicable for concentration measurements.

Limiting currents for Ag^+ ion reduction allow determination
of the solubility of Ag_2O in concentrated KOH solutions (the same
method was previously used to determine the diffusion coefficient
of silver ions [43]).

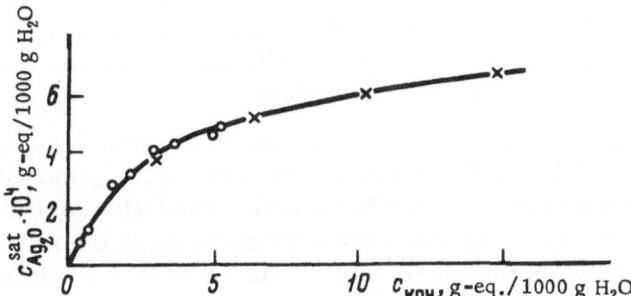

Fig. 2.9. Solubility of Ag_2O in KOH solutions. x) Determined using a rotating disc electrode [43]; O) determined by potentiometric titration [93].

The results of this study virtually coincide with the solubility value obtained by potentiometric titration [93] (Fig. 2.9). Similar values were obtained using radioactive indicators [94].

A platinum disc microelectrode was used by Bardin and Lalikov et al. [95] for analytical determination of gold, platinum and palladium, e.g., slimes and concentrates The same authors developed a method for determination of rhodium, osmium, and other metal ions using a platinum rotating microelectrode [96]. Kabanova proposed a method for determination of trace amounts of gold using gold and platinum electrodes [97].

Rotating platinum and nickel disc electrodes have been used in analytical determinations of a series of organic compounds [98], as well as in polarography in anhydrous liquid HF [99].

Finally, the rotating platinum disc microelectrode can be used to determine concentrations of Ag^+, Pb^{2+}, Cd^{2+}, Tl^+, Ni^{3+} ions in molten chlorides and of peroxide ions in molten nitrates [100].

If both the diffusion coefficient and concentration of the reacting species are known, Eqs. (2.37) and (2.38) can be used to determine the number of electrons n transferred in the heterogeneous reaction [101].

§ 2.8. Approximate Calculation
Methods for Convective Diffusion

Apart from the direct integration of the convective diffusion equation, approximate methods exist for calculating mass-transfer rates and are convenient for solving more complex problems of convective diffusion which do not easily lend themselves to direct integration of mass-transfer equations.

a. Nernst Model. We have discussed already in § 2.2 the basic physical concepts underlying the Nernst [1] model of heterogeneous processes. The mathematical formulation of this model applied to steady-state mass transfer consists in the following: within the diffusion layer δ_N mass transfer occurs by diffusion only, and the concentration distribution c of the transferred species is described by

$$D\frac{d^2c}{dz^2} = 0. \tag{2.39}$$

Intensive stirring maintains a constant concentration in the bulk:

$$c = c^* \quad \text{for} \quad z \gg \delta_N. \tag{2.40}$$

If, at the surface, a steady-state concentration c_S is established (the c_S value is determined by the specific rate of the heterogeneous reaction),

$$c = c_S \quad \text{for} \quad z = 0, \tag{2.41}$$

the concentration distribution within the layer is given by

$$c = c_S + (c^* - c_S)\, z/\delta_N,$$

while the dimensionless concentration C is

$$C = \frac{c^* - c}{c^* - c_S} = \begin{cases} 1 - z/\delta_N & \text{for} \quad 0 < z < \delta_N, \\ 0 & \text{for} \quad z > \delta_N. \end{cases} \tag{2.42}$$

Thus, within the diffusion layer δ_N, concentration varies linearly with distance from the surface.

According to (2.23), the diffusion flux density j is

$$j = -D(c^* - c_S)\left(\frac{dC}{dz}\right)_S = \frac{D(c^* - c_S)}{\delta_N}. \tag{2.43}$$

It should be mentioned again that the thickness of the diffusion layer δ_N cannot be calculated within the framework of the Nernst model.

Assuming that at a rotating disc $\delta_N = \delta_d$, and δ_d is described by expression (2.20), then the diffusion flux [Eq. (2.43)] based on the Nernst model is equal to j in Eq. (2.24).

The concentration distribution resulting from the Nernst model [Eq. (2.42)] is shown in Fig. 2.1 (curve 2). It can be seen that close to $z = \delta_d$ the Nernst concentration distribution differs considerably from the actual profile (curve 1).

b. Approximate Concentration Profile. Filinovskii and Kiryanov [102] developed an approximate method for solution of Eq. (2.14). It is used especially for solving nonstationary convective diffusion problems and is applicable to systems with high Sc values (Sc ≫ 1) only. It consists in the following.

The exact concentration distribution C(z) described by Eq. (2.21) is replaced by an approximate expression. The latter is chosen so that 1) the new distribution retains the character of changes of the former one and 2) both distributions coincide close to the surface of the disc. The latter requirement is of primary importance since the character of C(z) changes close to z = 0 and determines the value of the diffusional flux to the disc.

The choice of an approximate C(z) profile can obviously be made in many ways. We shall discuss that described in [102] and consider first, for simplicity, stationary diffusion.

The equation of convective diffusion is written below in dimensionless form, which will now be derived.

The dimensionless concentration C(z) was previously shown to be a function of a dimensionless ratio z/δ_d. The distance from the disc surface is therefore measured in units of thickness of the diffusion layer δ_d. It is convenient to assume for simplicity that the numerical value of δ_d is given by Eq. (2.20).† Thus, let

$$\xi = z/\delta_d. \qquad (2.44)$$

---

†Replacement of Eq. (2.20) by other expressions for δ_d [(2.34)–(2.36)] results only in a small change of numerical factors.

The concentration is again described by a dimensionless function

$$C(\xi) = \frac{c^* - c}{c^* - c_S}.$$

Substituting Eq. (2.44) and $C(\xi)$ into Eq. (2.14) leads to

$$\frac{d^2C}{d\xi^2} + a\xi^2 \frac{dC}{d\xi} = 0, \qquad (2.45)$$

where $a = 2.13$. Equation (2.45) was derived using Eq. (2.24) which describes δ_d, and only the first term (in z^2) was retained in the expansion representing the axial velocity component w (2.18a). The validity of this procedure for $Sc \gg 1$ has been discussed previously.

For concentrations c^* and c_S in the bulk and at the surface, respectively, the solution of Eq. (2.45) should satisfy the boundary conditions

$$C = 1 \quad \text{at} \quad \xi = 0, \qquad (2.46a)$$

and

$$C \to 0 \quad \text{as} \quad \xi \to \infty. \qquad (2.46b)$$

It has been shown in § 2.3 that the exact solution of Eq. (2.45) with the boundary conditions (2.46a) and (2.46b) has the following form:

$$C(\xi) = 1 - \frac{\int_0^\xi \exp\left(-at^3/3\right)dt}{\int_0^\infty \exp\left(-at^3/3\right)dt}.$$

It is easy to show that owing to the choice of the unit distance (δ_d) the integral in the denominator of the above expression is

$$\int_0^\infty \exp\left(-at^3/3\right)dt = 1. \qquad (2.47)$$

This simplifies the function $C(\xi)$ to

$$C(\xi) = 1 - \int_0^\xi \exp(-at^3/3)\, dt, \tag{2.48}$$

and the concentration distribution can be described by an incomplete gamma-function (2.21).

In order to obtain an approximate solution of Eq. (2.45), the function $C(\xi)$ will be sought in the form

$$C(\xi) = f(\xi) \exp(-a\xi^3/6). \tag{2.49}$$

Substituting Eq. (2.49) into (2.45), we obtain an equation which should be satisfied by the function $f(\xi)$:

$$\frac{d^2f}{d\xi^2} - \left(a\xi + \frac{a^2\xi^4}{4}\right)f = 0. \tag{2.50}$$

The boundary conditions (2.46a) and (2.46b) retain their original form.

At the disc surface (at $\xi = 0$) the solution of Eq. (2.50) is close to the solution of the equation

$$\frac{d^2f_0}{d\xi^2} - a\xi f_0 = 0. \tag{2.51}$$

The latter, being Airy's equation, has a solution in the form of Airy functions of the first and second kind [103]:

$$f_0 = A_1 Ai(a^{1/3}\xi) + A_2 Bi(a^{1/3}\xi). \tag{2.52}$$

With boundary conditions (2.46a) and (2.46b),† we have

$$C(\xi) \approx \frac{Ai(a^{1/3}\xi)}{Ai(0)} \exp(-a\xi^3/6). \tag{2.53}$$

---

†It should be mentioned that on account of the factor $\exp(-a\xi^3/6)$ both solutions of Airy's equation satisfactorily describe the behavior of $C(\xi)$ at $\xi \to \infty$. However, $Bi(a^{1/3}\xi)$ does not ensure the necessary behavior of $dC/d\xi$ at $\xi \to 0$.

The concentration distribution described by Eq. (2.53) is shown in Fig. 2.1 (curve 3). Comparison of expressions (2.48) [exact solution of Eq. (2.45) — curve 1] and (2.53) [approximate solution — curve 3] shows that when $\xi = 0$ both solutions coincide. Far from the disc, the concentration described by Ea. (2.53) decreases somewhat faster than that corresponding to Eq. (2.48).

The distribution equation (2.53) enables one to calculate the diffusional flux to the disc surface:

$$j = -\frac{D(c^* - c_S)}{\delta_d} \left(\frac{dC}{d\xi}\right)_S = -\frac{D(c^* - c_S)}{\delta_d} a^{1/s} \frac{Ai'(0)}{Ai(0)}. \qquad (2.54)$$

Substituting numerical values of $Ai'(0)$ and $Ai(0)$, we obtain

$$j = 0.94 \frac{D(c^* - c_S)}{\delta_d}. \qquad (2.55)$$

Thus, use of the approximate concentration profile results in a diffusional flux to the rotating disc which differs by 6% from the exact value (2.24). This is a rather large error in the description of the steady-state flux to the rotating disc. However, it will be shown later that the distribution (2.53) can be conveniently used in describing nonstationary mass-transfer processes.

It should be mentioned that taking into account further terms of expansion (2.18a) results in the change of a and in the appearance in Eqs. (2.45) and (2.50) of additional terms. In particular, $a = 2.25$ if δ_d is described by Eq. (2.28). The appearance of new terms in Eq. (2.50) has, however, little effect on the final result.[†]

Substituting the new value into Eq. (2.54), we obtain

$$j = 0.96 \frac{D(c^* - c_S)}{\delta_d}.$$

In this way, the difference between the approximate and exact solution is reduced to 4%.

This is apparently connected with the fact that the approximate concentration distribution described by Eq. (2.53) is somewhat closer to the actual distribution near a rotating disc.

[†]Terms containing higher powers of ξ were neglected in Eq. (2.51).

c. Integral Method of von Kármán–Pohlhausen. The integral method of mass-transfer calculations is a natural generalization of the integral method of von Kármán and Pohlhausen (cf. e.g., [104, 105]) developed originally for calculations of liquid flow within the boundary layer.

The basic concept consists in a replacement of the exact velocity distribution of liquid inside the boundary layer by an approximate distribution. The latter depends on a number of parameters chosen during the process of solving the problem. The von Kármán–Pohlhausen method has proved to be very fruitful for the solution of many problems connected with the theory of boundary layers. In particular, a similar method was used to calculate heat and mass transfer [104, 105]. In a subsequent section, results obtained by Olander for mass-transfer processes in the vicinity of a rotating disc surface [106] will be described.

It is convenient to consider once more the equation for convective diffusion near the surface of a rotating disc[†]:

$$u \frac{\partial c}{\partial r} + w \frac{\partial c}{\partial z} = D \frac{\partial^2 c}{\partial z^2}.$$ (2.56)

It has been shown in §1.2 that the radial (u) and normal (w) liquid velocity components can be represented by

$$u = r\omega F(\zeta) \quad \text{and} \quad w = \sqrt{\nu\omega} H(\zeta),$$ (2.57)

where

$$\zeta = z \sqrt{\frac{\omega}{\nu}}.$$ (2.58)

The u and w components are related by the continuity equation:

$$\frac{\partial u}{\partial r} + \frac{\partial w}{\partial z} = -\frac{u}{r}.$$ (2.59)

To be more definite, the problem of disc dissolution can be considered. It will be assumed that the central part of the disc $(r < r_{10})$ is "coated" and does not dissolve, whereas the peripheral

---

[†]Here the steady-state mass-transfer process is discussed and, as usually, radial diffusion is neglected.

part $(r > r_{10})$ can dissolve upon rotation; it is further assumed that the hydrodynamics of the process are not significantly affected by dissolution of the disc material.

The boundary conditions necessary for solution of Eq. (2.56) are chosen as

$$c \to 0 \quad \text{as} \quad z \to \infty \quad \text{at any } r, \tag{2.60a}$$

$$c = c_S \quad \text{at} \quad z = 0, \; r > r_{10}, \tag{2.60b}$$

$$c = 0 \quad \text{at} \quad r < r_{10} \text{ and at any } z. \tag{2.60c}$$

The boundary condition (2.60b) indicates that the dissolving part of the disc is exposed to the flow of pure solution.

Integration of Eq. (2.56) is carried out between the limits $z = 0$ and $z = z^*$, point z^* being chosen sufficiently far from the disc surface so that it is in the region of pure solution. Then, at $z = z^*$, $c_{z=z^*} = 0$ and $(\partial c/\partial z)_{z=z^*} = 0$. Integration of Eq. (2.56) gives

$$\frac{d}{dr}\int_0^{z^*} (uc)\, dz - \int_0^{z^*} c\left(\frac{\partial u}{\partial r} + \frac{\partial w}{\partial z}\right) dz - (wc)_S = -D\left(\frac{\partial c}{\partial z}\right)_S. \tag{2.61}$$

The right side of Eq. (2.61) includes the unknown diffusion flux to the surface of the rotating disc, $-D(\partial c/\partial z)_S$.

Using Eq. (2.61) together with the continuity equation (2.59) and the condition $w_{z=0} = 0$ (the latter condition states that the disc surface is impermeable to the liquid, and that convective flux does not affect hydrodynamics), we obtain the result

$$\frac{d}{dr}\int_0^{z^*} (uc)\, dz + \int_0^{z^*} \left(\frac{uc}{r}\right) dz = -D\left(\frac{\partial c}{\partial z}\right)_S. \tag{2.62}$$

Equation (2.62) can be transformed into

$$r\frac{d}{dr}\int_0^{z^*} \left(\frac{uc}{r}\right) dz + 2\int_0^{z^*} \left(\frac{uc}{r}\right) dz = -D\left(\frac{\partial c}{\partial z}\right)_S. \tag{2.63}$$

Replacing the variable z in Eq. (2.63) by ζ using Eq. (2.58), we obtain the upper integration limit

$$\zeta^* = z^* \sqrt{\omega/\nu}$$

and transform Eq. (2.63) into

$$r \frac{d}{dr} \int_0^{\zeta^*} \left(\frac{uc}{r} \right) d\zeta + 2 \int_0^{\zeta^*} \left(\frac{uc}{r} \right) d\zeta = - \frac{D\omega}{\nu} \left(\frac{\partial c}{\partial \zeta} \right)_s. \qquad (2.64)$$

The radial velocity component u in Eq. (2.64) is known from the solution of the hydrodynamic problem. According to § 1.2, u is described by

$$u = 0.51 r \omega \zeta. \qquad (2.65)$$

Substituting Eq. (2.65) into Eq. (2.64) gives

$$r \frac{d}{dr} \int_0^{\zeta^*} c \zeta d\zeta + 2 \int_0^{\zeta^*} c \zeta d\zeta = - \frac{1}{0.51 \, \mathrm{Sc}} \left(\frac{\partial c}{\partial \zeta} \right)_s. \qquad (2.66)$$

The diffusion process occurs at the active part of the disc only ($r > r_{10}$), and it is convenient therefore to introduce a dimensionless distance:

$$R = \frac{r}{r_{10}} - 1. \qquad (2.67)$$

From Eqs. (2.66) and (2.67), we obtain

$$(1 + R) \frac{d}{dR} \int_0^{\zeta^*} c \zeta d\zeta + 2 \int_0^{\zeta^*} c \zeta d\zeta = - \frac{1}{0.51 \, \mathrm{Sc}} \left(\frac{\partial c}{\partial \zeta} \right)_s. \qquad (2.68)$$

Equation (2.68) is equivalent to the original equation (2.56). Both expressions can be used to calculate concentration c(r, z) in the vicinity of the rotating disc. However, using Eq. (2.56), it is necessary to solve a differential equation containing partial derivatives. This requires a rather complex method which will be described in § 8.1.

Equation (2.68) offers the possibility of finding the concentration distribution c(r, z) in another way. It is assumed initially that any arbitrary concentration distribution obtains but that this distribution depends on some parameter. This parameter will be chosen in such a way that the postulated concentration distribution satisfies the equation and the boundary conditions (2.60a)–(2.60c) (for details, see [105]).

It is also assumed that mass transfer occurs within a thin liquid layer immediately adjacent to the surface. Its thickness ζ_d, can vary with distance from the rotation axis. The concentration profile of the transferred species within the ζ_d layer will be presented as a polynomial

$$c\,(r,\ z) = a_1 + a_2\,(\zeta/\zeta_d) + a_3\,(\zeta/\zeta_d)^2 + a_4\,(\zeta/\zeta_d)^3. \qquad (2.69)$$

Using the boundary conditions (2.60a)–(2.60c) and additional considerations concerning the functional dependence of c near $\zeta = 0$ and $\zeta = \zeta_d$ [106], it is possible to find coefficients $a_1, a_2, a_3,$ and a_4 and obtain the following expression:

$$c(r, \zeta) = c_S\left[1 - \frac{3}{2}\left(\frac{\zeta}{\zeta_d}\right) + \frac{1}{2}\left(\frac{\zeta}{\zeta_d}\right)^3\right]. \qquad (2.70)$$

After substitution of Eq. (2.70) into Eq. (2.68) and integration (replacing the upper limit $\zeta*$ by ζ_d†), an ordinary differential equation for determining the parameter ζ_d is obtained:

$$(1 + R)\,\zeta_d\,\frac{d}{dR}\,(\zeta_d^2) + 2\zeta_d^3 = \frac{15}{0.51\,Sc}. \qquad (2.71)$$

Solving this equation [106] using the additional condition that $\zeta_d = 0$ at $R = 0$ leads to

$$\zeta_d^3 = \frac{15}{2 \cdot 0.51\,Sc}\left[1 - \frac{1}{(1 + R)^3}\right]. \qquad (2.72)$$

It is easy now [using the concentration profile (2.70)] to find the expression describing the diffusional flux to the dissolving part of the disc:

$$j = -\,D\left(\frac{\partial c}{\partial z}\right)_S = \begin{cases} 0 & \text{for } r < r_{10} \\ \dfrac{Dc_S}{\zeta_d}\left(\dfrac{3}{2}\sqrt{\dfrac{\omega}{\nu}}\right) = \dfrac{Dc_S}{\delta_d} & \text{for } r \geqslant r_{10}. \end{cases} \qquad (2.73)$$

The thickness of the diffusion boundary layer in Eq. (2.73) is

$$\delta_d = \frac{2}{3}\left(\frac{\nu}{\omega}\right)^{1/2} \zeta_d. \qquad (2.74)$$

---

†According to the assumptions made, c = 0 beyond the limits of the ζ_d layer.

Substitution of ζ_d from Eq. (2.72) gives

$$\delta_d = 1.63 \left(\frac{\nu}{\omega}\right)^{1/2} \mathrm{Sc}^{-1/3} \left(1 - \frac{r_{10}^3}{r^3}\right)^{1/3}. \tag{2.75}$$

According to Eq. (2.75), the thickness of the diffusion boundary layer increases gradually beyond the limits of the "coated" part of the disc. The active part turns out in this case to be nonuniformly accessible. This results from the fact that only pure solution flows to the active sections of the disc surface closest to the "coated" part. Therefore, diffusion-controlled dissolution of the disc material is faster in the vicinity of the "coated part." At the edges (at $r \gg r_{10}$) the thickness of the diffusion boundary layer tends to a constant value

$$\delta_{d\infty} = 1.63 \left(\frac{\nu}{\omega}\right)^{1/2} \mathrm{Sc}^{-1/3}. \tag{2.76}$$

The above value is in good agreement with the thickness of the diffusion boundary layer at a uniform disc calculated previously (§ 2.3). It can be assumed that at distances $r \approx 2r_{10}$, $\delta_d \approx \delta_{d\infty}$. Thus, if the uniformity condition is withdrawn, effects of nonuniformity extend over the disc surface to distances commensurate with the dimensions of the nonuniform surface.

If reactivity of the disc is uniform over the whole surface ($r_{10} = 0$), the thickness of the diffusion boundary layer is also uniform and equal to $\delta_{d\infty}$.

Consider now dissolution of a ring of radius r_{10} and width Δr. The density of the diffusion flux is again given by Eq. (2.73). The total flux of the substance diffusing away from the dissolving ring is

$$\bar{j}_{\mathcal{R}} = j_d \int_{r_{10}}^{r_{10}+\Delta r} \frac{2\pi r\, dr}{(1 - r_{10}^3/r^3)^{1/3}}, \tag{2.77}$$

where

$$j_d = Dc_S/\delta_{d\infty}. \tag{2.78}$$

After integration, we obtain

$$\bar{j}_{\mathcal{R}} = j_d\, \pi r_{10}^2 \left[\left(\frac{r_{10} + \Delta r}{r_{10}}\right)^3 - 1\right]^{2/3}. \tag{2.79}$$

For sufficiently narrow rings ($\Delta r \ll r_{10}$), Eq. (2.79) can be simplified:

$$\overline{j}_{\mathscr{R}} \approx j_d 3^{3/2} \pi r_{10}^{4/3} (\Delta r)^{2/3}. \qquad (2.80)$$

Thus, the total flux is proportional to the 4/3 power of the ring radius r_{10} and to the 2/3 power of the ring width Δr. A ring-shaped active surface results in higher diffusion flux densities than those for a uniform disc. In the latter case, the total flux from an identical ring is

$$\overline{j}_{\mathscr{D}} = j_d \cdot 2\pi r_{10} \Delta r.$$

The ratio

$$\overline{j}_{\mathscr{R}}/\overline{j}_{\mathscr{D}} = \frac{3}{2}^{2/3} \left(\frac{r_{10}}{\Delta r}\right)^{1/3} \qquad (2.81)$$

characterizes the "enhanced effectivity" of a ring electrode. Since $\Delta r \ll r_{10}$, $\overline{j}_{\mathscr{R}} \gg \overline{j}_{\mathscr{D}}$. It follows also from Eq. (2.81) that the ratio $\overline{j}_{\mathscr{R}}/\overline{j}_{\mathscr{D}}$ increases with increasing distance of a ring (of constant width) from the disc's center.

The following peculiarity of the diffusion flux equation (2.73) will now be considered. At the limits of the active part of the disc surface (near $r = r_{10}$), the diffusion flux described by Eq. (2.73) becomes infinite. This is connected with the absence of diffusional limitations near $r = r_{10}$. The real value of the flux in this region can be obtained only after the kinetics of the heterogeneous reaction itself have been taken into account. The effect of the latter on the diffusion flux to a partly blocked disc surface has been discussed by Rosner [107].

Finally, it should be mentioned that the solution of the above problem using Eq. (2.56) is given in Levich's monograph [2]. Results obtained by the integral method are in good agreement with Levich's calculations.

The experimental verification of Eq. (2.79) has been carried out in several systems using a rotating ring (or equivalently, a disc with a "coated" central part). Kabanov and Nikiforova [2, p. 316] studied oxygen reduction at a platinum ring; Bruckenstein et al. [108], reduction of copper ions at a copper ring; Gregory

TABLE 2.7. Reduction of I_3^- Ions in 0.1 N KI at a Platinum
Ring Electrode [110]

$r_{10}/(r_{10} + \Delta r)$	0	0.1	0.2	0.3	0.4
$\dfrac{\bar{i}_{\mathscr{R}}}{i_d \pi r_{10}^2}$	1.00	1.00 (1.00) [1]	0.99 (1.00)	0.97 (0.96)	0.96 (0.96)

$r_{10}/(r_{10} + \Delta r)$	0.5	0.6	0.7	0.8	0.9
$\dfrac{\bar{i}_{\mathscr{R}}}{i_d \pi r_{10}^2}$	0.92 (0.92)	0.85 (0.85)	0.74 (0.76)	0.60 (0.62)	0.42 (0.42)

and Riddiford [14, 109], dissolution of a copper ring in sulfuric
acid solutions of potassium dichromate; and Daguenet and Robert
[110, 111], iodine reduction at a platinum ring; the ratio $r_{10}/\Delta r$
was varied in these experiments. In [14], viscosity and diffusion
coefficients were additionally varied by change of temperature.
In all cases, good agreement of the measured limiting currents
with the values calculated according to Eq. (2.79) was reported,
as exemplified by the data given in Table 2.7.

§ 2.9. Diffusion Flux under
Turbulent Conditions

The theory of mass transfer in a turbulent flux follows from
the same concepts which underlie the semiempirical Prandtl −
von Kármán theory and which were briefly discussed in § 1.4. The
analogy between mass and momentum transfer allows one to assume that turbulent velocity fluctuations result in additional mass
transfer. The intensity of this process can be characterized by
a coefficient of turbulent diffusion D_{turb}. The total mass−transfer
rate in turbulent flux is thus described by

$$j = -(D + D_{turb}) \frac{dc}{dz}. \qquad (2.82)$$

Since the D_{turb} value depends on the intensity of turbulent stirring, i.e., on the same mechanism which results in the appearance
of turbulent viscosity ν_{turb}, it can be assumed that $D_{turb} = \nu_{turb}$.

The functional dependence of ν_{turb} on distance z from the solid
wall depends, in turn, on the rate of damping of turbulent eddies

as discussed earlier in § 1.4. It has been mentioned previously that fully satisfactory agreement with experimental data can be reached using the four-layer Landau–Levich model for turbulent flux near a solid wall.† A discussion of other functional dependences of D_{turb} can be found in the paper by Loitsyanskii [112].

Integration of expressions (1.28) and (2.82) between the limits z = 0 to a point z = ∞ located in the developed flux region gives

$$\omega r = \frac{\bar{\tau}_{z\varphi}}{\rho \nu} \int_0^\infty \frac{dz}{1 + A(z)} , \qquad (2.83a)$$

$$c^* - c_S = -\frac{j}{D} \int_0^\infty \frac{dz}{1 + Sc \cdot A(z)} . \qquad (2.83b)$$

Integration was carried out assuming v = rω at z = 0 and v → 0 as z → ∞, and c = c_S at z = 0 and c = c* as z → ∞. The function A(z) = ν_{turb}/ν describes changes of turbulent viscosity near a solid wall.

From Eqs. (2.83a) and (2.83b), the relation between the mass-transfer rate and one of the components of shear stress can be easily obtained:

$$j = -\frac{D(c^* - c_S)}{\rho \nu \omega r} \, \bar{\tau}_{z\varphi} \, \frac{\int_0^\infty [1 + A(z)]^{-1}dz}{\int_0^\infty [1 + Sc \, A(z)]^{-1}dz} . \qquad (2.84)$$

It has been mentioned in § 1.4 that generally function A(z) has a rather complex form. It can be assumed that close to the solid surface the function can be expressed by

$$A(z) = A_0 (z/\delta_0)^{1/\gamma}. \qquad (2.85)$$

where A_0 is a numerical factor and δ_0 is the thickness of the hydrodynamic boundary layer.

---

†A detailed exposition of the four-layer model of turbulent diffusion can be found in Levich's [2] and Frank-Kamenetskii's [3] monographs.

After substitution of Eq. (2.85) into Eq. (2.84) and change of variables in the integrands, Eq. (2.84) is transformed into

$$j = - \frac{D(c^* - c_S)}{\rho v \omega r} \bar{\tau}_{z\varphi} Sc^\gamma. \tag{2.86}$$

The quantity $\bar{\tau}_{z\varphi}$ can be calculated in the process of solving the hydrodynamic problem. Von Kármán [113, 114] obtained the following expression:

$$\bar{\tau}_{z\varphi} \approx - 0.0227\rho \, (r\omega)^{7/4}(v/\delta_0)^{1/4}, \tag{2.87}$$

where δ_0 is given by Eq. (1.30).

Combining Eqs. (1.30), (2.87), and (2.86), we obtain

$$j \approx 0.024 \frac{D(c^* - c_S)}{r} \left(\frac{r^2\omega}{v}\right)^{4/5} Sc^\gamma. \tag{2.88}$$

The flux of the solute, $\bar{j}$, to the rotating disc of radius r_0 is derived by integration of Eq. (2.88):

$$\bar{j} = 2\pi \int_0^{r_0} jr\,dr \approx 0.02\pi D(c^* - c_S)\, r_0 \, Re^{4/5} Sc^\gamma, \tag{2.89}$$

where $Re = r_0^2 \omega/v$ is the Reynolds number.

Results of experimental studies of mass-transfer rates in a turbulent flow regime are usually presented in terms of characteristic numbers. Equation (2.89) can be easily expressed in this form:

$$Nu_d = \frac{\bar{j}r_0}{\pi r_0^2 D(c^* - c_S)} \approx 0.02\, Re^{4/5} Sc^\gamma. \tag{2.90}$$

Following the assumption of Levich and Landau that damping of turbulent eddies in a narrow sublayer occurs according to a 4th power law [i.e., $1/\gamma = 4$ in Eq. (2.85)], the diffusional Nusselt number may be represented as

$$Nu_d \approx 0.02 \, Re^{0.8} Sc^{0.25}. \tag{2.91}$$

Equation (2.91) was derived with several simplifying assumptions. Therefore, the equation has basically only a qualitative character. It should be mentioned that relations similar to (2.91) have been derived by several authors using various assumptions and approximations. Summarizing their results, Eq. (2.91) can be presented in the general form

$$Nu_d \approx M\, Re^\beta\, Sc^\gamma. \tag{2.92}$$

The quantities M, β, and γ will now be discussed in more detail.

It has been shown above that γ characterizes the rate of damping of turbulent eddies in the viscous sublayer. According to the Landau–Levich theory, confirmed by the conclusions of other authors [3, 112], the value $\gamma = 0.25$ is based on the most solid physical background (especially for high Sc values). Nevertheless, the value $\gamma = 1/3$ (and consequently a cubic law of damping of turbulent eddies in the viscous sublayer) is often used (e.g., [110, 115]). It should be stressed that experimental determinations of γ are extremely useful for establishing the detailed structure of the turbulent boundary layer. Thus, mass-transfer data can throw additional light on hydrodynamics of turbulent fluxes.

The Landau–Levich theory results in $\gamma \approx 1$ for low Sc (Sc ~ 1) values. Similar values are obtained in calculations of heat transfer to a rotating disc [116, 117].

The following values of β were reported. The majority of authors [114, 116-118] obtained $\beta = 0.8$. Levich's [2] assumption of a logarithmic concentration profile in the turbulent boundary layer resulted in $\beta = 0.9$. Kaufmann [115] calculated $\beta = 0.78$; Kishinevskii [119], $\beta = 0.92$. The value of the numerical factor M depends on the properties of the hydrodynamic system and varies over a wide range.

Relatively few papers of an electrochemical nature involve experimental studies of mass transfer to a rotating disc in the turbulent flow regime. The first experimental verification of the equation of convective diffusion was carried out by Bagotskaya [120] who measured the limiting diffusion current for cathodic reduction of molecular oxygen and of hydrogen ions at a copper electrode. The limiting currents were approximately proportional to the angular velocity of the rotating electrode.

More accurate measurements were carried out by Daguenet [110, 121] who studied the reduction of I_3^- ions in KI solution at a platinum electrode. His experimental results can be presented in the form $Nu_d \sim Re^{0.9} Sc^{0.33}$.

According to Kishinevskii et al. [122, 123], $Nu_d \sim Sc^{0.5}$ (measurements of the dissolution rate in water of benzoic acid pressed in the form of disc). It should be mentioned, however, that the accuracy of the experimental measurements is hitherto insufficient for reliable determinations of the value of γ [cf. Eq. (2.92)].

The numerical factor M was measured by Krichevskii and Tsekhanskaya [31, 84]. They demonstrated the constancy of M over a wide range of Re and Sc numbers (dissolution of terephthalic acid in aqueous solutions of ammonia, triethylamine, and hexamethylenimine).

Transition from a laminar to a turbulent regime at a well-centered disc with a smooth surface begins at $Re_{cr} = 2.7 \cdot 10^5$ [124]. Initially, only the periphery of the disc is involved (a laminar flow regime is still maintained at the central part of the disc [34]). With increasing angular velocity, the flow regime becomes practically turbulent over the whole surface. The range of Re numbers corresponding to the transition from laminar to turbulent regime increases with decreasing Sc number (as can be seen from Fig. 2.10). As may be expected, the transition to the turbulent regime occurs more sharply at a ring than at a disc (the transition Re region is very narrow) [125, 126].

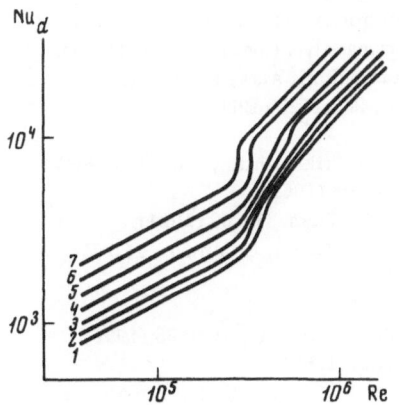

Fig. 2.10. Dependence of Nu_d on Re observed in the reduction of KI_3 in 0.1 N KI at a platinum disc [110]. Sc values: 1) 345; 2) 510; 3) 775; 4) 1212; 5) 1980; 6) 3440; 7) 6450.

Axial and radial wobbles of the disc cause additional stirring of the solution, and transition to a turbulent regime occurs at lower Reynolds numbers. The same effect results from any roughness of the disc surface: Re_{cr} decreases with increasing dimensions of microprotuberances [111].

Finally, experiments should be mentioned which involve a more complex character of liquid flow in the vicinity of the electrode, e.g., in the case of a disc rotating in suspensions of solid particles in aqueous solutions [127-129].

References

1.   W. Nernst, Z. Phys. Chem., 47:52 (1904).
2.   V. G. Levich, Physicochemical Hydromechanics, Prentice Hall, Inc. (1962).
2a.   V. G. Levich, Acta Physicochim. URSS, 17:257 (1942); Zh. Fiz. Khim., 18:335 (1944).
3.   D. A. Frank-Kamenetskii, Diffusion and Heat Transfer in Chemical Kinetics, Nauka, Moscow (1967).
4.   E. Jahnke, F. Emde, and F. Lösch, Special Functions, Nauka, Moscow (1964).
5.   D. P. Gregory and A. C. Riddiford, J. Chem. Soc., 1956:3756.
6.   J. Newman, J. Phys. Chem., 70:1327 (1966).
7.   T. F. Kassner, J. Electrochem. Soc., 114:689 (1967).
8.   A. S. Moshkevich, and A. A. Ravdel, Zh. Fiz. Khim., 43:71 (1970).
9.   D. B. Spalding, Convective Mass Transfer, Energiya, Moscow—Leningrad (1965), p. 59.
10.   S. W. Feldberg, in: Electroanalytical Chemistry, Vol. 3, A. J. Bard, ed., Marcel Dekker, New York (1969), p. 199.
11.   J. D. Newson and A. C. Riddiford, J. Electrochem. Soc., 108:695 (1961).
12.   Yu. G. Siver and B. N. Kabanov, Zh. Prikl. Khim., 22:53 (1948); 23:428 (1949).
13.   S. Azim and A. C. Riddiford, Analyt. Chem., 34:1023 (1962).
14.   D. P. Gregory and A. C. Riddiford, J. Electrochem. Soc., 107:950 (1960).
15.   K. F. Blurton and A. C. Riddiford, J. Electroanalyt. Chem., 10:457 (1965).
16.   C. A. Emergy and H. E. Hintermann, Electrochim. Acta, 13:127 (1968); M. Daguenet and J. Robert, Compt. Rend., 262:1125 (1966).
17.   D. Jahn, Dissertation, Bohn (1961).
18.   M. Kh. Kishinevskii and G. B. Denisova, Zh. Prikl. Khim., 37:1544 (1964).
19.   W. Geibler and R. Landsberg, Z. Chem., 1:308 (1961).
20.   R. Landsberg, W. Geissler, and S. Müller, Z. Chem., 1:169 (1961).
21.   M. Stackelberg and M. Pilgram, Coll. Czech. Chem. Commun., 25:2974 (1960).
22.   V.S. Bagotskii, Zh. Fiz. Khim., 22:1466 (1948).
23.   S. E. Beacom, and N. R. Hollier, J. Electrochem. Soc., 109:495 (1962).
24.   R. N. O'Brien, J. Electrochem. Soc., 114:710 (1967).
25.   S. V. Gorbachev and I. I. Aryamova, Trudy MKhTI im. D. I. Mendeleeva, 32:5 (1961); S. V. Gorbachev and V. A. Belyaeva, Zh. Fiz. Khim., 35:2158

(1961); 36:229 (1962); L. P. Kholpanov and S. V. Gorbachev, Zh. Fiz. Khim.,
36:855, 859, 1074 (1962); 38:3016 (1964); Yu. A. Korostelin and S. V.
Gorbachev, Zh. Fiz. Khim., 40:2324 (1966); L. I. Ivanovskaya and S. V.
Gorbachev, Zh. Fiz. Khim., 37:2305 (1963); Trudy MKhITI im. D. I. Men-
deleeva, 44:59 (1963); K. I. Yuodkaizis, V. L. Yasinskaite, and R. M.
Vishomirskis, Tr. Akad. Nauk LitSSR, Ser. B, 3(54):3 (1968).

26.   L. A. Poluboyartseva, P. I. Zarubin, and V. M. Novakovskii, Zh. Prikl.
       Khim., 36:1264 (1963).

27.   V. N. Boronenkov, O. A. Esin, and F. M. Shurygin, Élektrokhimiya, 1:592 (1965).

28.   G. P. Dezider'ev and S. I. Berezina, Dokl. Akad. Nauk SSSR, 130:1270 (1960);
       Izv. Kazan. Fil. Akad. Nauk SSSR, Ser. Khim. Nauk, 3:41 (1957); Proceedings
       of the Fourth Conference on Electrochemistry, Izd. Akad. Nauk SSSR, Moscow
       (1959), p. 672.

29.   N. Ibl, Galvanotechnik Oberflächenschutz, 4:265 (1963).

30.   Z. A. Solov'eva, L. N. Solobkova, and A. T. Vagramyan, Élektrokhimiya,
       6:590 (1970); R. Piontelli, B. Rivolta, and G. C. Razzini, Electrochim. Metal,
       4:218 (1969).

31.   I. R. Krichevskii and Yu. V. Tsekhanskaya, Dokl. Akad. Nauk SSSR, 122:258
       (1958).

32.   G. T. Rogers and K. J. Taylor, Nature, 200:1062 (1963).

33.   A. A. Ravdel and G. N. Gorelik, Zh. Prikl. Khim., 37:275 (1964).

34.   V. I. Eremenko and Ya. V. Natanzon, Zh. Fiz. Khim., 42:398 (1968).

35.   N. Gregory, J. T. Stuart, and W. S. Walker, Phil. Trans. Roy. Soc., London,
       Ser. A. Math. Phys. Sci., 248:155 (1955).

36.   F. G. Cottrell, Z. Phys. Chem., 42:385 (1902).

37.   Ya. N. Gokhshtein, Zh. Fiz. Khim., 30:1584 (1956); 31:404 (1957).

38.   V. D. Yukhtanova, Dokl. Akad. Nauk SSSR, 120:137 (1958); Zh. Fiz. Khim.,
       35:2778 (1961).

39.   A. I. Fedorova, G. L. Vidovich, L. I. Boguslavskii, and V. D. Yukhtanova,
       Proceedings of the Fourth Conference on Electrochemistry, Izd. Akad. Nauk
       SSSR, Moscow (1959), p. 665.

40.   M. Stackelberg, Rev. of Polarography, 5:133 (1957).

41.   M. Stackelberg, H. Pilgram, and V. Toome, Z. Elektrochem., 57:342 (1953).

42.   M. Kraichman and E. Hogge, J. Phys. Chem., 59:986 (1955).

43.   Yu. V. Pleskov and B. N. Kabanov, Zh. Neorg. Khim., 2:1807 (1957).

44.   R. R. M. Johnston and M.Spiro, J. Phys. Chem., 71:3784 (1967).

45.   L. N. Nekrasov, Dissertation, Moscow State University (1961).

46.   D. I. Macero and C. L. Rulfs, J. Amer. Chem. Soc., 81:2942 (1959).

47.   M. B. Barbin, Tu. S. Lyalikov, and V. S. Temyanko, Uch. Zap. Kishinev,
       Univ., 56:99 (1960); Yu. S. Lyalikov, M. B. Barbin, R. M. Novik, and V. S.
       Temyanko, Analele Romino-Sovietice, Ser. Khim., 14(3):30:193 (1959).

48.   C. E. Plock, J. Inorg. Nucl. Chem., 30:3023 (1968).

49.   R. Greef and H. Aulich, J. Electroanalyt. Chem., 18:295 (1968).

50.   E. Budevskii and S. Toshev, Izv. Inst. Fiz. Chem. Bulgar. Akad. Nauk,
       Otdel. Geol., Geograf. i Khim. Nauk, 1:183 (1960).

51.   M. D. Zholudev and V. V. Stender, Ukr. Khim. Zh., 23:200 (1957).

52.   M. Stackelberg, W. Vielstich, and D. Jahn, Anal. Real. Espanol. Fis. Quim.
       B56:475 (1960).

53.   W. Vielstich and D. Jahn, in: Advances in Polarography, I. Langmuir, ed., Oxford, Pergamon Press (1960), p. 281.

54.   D. Glietenberg, A. Kutschker, and M. Stackelberg, Ber. Bunsenges, 72:562 (1968).

55.   J. A. Olabe and A. J. Arvia, Electrochim. Acta, 14:785 (1969).

56.   R. M. Vishomirskis and Yu. Yu. Matulis, Proceedings of the Fourth Conference on Electrochemistry, Izd. Akad. Nauk SSSR, Moscow (1959), p. 86.

57.   E. A. Aykazyan and A. I. Fedorova, Dokl. Akad. Nauk SSSR, 86:1137 (1952).

58.   E. A. Hogge and M. B. Kraichman, J. Amer. Chem. Soc., 76:1431 (1954).

59.   R. Landsberg and S. Müller, Wiss. Z. Hochschule für Chemie Leuna – Merseburg, 3:319 (1960-1961).

60.   S. Minc, J. Sobkowski, and K. Kijowska, Roczn. Chem., 40:1575 (1966).

61.   M. C. Giordano and A. J. Arvia, Extended Abstracts, 19th CITCE Meeting, Detroit (1968), p. 149.

62.   V. A. Macagno, M. C. Giordano, and A. J. Arvia, Electrochim. Acta, 14:335 (1969).

63.   P. Beran and S. Bruckenstein, Anal. Chem., 40:1044 (1968).

64.   T. Iwasita and M. C. Giordano, Electrochim., Acta, 14:1045 (1969).

65.   G. A. Tedoradze, Zh. Fiz. Khim., 33:129 (1959).

66.   R. Landsberg, S. Müller, and J. Hendel, J. Electroanalyt. Chem., 2:400, 484 (1961); R. Landsberg and R. Thiele, Chem. Techn., 15:627 (1963).

67.   D. Jahn and W. Vielstich, J. Electrochem. Soc., 109:849 (1962).

68.   J. C. Bazan and A. J. Arvia, Electrochim. Acta, 10:1025 (1965).

69.   A. J. Arvia, S. L. Marchiano, and J. J. Podesta, Electrochim. Acta, 12:259 (1967).

70.   A. J. Arvia, J. C. Bazan, and J. S. W. Carrozza, Electrochim. Acta, 13:81 (1968).

71.   F. Lohmann and W. Mehl, Ber. Bunsenges., 71:493 (1967).

72.   M. Breiter and K. Hoffmann, Z. Elektrochem., 64:462 (1960).

73.   G. P. Lewis and P. Ruetschi, J. Phys. Chem., 67:65 (1963).

74.   R. E. Davis, G. L. Horwatch, and C. W. Tobias, Electrochim. Acta, 12:287 (1967).

75.   Z. Zembura and H. Kolny, Roczn. Chem., 39:1921 (1965); 41:1629 (1967).

76.   P. Stonehart, J. Electroanalyt. Chem., 15:239 (1967).

77.   J. C. W. Carrozza, H. A. Garrera, and A. J. Arvia, Electrochim. Acta, 14:205 (1969).

78.   E. A. Aykazyan and Yu. V. Pleskov, Zh. Fiz. Khim., 31:205 (1957).

79.   L. S. Reishakhrit et al., Vestn. LGU, Ser. Fiz. Khim., No. 22, p. 131 (1963); No. 16, pp. 124, 158 (1964); No. 22, p. 127 (1964); No. 4, p. 146 (1965); No. 22, p. 150 (1966); Élektrokhimiya, 5:721 (1969).

80.   I. A. Tserkovnitskaya, N. S. Borovaya, and M. M. Ankubinova, Vestn. LGU, 16(3):149 (1968).

81.   T. A. Miller, B. Lamb, K. Prater, J. K. Lee, and R. N. Adams, Anal. Chem., 36:418 (1964).

82.   Z. Galus, C. Olson, H. Y. Lee, and R. N. Adams. Anal. Chem., 34:164 (1962).

83.   I. R. Křichevskii and Yu. V. Tsekhanskaya, Zh. Fiz. Khim., 30:2315 (1956).

84.   I. R. Krichevskii and Yu. V. Tsekhanskaya, Zh. Fiz. Khim., 33:2331 (1959).

85.   T. B. Denisova and M. Kh. Kishinevskii, Tr. Kishinev, Politekhn. Inst., 5:13 (1966).

86.   P. M. Shurygin and V. D. Shantarin, Zh. Fiz. Khim., 42:463 (1968).

87.   N. G. Chovnyk and V. V. Vashchenko, Zh. Fiz. Khim., 35:580 (1961); J. E. Bowcott and B. A. Plunkett, Electrochim. Acta, 14:883 (1969).

88.   Yu. K. Delimarskii and G. V. Shilina, Dokl. Akad. Nauk Ukr. SSR, 1964:770; Élektrokhimiya, 1:532 (1965); Ukr. Khim. Zh., 30:1045 (1964); Electrochim. Acta, 10:973 (1975).

89.   I. D. Panchenko et al., Zh. Anal. Khim., 18:920 (1963); Ukr. Khim. Zh., 29:1164 (1963); 31:1203 (1865); 32:1296 (1966); Élektrokhimiya, 2:529 (1966).

90.   V. N. Boronenkov, O. A. Esin, and P. M. Shurygin, Dokl. Akad. Nauk SSSR, 151:872 (1963); Zh. Fiz. Khim., 38:1148 (1964); Yu. M. Sizov and O. A. Esin, Élektrokhimiya, 2:974 (1966).

91.   R. Landsberg and P. Örgel, Chem. Techn., 13:665 (1961).

92.   R. Landsberg and W. Geissler, Mitteilungsblatt der Chem. Ges. DDR, Sonderheft, 1960, Anal. Chem. S. 301; S. Schwarzer and R. Landsberg, J. Electroanal. Chem., 14:339 (1967).

93.   H. L. Johnston, F. Cuta, and A. B. Garrett, J. Amer. Chem. Soc., 55:2311 (1933).

94.   L. D. Kovba, N. A. Balashova, Zh. Neorgan. Khim., 4:225 (1959).

95.   M. B. Barbin, Yu. S. Lyalikov, and V. S. Temyanko, Zh. Anal. Khim., 14:24, 677 (1959); Izv. Vuzov. Khim. i Khim. Tekhnol., No. 4, p. 503 (1959); M. B. Barbin, V. S. Temyanko, M. M. Mikhailova, and T. A. Manuilova, in: Theory and Practice of Polarographic Analysis, Kishinev (1962), p. 212.

96.   Yu. S. Lyalikov and O. M. Mukhamednazarova, Izv. Akad. Nauk Turkm. SSR, Ser. Fiz.-Tekh., Khim. i Geol. Nauk, No. 5, p. 45 (1963); Izv. Akad. Nauk Mold. SSR, Ser. Biol. i Khim. Nauk, 1964:38; M. B. Barbin and V. P. Goncharenko, Zh. Fiz. Khim., 38:2626 (1964); M. B. Barbin and V. I. Shapiro, Uch. Zap. Kishinev, Univ., 68:64 (1964).

97.   O. L. Kabanova, Zh. Anal. Khim., 17:796 (1962).

98.   Yu. I. Veinstein, I. N. Palant, L. N. Yakhontov, D. M. Krasnokutskaya, and M. V. Rubtsov, Tr. IREA, 30:329 (1967); L. S. Marcoux and R. N. Adams, Anal. Chem., 39:1898 (1967); N. E. Khomutov, C. N. Skomayakova, and I. O. Poluyanova, Tr. MKhTI im. D. I. Mendeleeva, 48:17 (1965); V. D. Bezuglyi and Yu. I. Beilis, Zh. Anal. Chem., 20:1000 (1965).

99.   E. A. Aykasyan, N. M. Arakelyan, and S. E. Isavekyan, Izv. Akad. Nauk Arm. SSR, Khim. Nauk, 17:131 (1964).

100.   Yu. K. Delimarskii, I.D. Panchenko, and G. Ya. Shilina, Coll. Czech. Chem. Commun., 25:3061 (1960); N. G. Chovnyk et al., Ukr. Khim. Zh., 32:454, 595 (1966); P. G. Zambonin and J. Jordan, J. Amer. Chem. Soc., 91:2225 (1969); P. G. Zambonin, J. Electroanalyt. Chem., 24:365 (1970).

101.   W. T. Tiedemann and D. N. Bennion, J. Electrochem. Soc., 117:203 (1970).

102.   V. Yu. Filinovskii and V. A. Kiryanov, Dokl. Akad. Nauk SSSR, 156:1412 (1964).

103.   A. Erdelyi, Asymptotic Expansions, Dover, New York (1961).

104.   H. Schlichting, Boundary Layer Theory, McGraw-Hill, New York (1968).
105.   L. G. Loitsyanskii, in: Transactions of an All-Union Conference on Theoretical
       and Applied Mechanics, Izd. Akad. Nauk SSSR, Moscow—Leningrad (1962), p. 145.
106.   D. Olander, Chem. Eng. Sci., 18:123 (1963); 19:275 (1964).
107.   D. Rosner, J. Electrochem. Soc., 113:624 (1966).
108.   D. T. Napp, D. C. Johnson, and S. Bruckenstein, Anal. Chem., 39:481 (1967).
109.   A. C. Riddiford, J. Electrochem. Soc., 108:610 (1961); N. Ibl, J. Electro-
       chem. Soc., 108:610 (1961).
110.   M. Daguenet, J. Heat Mass. Transfer, 11:1581 (1968).
111.   M. Daguenet and J. Robert, J. Chim. Phys. Phys.-Chim. Biol., 65:1668 (1968).
112.   L. G. Loitsyanskii, in: Transaction of an All-Union Conference on Theotetical
       and Applied Mechanics, Izd. Akad. Nauk SSSR, Moscow—Leningrad (1962),
       p. 145.
113.   Th. v. Kármán, Z. Angew. Math. Mech., 1:224 (1921).
114.   L. A. Dorfman, Hydrodynamic Drag and Heat Transfer of Rotating Bodies,
       Fizmatgiz, Moscow (1960).
115.   T. G. Kaufmann and E. F. Leonard, AIChE Journal, 14:421 (1968).
116.   E. C. Cobb and O. A. Saunders, Proc. Roy. Soc., 236:343 (1956).
117.   J. P. Hartnett, S. Tsai, and H. N. Jantscher, Teploperedacha, No. 3, p. 47
       (1965).
118.   D. R. Davies, Quart. J. Mech. Appl. Math., 12(2):151 (1959).
119.   M. Kh. Kishinevskii, in: Processes of Chemical Technology, Nauka, Moscow—
       Leningrad (1965), p. 160.
120.   I. A. Bagotskaya, Dokl. Akad. Nauk SSSR, 85:1057 (1952).
121.   M. Daguenet, I. Epelboin, and M. Froment, Compt. Rend., 258:3694 (1964).
122.   G. B. Denisova and M. Kh. Kishinevskii, in: Processes of Chemical Tech-
       nology, Nauka, Moscow (1965), p. 165.
123.   V. P. Popovich and M. Kh. Kishinovskii, Mass Transfer Processes in Chemical
       Technology, 3rd ed., Khimiya, Moscow (1968), p. 11.
124.   M. Daguent and J. Robert, J. Chim. Phys., 1967:395.
125.   M. Daguenet and J. Robert, Compt. Rend., Ser. C, 264:161 (1967).
126.   M. Daguenet, I. Epelboin, and J. Vanhaecht, Compt. Rend., Ser. A, 265:319
       (1967).
127.   K. D. N. Brummer, J. Catalysis, 9:207 (1967).
128.   M. R. Dausheva, O. A. Songina, and S. I. Zhdanov, Élektrokhimiya, 6:285
       (1970).
129.   F. Vydra, J. Electroanal. Chem., 25:App. 13 (1970).

Chapter 3

Mixed Kinetics of Heterogeneous Reactions

In the discussion of reactant transport phenomena we have hither-to neglected the kinetics of the heterogeneous process which the reactants undergo at the electrode surface. It has simply been assumed that the concentration close to the surface is constant and equal to some arbitrary value c_S without any further analysis of what determines this value.

The factors which determine this concentration will now be discussed. This requires an analysis of convective diffusion processes, of the kinetics of the heterogeneous reaction itself, and of their coupling.

§ 3.1. The Rate of the Heterogeneous Reaction

The kinetic equations characterizing the rate of a hetero-geneous reaction depend, on the one hand, on the concentrations of the reaction components, and on the other hand, on the reactivity of the surface. Generally, the rate of a heterogeneous reaction can be represented by

$$q = kf(c_1, c_2, \ldots). \tag{3.1}$$

The function f, depending on the concentrations of the reaction components $c_1, c_2, \ldots$, expresses the kinetics of transformations occurring in the reaction process. The function often is of the form

$$q = k c_1^{\mu_1} c_2^{\mu_2} \ldots \tag{3.2}$$

The quantity k, which depends on temperature and on surface properties, is called the rate constant, while the quantities $\mu_1, \mu_2, \ldots$ characterize the reaction orders with respect to the given indicated components.

The temperature dependence of k generally has the form

$$k = k_0 \exp\left(-\Delta U / RT\right),\tag{3.3}$$

where ΔU is the activation energy of the process.

The rate of electrochemical reactions occurring at the electrode surface depends markedly (normally exponentially) on the potential difference at the electrode−solution interface.

Consider a simple first-order redox reaction:

$$\text{Ox} + ne^- \underset{k_b}{\overset{k_f}{\rightleftarrows}} \text{Red},\tag{3.4}$$

where Ox is the oxidized form and Red the reduced form of the reacting species; k_f and k_d are the formal rate constants of the respective reactions.

Denoting the concentrations of the oxidized and reduced forms by c_{OS} and c_{RS}, respectively, we can express the rates of the forward reaction (reduction of Ox), q_f, and of the back reaction, q_b, by

$$q_f = k_f c_{OS} \quad \text{and} \quad q_b = k_b c_{RS}.\tag{3.5}$$

The net rate q of reaction (3.4) referred to a unit electrode surface is then the difference of the rates of the simultaneous forward and backward processes, i.e.,

$$q = q_f - q_b = k_f c_{OS} - k_b c_{RS}.\tag{3.6}$$

The electrochemical reaction (3.4) is accompanied by charge transfer and thus results in a flow of current across the electrode−solution interface. The current density i is related to the rate of the surface reaction as follows:

$$i = nFq = nF\left(k_f c_{OS} - k_b c_{RS}\right).\tag{3.7}$$

According to the theory of rate-controlling discharge [1, 2], the rate constants k_f and k_b are related to the electrode potential by

$$k_f = k_f^0 \exp\left[-\alpha nF\varphi/RT\right], \tag{3.8a}$$

and

$$k_b = k_b^0 \exp\left[(1-\alpha)nF\varphi/RT\right]. \tag{3.8b}$$

Here, k_f^0 and k_b^0 are constant, potential-independent coefficients (the "standard" rate constants), α is the transfer coefficient, and the potential φ varies with respect to some reference electrode.

The net current is equal to zero at the equilibrium potential φ_e, and therefore

$$nFk_f^0 c_O^* \exp\left[-\alpha nF\varphi_e/RT\right] = nFk_b^0 c_R^* \exp\left[(1-\alpha)nF\varphi_e/RT\right]. \tag{3.9}$$

Equation (3.9) assumes the concentration gradients at the interface to be zero at equilibrium; in other words, the surface and bulk concentrations of Ox and Red are assumed equal; i.e., $c_{OS} = c_O^*$ and $c_{RS} = c_R^*$. The oxidation and reduction currents at the equilibrium potential are equal, and their value is called the exchange current i_0, which can be represented by:

$$i_0 = nFk_f^0 c_O^* \exp\left[-\alpha nF\varphi_e/RT\right] = nFk_b^0 c_R^* \exp\left[(1-\alpha)nF\varphi_e/RT\right]. \tag{3.10}$$

The net reaction current can be expressed in terms of the exchange current and of the departure of the electrode potential from the φ_e value; thus,

$$i = i_0 \left\{ \frac{c_{OS}}{c_O^*} \exp\left[-\alpha nF(\varphi-\varphi_e)/RT\right] - \frac{c_{RS}}{c_R^*} \exp\left[(1-\alpha)nF(\varphi-\varphi_e)/RT\right] \right\}. \tag{3.11}$$

The difference $(\varphi - \varphi_e)$ is called the overpotential.

At sufficiently high overpotentials, one of the partial currents (of oxidation or reduction) becomes negligible in comparison with the other, and consequently one of the exponential terms in Eq. (3.11) disappears.

§ 3.2.   Basic Equation of Mixed Kinetics

Under steady-state conditions, the entire substance supplied by convective diffusion to the disc surface takes part in the heterogeneous process. Therefore, the material-balance condition

$$j = q. \tag{3.12}$$

obtains at the disc's surface. Here $j = D(dc/dz)_S$ is the density of the diffusional flux to the surface, i.e., the number of moles of the substance supplied by diffusion in unit time to unit surface of the disc, and q characterizes the rate of the transformation.

It will be assumed, following § 3.1, that the rate of the heterogeneous reaction is given by

$$q = kc_S^\nu, \tag{3.13}$$

where c_S is the concentration of the substance at the disc surface.

Combining Eqs. (3.12), (2.23), and (3.13), we obtain

$$D \left(\frac{dc}{dz}\right)_S = kc_S^\nu. \tag{3.14}$$

Equation (3.14) represents the material-balance condition which takes into account the finite rate of the heterogeneous reaction. It should be added to the equation for convective diffusion (2.10) as a boundary condition replacing Eq. (2.12). It should be noted that such replacement makes the surface concentration c_S an unknown variable, which can be determined by solution of the equation of convective diffusion only.

For the time being, it can be assumed that c_S is known. Then relation (2.24),

$$j = D(c^* - c_S)/\delta_d,$$

can still be used to express the diffusion flux. Introducing Eq. (2.24) into Eq. (3.14), we obtain the following algebraic equation for the surface concentration c_S:

$$\frac{D(c^* - c_S)}{\delta_d} = k c_S^{\mu}. \tag{3.15}$$

The solution of Eq. (3.15) is conveniently obtained by the following graphical method. Introducing the dimensionless concentration $C_S = c_S/c^*$ and a dimensionless parameter $K = k\delta_d c^{*(\mu-1)}/D$, we transform Eq. (3.15) into

$$C_S^{\mu} = (1 - C_S)/K. \tag{3.16}$$

Figure 3.1, reprinted from the monograph by Frank-Kamenetskii [3], shows the functions C_S^{μ} and $(1 - C_S)/K$ plotted against the dimensionless concentration C_S. The family of curves passing through the point $C_S = 0$ represents functions C_S^{μ} for various values of the power μ. The family of straight lines through the point $C_S = 1$ represents functions $(1 - C_S)/K$ for various values of the coefficient $1/K$. The abscissas of points of intersection of the curves with straight lines correspond to the concentrations C_S

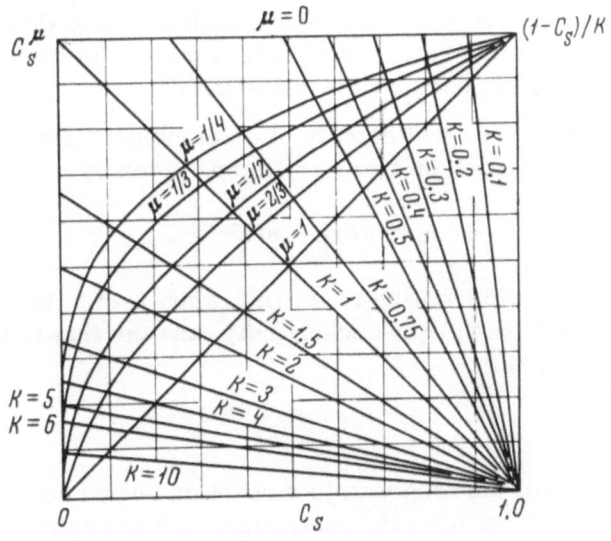

Fig. 3.1. Diffusional kinetics of a reaction of fractional order [3].

which represent solutions of Eq. (3.16). It can be seen from Eq. (3.1) that the intersection points and, consequently, the concentration C_S vary with the parameters μ and $1/K$.

Having determined the value of the surface concentration C_S, it is possible to find the mass flux to the disc surface:

$$j = D\left(\frac{dc}{dz}\right)_S = \frac{Dc^*}{\delta_d}(1 - C_S) = \frac{Dc^*}{\delta_d} KC_S^\mu. \tag{3.17}$$

It can be easily seen from Fig. 3.1 that two limiting cases exist, depending on the value of the parameter K.

At low $1/K$ values $(K \gg 1)$, all intersection points tend to the value $C_S = 0$. It follows from Eq. (3.17) that the current density is then given by

$$j = j_d = \frac{Dc^*}{\delta_d}. \tag{3.18}$$

The above result has a simple physical significance. When the surface reaction is fast $(k \gg D/\delta_d c^{*(\mu-1)})$, the whole amount of the substance supplied to the surface of the disc disappears immediately in the reaction process. The reaction rate is then completely determined by the rate of the mass transport to the surface, i.e., the reaction occurs under "diffusion control."

The limiting current density of an electrochemical reaction accompanied by electron transfer is then given by

$$i_d = nFj_d = nFDc^*/\delta_d. \tag{3.18a}$$

In the opposite limiting case, i.e., when $K \ll 1$, the dimensionless concentration C_S approaches unity, and the flux to the rotating disc is given by

$$j_k = kc^{*\mu}. \tag{3.19}$$

The latter result corresponds to a reaction occurring under "kinetic control" with a rate determined solely by the rate of the surface process $(k \ll D/\delta_d c^{*(\mu-1)})$.

The limiting kinetic current is

$$i_k = nFj_k = nFkc^{*\mu}. \tag{3.19a}$$

When the rates of the diffusional and kinetic steps are commensurate, the reaction proceeds under "mixed control."

Since the diffusion rate depends on the thickness of the diffusion layer, and the rate of the electrochemical surface process proper depends on the overpotential, a suitable choice of the rate of stirring and of overpotential allows, in principle, the rate of a process to be shifted into the region of mixed kinetic control.

In the case of gaseous (nonelectrochemical) reactions the rate constants of which cannot be changed by potential, the reaction can be conveniently shifted from diffusional to the kinetic control region by a temperature decrease. Thus, for example, iodine etching of a rotating germanium disc shifts from diffusional to mixed control below 360°C [4].

A characteristic indication of mixed kinetics is a more complex dependence of the current on the rotational speed of the electrode than that expected in purely hydrodynamic mass transfer. This is illustrated in Fig. 3.2, where the ionization current of chlorine at a Pt electrode is plotted as a function of the square root of the number of revolutions per second of the disc [5]. At low angular velocities, when the rate of supply of chlorine to the surface is low and the reaction occurs under diffusion control, the current is proportional to $\sqrt{\omega}$, in agreement with Eq. (3.18a). Conversely, at high angular velocities, when kinetic control obtains,

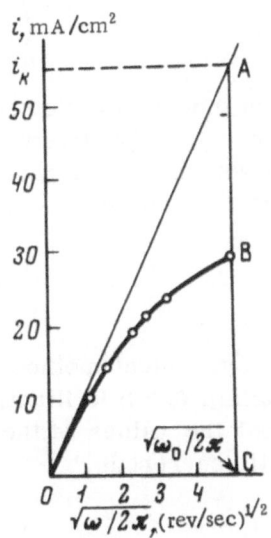

Fig. 3.2. Dependence of the limiting current density of chlorine ionization at a Pt electrode on the square root of the angular velocity [5]. Solution: 0.1 N HCl + 1.9 N H$_2$SO$_4$. Overpotential: 0.04 V.

the current becomes independent of the angular velocity, in agreement with Eq. (3.19a).

§ 3.3. Determination of the Reaction
Order and of the Rate Constant

The reaction order μ and the rate constant k are important kinetic characteristics of a reaction.

For the determination of the reaction order, Eq. (3.17) is considered. The dimensionless surface concentration C_S can be expressed by the measurable quantities j and j_d:

$$C_S = \left(\frac{j}{Kj_d}\right)^{1/\mu}. \tag{3.20}$$

Combining the above expression with (3.16) gives

$$\left(\frac{j}{Kj_d}\right)^{1/\mu} = 1 - \frac{j}{j_d}. \tag{3.21}$$

For two different angular velocities ω_1 and ω_2, we have

$$\left(\frac{j^{(1)}}{K^{(1)}j_d^{(1)}}\right)^{1/\mu} = 1 - \frac{j^{(1)}}{j_d^{(1)}}, \tag{3.22a}$$

$$\left(\frac{j^{(2)}}{K^{(2)}j_d^{(2)}}\right)^{1/\mu} = 1 - \frac{j^{(2)}}{j_d^{(2)}}, \tag{3.22b}$$

where the superscripts (1) and (2) denote values corresponding to ω_1 and ω_2, respectively. Substituting into Eqs. (3.22a) and (3.22b) expressions for the parameters $K^{(1)}$ and $K^{(2)}$, and taking logarithms, we obtain

$$\mu \ln \frac{1 - j^{(1)}/j_d^{(1)}}{1 - j^{(2)}/j_d^{(2)}} = \ln \frac{j^{(1)}}{j^{(2)}}. \tag{3.23}$$

A graphical method for determination of the reaction order μ using Eq. (3.23) is illustrated in Fig. 3.3. The dashed lines represent the values of the limiting diffusion current i_d and of the kinetic current i_k.

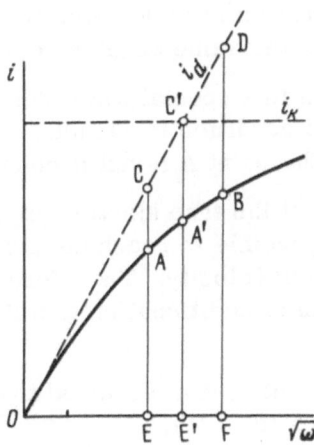

Fig. 3.3. Determination of the reaction order from the dependence of the limiting current on the square root of the angular velocity of the electrode.

Points A and B on the curve correspond to currents registered for two different angular velocities ω_1 and ω_2. It can be easily seen that Eq. (3.23) can be replaced by

$$\mu = \frac{\ln BF - \ln AE}{(\ln CE - \ln AC) - (\ln DF - \ln BD)}. \tag{3.24}$$

The reaction order μ can be determined by substituting the lengths of the respective sections into Eq. (3.24).

Analysis of Eq. (3.24) shows that determination of μ requires results obtained in the region of mixed kinetics. Indeed, if points A and B belong to the diffusional region, then AC = BD = 0 and infinite terms appear in the denominator of Eq. (3.24). If both points A and B lie in the kinetic region BF $\approx$ AE, again μ cannot be determined.

Frumkin, Aikazyan, and Tedoradze [5, 6] and Riddiford in his review [7] propose a somewhat different method of determination of μ, which, however, is one of the variants of the general method described above.

If the point B is chosen far in the kinetic region, the section BF = i_k (where i_k is the true value of the kinetic current). Then BD $\approx$ DF, and Eq. (3.24) can be transformed into

$$\mu = \frac{\ln i_k - \ln i}{\ln i_d - \ln (i_d - i)}. \tag{3.25}$$

Here i is the value of the current at a certain angular velocity ω, and i_d is the limiting diffusion current at the same angular velocity.

In [5, 6] the point A was also chosen in a special way: the angular velocity ω^0 was chosen from the condition $i_d^0 = i_k$ (cf. Fig. 3.2). However, such a location of the point A is not necessary.

Equation (3.25) requires knowledge of the true kinetic current i_k. In practice, however, it is often impossible to reach the purely kinetic region by increasing the angular velocity. Therefore, determination of i_k requires construction of additional plots which will be described in § 3.4.

In order to determine the rate constant k, it is assumed that the reaction order μ is known. According to (3.17),

$$C_S = 1 - j/j_d \quad \text{and} \quad C_S^\mu = j/Kj_d ,$$

wherefrom

$$K = j [j_d (1 - j/j_d)^\mu]^{-1}$$

or

$$k = j [c^{*\mu} (1 - j/j_d)^\mu]^{-1} \tag{3.26}$$

is obtained.

In terms of electrical units, Eq. (3.26) becomes

$$k = i [nFc^{*\mu} (1 - i/i_d)^\mu]^{-1}. \tag{3.26a}$$

Substituting the experimental current density i and the theoretically calculated value of the limiting current density i_d for the same angular velocity into Eq. (3.26a), we can easily obtain the rate constant k.

§ 3.4.   First-Order Reactions

The general relations described in the former sections are considerably simplified in the case of first-order reactions.

For $\mu = 1$, the solution of Eq. (3.16) has the form
$$C_S = 1/(1 + K), \tag{3.27}$$

where

$$K = k\delta_d/D. \tag{3.28}$$

Substituting the expression for C_S into Eq. (3.17), we find

$$j = j_d \frac{K}{1+K} = \frac{Dc^*}{\delta_d + D/k}. \tag{3.29}$$

Expression (3.29) describes the flux to the rotating disc whose surface is the site of a first-order reaction.

The most interesting consequence of Eq. (3.29) is the dependence of $1/j$ on $1/\sqrt{\omega}$. Substituting the value of δ_d (2.20) into (3.29), we easily obtain the result

$$\frac{1}{i} = \frac{1}{nFkc^*} + \frac{1.61\nu^{1/6}}{nFD^{2/3}c^*}\frac{1}{\omega^{1/2}}. \tag{3.30}$$

(Here electrical variables have been used.)

A graphical representation of Eq. (3.30) is shown in Fig. 3.4 reprinted from an experimental paper of Frumkin and Tedoradze [5]. A straight line is obtained in the coordinate system i^{-1} vs $(\omega/2\pi)^{-1/2}$.

Extrapolation of the straight line to $\omega \to \infty$ [i.e., to $(\omega/2\pi)^{-1/2} \to 0$] makes the determination of the true kinetic current, $i_k = nFkc^*$, possible, and it can be used as a direct method for determination of the rate constant k.

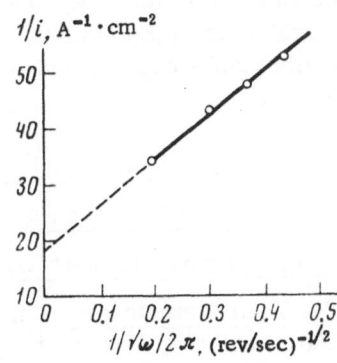

Fig. 3.4. Dependence of i^{-1} on $(\omega/2\pi)^{-1/2}$ for chlorine evolution at a Pt electrode (from Fig. 3.2) [5].

Determination of the rate constants k_f^0 and k_d^0 of an electrochemical redox reaction of the type of Eq. (3.4) close to the equilibrium potential will now be considered.

The transport of the oxidized and reduced forms to the surface of a rotating disc electrode is described by

$$i = nFD_O \frac{(c_O^* - c_{OS})}{\delta_{dO}} \,, \tag{3.31a}$$

$$i = - nFD_R \frac{(c_R^* - c_{RS})}{\delta_{dR}} \,, \tag{3.31b}$$

where D_O and D_R are the diffusion coefficients of the respective species, and δ_{dO} and δ_{dR} are the respective diffusion layer thicknesses.

Substituting the values of c_{OS}/c_O^* and c_{RS}/c_R^* from (3.31a) and (3.31b) into (3.11), and after a few transformations, we find

$$\frac{1}{i} = \frac{b^\alpha}{(1-b)\,i_0} + \left\{ \frac{1}{D_O^{2/3}c_O^*} + \frac{b}{D_R^{2/3}c_R^*} \right\} \frac{1,61\nu^{1/6}}{(1-b)\,nF\,\sqrt{\omega}} \,, \tag{3.32}$$

where

$$b = b(\varphi) = \exp\left[nF(\varphi - \varphi_e)/RT\right]. \tag{3.33}$$

In deriving Eq. (3.32) the expression (2.27) was used for the limiting diffusion current density, and expression (2.20) was used for δ_A. It can be seen from Eq. (3.32) that at a fixed potential φ the plot i^{-1} vs $\omega^{-1/2}$ is a straight line. The latter can be constructed from experimental data obtained at various angular velocities. The intercept (at $\omega \to \infty$) represents the first term on the right side of Eq. (3.32). Its value depends on the potential φ. Measurements carried out for different φ values allow i_0 and α to be determined.

§ 3.5. Experimental Determination of the Kinetic Characteristics of an Electrode Reaction

a. Reaction Order. The determination of the electrochemical reaction order by the rotating-electrode method was first made by

Frumkin, Tedoradze, and Aikazyan [5, 6] for chlorine and hydrogen evolution on platinum. Chlorine ionization proceeds in two steps:

$$Cl_2 + e^- \rightarrow Cl_{ads} + Cl^-,$$
$$Cl_{ads} + e^- \rightarrow Cl^-,$$

the first of which is the rate-determining one.

The dependence of the chlorine ionization rate at constant potential on the angular velocity of the electrode is shown in Fig. 3.2. Extrapolation to $\omega \rightarrow \infty$ (in the coordinate system i^{-1} vs $\omega^{-1/2}$) makes determination of the true kinetic current possible (cf. Fig. 3.4). Substituting this value of i_k into Eq. (3.25), the authors found the order of the rate-determining step to be unity ($\mu = 1$).† This was confirmed by independent measurements of the kinetic current i_k, which was found to be proportional to the chlorine concentration in the solution. The same result was later obtained by Wynne-Jones et al. [9].

Hydrogen ionization [6, 10] also involves several consecutive steps. The first is hydrogen adsorption ($H_2 \rightarrow 2H_{ads}$). The coverage of Pt with adsorbed hydrogen is close to unity. However, the electrochemical reaction itself (ionization of adsorbed hydrogen) involves a part of the electrode only, i.e., its active sites, the total area of which is apparently rather small. Therefore, the adsorption step is followed by surface diffusion of adsorbed hydrogen from the adsorption sites to the "active centers" at which the reaction $H_{ads} \rightarrow H^+ + e^-$ occurs. As usual, the kinetics of the whole process is determined by the step with the lowest specific rate.

It was found that at an active surface‡ the rate-determining step is hydrogen ionization. Its order with respect to dissolved hydrogen is unity [10]. This result, obtained using a rotating disc electrode, was repeatedly confirmed by direct measurements of the dependence of the kinetic current on hydrogen concentration.

Surface diffusion of hydrogen becomes apparently inhibited upon cathodic polarization of the platinum surface and eventually

---

† It was to be expected already on the basis of the linear dependence of i^{-1} on $\omega^{-1/2}$ (cf. [8]).

‡ The platinum electrode was activated by alternating cathodic and anodic polarization.

becomes the rate-determining step. The diffusion rate, however, is independent of the concentration of dissolved hydrogen, since at all concentrations studied the limiting coverage of the surface with adsorbed hydrogen has been attained. Therefore, the ionization process is concentration independent, i.e., its order with respect to hydrogen concentration is zero. The form of the current vs square root of angular velocity curve is very characteristic: the section corresponding to diffusion control ($i \sim \sqrt{\omega}$) sharply passes into one of kinetic control where current is independent of the angular velocity (Fig. 3.5). It is difficult even to refer here to "mixed kinetics"; the corresponding part of the experimental curve is absent.

When the relation between the specific rates of consecutive reaction steps is different from the case considered above, the transition from diffusional to kinetic-control regions occurs less sharply. In [6] curves of i vs $\sqrt{\omega}$ have been calculated corresponding to values of the reaction order in the range $0 \leq \mu \leq 1$ (for various assumptions of the mechanism of the process and of the rates of individual steps). The smoothest transition from diffusional to kinetic control corresponds to $\mu = 1$ (cf. Fig. 3.2).

This approach was utilized by Müller [11] who calculated i vs $\sqrt{\omega}$ curves and the formal μ values for the case of preceding adsorption of the reacting species at the electrode. It was found that the formal reaction order μ depends on the type of adsorption isotherm (Langmuir and Temkin isotherms were considered) and on the surface coverage. Thus, an analysis based on the reaction

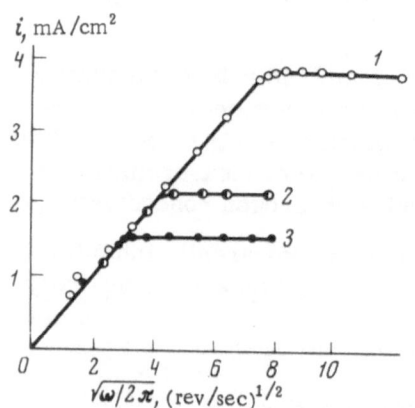

Fig. 3.5. Dependence of the ionization current of hydrogen (at overpotential 0.045 V) on the square root of the angular velocity: 1) 1 N H_2SO_4; 2) 1 N HCl; 3) 1 N HBr [6].

orders obtained using a rotating disc electrode allows one to gain an insight into the adsorption process preceding electrochemical reactions.†

Below, papers are listed which involve determination of reaction orders of electrode reactions using a rotating disc electrode.

Nekrasov and Zolotova [13] determined the order of the oxygen ionization reaction at Pt utilizing Eq. (3.25) and a more general formula of the type (3.24). Both methods gave the same result, $\mu = 1 \pm 0.04$. Oxygen ionization on pyrolytic graphite is also of the first order [14].

Determination of μ in other papers was made usually by the method of Frumkin, Tedoradze and Aikazyan, i.e., using the value of the true kinetic current i_k obtained by extrapolation to $\omega \to \infty$. It follows from the above that the highest accuracy is achieved in the case of first-order reactions.

The iodine–iodide system was studied using a rotating disc electrode in aqueous and organic solutions. Both in dimethyl sulfoxide and in dimethylformamide [15], a cathodic wave of I_3^- reduction and two anodic waves of I^- oxidation were observed at Pt electrodes. In all cases, the reaction order was found to be unity.

Reduction of I_3^- from aqueous solutions at Pt [16] and Ge [17] electrodes was found to proceed with a reaction order $\mu = 0.5$; oxidation of I^- to IO_3^- at a graphite electrode [18] is a first-order reaction.

The rotating disc electrode was also used to establish the reaction order of ferricyanide reduction at gold and platinum electrodes [19], hypochlorite reduction at platinum [20, 21], manganese ion oxidation at electrodes of platinum and azobenzene (cf. § 9.1) in alkaline solution [22], oxidation of N-methylaniline at electrodes made of platinum and carbon paste [23], reduction of bromine at platinum in hydrofluoric acid [24], and reduction of persulfate ions at a lead electrode in the region of decreasing current (i.e.,

---

†Analysis of kinetic equations, particularly of i vs $\sqrt{\omega}$ curves for the case of a preceding adsorption step, as well as for a catalytic surface reaction, has been presented in [12].

in the vicinity of the point of zero charge the kinetics acquire a mixed character) [25]. The reaction orders in all the above cases were unity.

The reaction order (with respect to HF) of dissolution of a zirconium disc in $HF-HNO_3$ and $HF-H_2SO_4$ mixtures is close to 1 and 2, respectively [26]. Iron corrosion in oxygen-saturated acidic Na_2SO_4 solutions proceeds in the mixed-kinetics region with respect to oxygen, and the reaction order is 0.4 [27].

Reaction orders close to zero were observed for dissolution of zinc oxide (pressed into a disc) in NaOH and KOH solutions [28], for dissolution of naphthalene, diphenyl, and diphenylamide in heptane [29], for oxidation of trivalent chromium at an oxidized lead anode [30], for oxidation of chloride ions at an iridium — platinum alloy electrode [31], and for reduction of iodate ions at Pt in acid solutions and of persulfate ions at silver electrodes [32]. Anodic oxidation of hydrazine at bright nickel also proceeds with a reaction order of zero, as shown (apart from the characteristic form of the current vs square root of angular velocity curve) by the fact that the kinetic currents are independent of the bulk concentration of hydrazine [33].

It should be mentioned that a zero, or close to zero reaction order, is most conveniently measured using an electrode uniformly accessible for the diffusing substance, i.e., a rotating disc electrode. Under conditions corresponding to a nonuniformly accessible electrode, the measured reaction rate is a value averaged over sections with different current densities; consequently, the discontinuity on the i vs $\sqrt{\omega}$ curve becomes less sharply defined.

b. Determination of the Rate Constant, the Exchange Current, and the Transfer Coefficient. Various methods of analyzing experimental results to obtain the rate constants reduce to one or another way of extrapolating the measured currents to infinite angular velocity of the rotating electrode ($\omega \rightarrow \infty$) and thus remove the diffusional limitation. The true kinetic currents i_k thus obtained correspond to a situation in which the surface and bulk concentrations are the same.

For first-order reactions, the extrapolation is easily carried out in the i^{-1} vs $\omega^{-1/2}$ coordinate system, according to Eq. (3.30). In this way the "true" polarization curve (i.e., uninfluenced by mass-transport processes) is obtained over the potential region corresponding to mixed kinetics, i.e., in the region for which the rates of the diffusion and electrode processes are commensurate. Polarization curves obtained at various concentrations of the reactants serve to determine the exchange current and the transfer coefficient.

Equation (3.30) can be utilized at potentials far from the reversible potential when the rate of the back reaction can be neglected (i.e., at $\varphi - \varphi_e \gg RT/\alpha nF$). However, in the case of very fast reactions, the rate of the electrochemical step becomes so high already at low overpotentials ($\varphi - \varphi_e \leqslant RT/\alpha nF$) that the reaction passes into the diffusion-control region. Then the mixed kinetic region is close to the reversible potential, and Eq. (3.33), which takes into account also the back reaction, must be used.

In this way Vielstich and Jahn [34] determined for the first time the exchange currents and transfer coefficient of a few fast reactions. They studied the $Fe^{2+}-Fe^{3+}$ and $Fe(CN)_6^{4-}-Fe(CN)_6^{3-}$ systems at bright platinum. Results obtained for the first couple extrapolated in i^{-1} vs $\omega^{-1/2}$ coordinates are shown in Fig. 3.6; the dependence of the calculated rate constants of the forward and back reactions on overpotential is shown in Fig. 3.7. The results

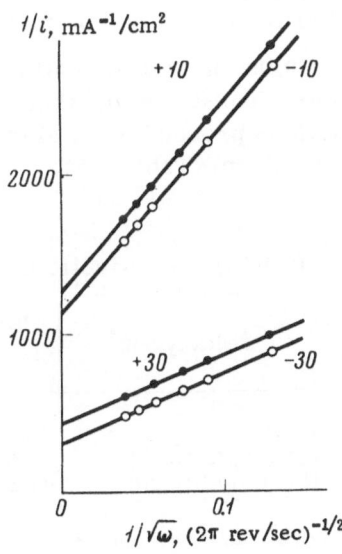

Fig. 3.6. Dependence of i^{-1} on $\omega^{-1/2}$ for the $Fe^{2+}-Fe^{3+}$ (10^{-2} M) couple in 1 M $HClO_4$ [34]. The values of overpotential (mV) are indicated on the curves.

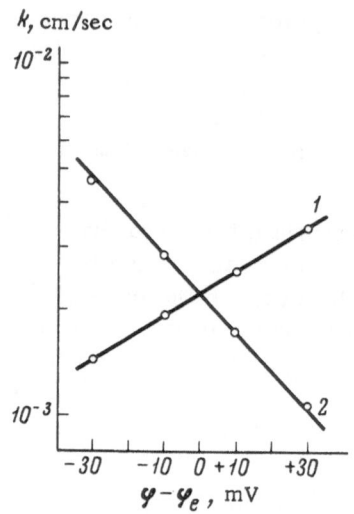

Fig. 3.7. Dependence of the rate constants of the forward (1) and backward (2) reactions (according to Fig. 3.6) on overpotential [34].

obtained using the rotating disc electrode are compared in Table 3.1 with the data of Randles and Somerton [35] obtained using the Faradaic impedance method. The agreement of the two sets of data is good. It should be mentioned that the superiority of the rotating disc method over AC methods consists in the absence of complications connected with the relaxation and capacitative properties of the double layer. According to [34] the rate constant at 120 rev $\cdot$ sec^{-1} can be measured with an accuracy of 0.1 cm $\cdot$ sec^{-1}.

Galus and Adams [36] utilized Eq. (3.30) to determine the reaction constants and their dependence on overpotential for a few reactions proceeding at electrodes constructed of carbon paste and platinum (Table 3.2).

TABLE 3.1. Kinetic Characteristics of Certain Reactions

Redox system	Inert electrolyte	α	i_0, A/cm^2 *	
			[34]	[35]
Fe^{2+}/Fe^{3+}	1 M HClO$_4$	0.63	0.23	0.5
K$_4$Fe(CN)$_6$/K$_3$Fe(CN)$_6$	1 M KCl	0.61	5	9
K$_4$Fe(CN)$_6$/K$_3$Fe(CN)$_6$	0.5 M K$_2$SO$_4$	0.56	7	13

*Reactant concentrations 1 mole/liter.

TABLE 3.2. Rate Constants of Certain Reactions

Redox system	Inert electrolyte (1 M)	k_b^0 at $\varphi = \varphi_e$, cm $\cdot$ sec^{-1}	
		Platinum	Carbon paste
Fe^{2+}/Fe^{3+}	KCl	$4.3 \cdot 10^{-3}$	$5.4 \cdot 10^{-5}$
$K_4Fe(CN)_6/K_3Fe(CN)_6$	KCl	—	$7.1 \cdot 10^{-3}$
Ce^{3+}/Ce^{4+}	H_2SO_4	$3.7 \cdot 10^{-4}$	$3.8 \cdot 10^{-4}$

With the exception of the $Ce^{3+} - Ce^{4+}$ couple, the rate constants were 1-2 orders of magnitude lower on carbon than on platinum electrodes. The accuracy of determination of rate constants was estimated to be within ±5% for the upper limit $1.5 \cdot 10^{-2}$ cm $\cdot$ sec^{-1} (at 120 rev $\cdot$ sec^{-1}).

In a similar way, the kinetic characteristics were studied for oxidation of p-nitrobenzene and p-nitrotoluene in an aqueous ethyl alcohol solution of NaOH at gold, silver, platinum, and steel electrodes. The reaction proceeds in two steps: reversible formation of an ion-radical followed by its slow reduction. The rate constant at the reversible potential of the second step was found to be lower on noble metals (10^{-4} cm $\cdot$ sec^{-1}) than on mercury, even in the presence of inhibitors (10^{-3} cm $\cdot$ sec^{-1}) [37].

The "true" polarization curves obtained using the above method for the iodine − iodide and bromine − bromide systems in dimethyl sulfoxide and acetonitrile were used to calculate the exchange currents and transfer coefficients for anodic and cathodic processes at platinum electrodes [15, 38, 39]. The same method was also used in the analysis of oxygen ionization kinetics at silver [40] and iron [41], Ce^{4+} reduction at anthracene [42], reactions in the ferriferrocyanide system at gold [43], and oxidation of nitrite in melts [44].

§ 3.6. The State of the Electrode Surface and the Transition from Diffusion to Mixed Kinetic Control

The rate constant usually increases with increasing overpotential, and chemical or mixed kinetic control changes into dif-

fusion control.† However, if the rate constant decreases with
increasing overpotential, the opposite transition takes place. The
latter can be exemplified by the reduction of persulfate anions at
mercury, which has been studied extensively at a dropping mercury electrode. The electrode surface becomes negatively charged
upon cathodic polarization, thus inhibiting the discharge of anions
so that the polarization curves show a sharp current decrease in
the vicinity of the potential of zero charge at mercury. Curves
of this type were obtained by Florianovitch and Frumkin [46] at an
amalgamated rotating disc electrode (cf. Fig. 3.8). In the limiting-
current region, the rate is diffusion controlled (current proportional
to square root of the angular velocity of the electrode), and in the
region of minimum current, the rate is kinetically controlled
(current independent of the angular velocity). Current decrease
is also observed during cathodic reduction of persulfate ions at
rotating electrodes of lead and other metals [25].

A decrease of current with increasing overpotential is also
observed in the case of electrode passivation. Passivation, i.e.,
inhibition of an electrochemical process due to, for example,
changes of the electrode surface properties, is very often accompanied by a transition of the process from diffusion into mixed or
chemical kinetic control. This section concerns results of kinetic
studies of certain reactions proceeding at passivating rotating disc
electrodes.

One of the processes which was extensively studied in this
way is the previously discussed (§ 3.4) ionization of molecular
hydrogen on platinum [6, 47]. A polarization curve for this reaction in H_2SO_4 is shown in Fig. 3.9. At lower angular velocities in
the potential region marked a of a sufficiently active electrode,
the kinetics is controlled by diffusion of H_2 molecules to the electrode surface; the limiting current is proportional to the square
root of angular velocity. The diffusional nature of the limiting
current is confirmed by the same value of its density on bright
and platinized platinum electrodes.

At sufficiently positive potentials, platinum undergoes passivation due to anionic adsorption, and at still higher potentials
it undergoes passivation due to the adsorption of oxygen (region

---

†Gorbachev et al. [45] studied this transition for a series of redox systems by measuring the i vs $\sqrt{\omega}$ curves obtained at a rotating disc electrode.

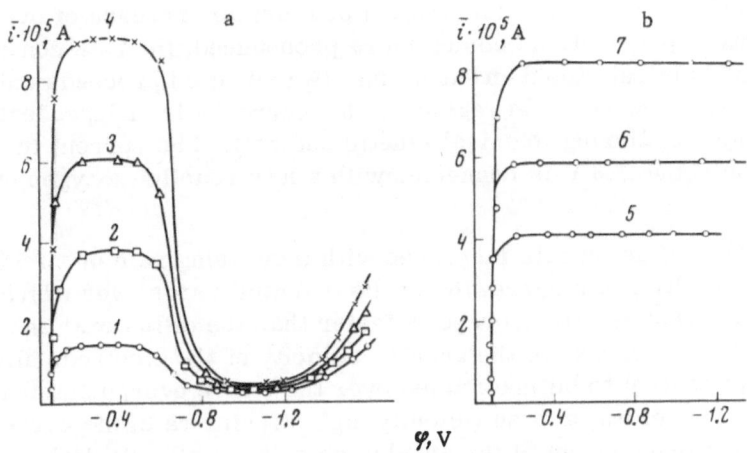

Fig. 3.8. Polarization curves for persulfate-ion reduction at an amalgamated copper electrode [46]. a) $2 \cdot 10^{-3}$ N $K_2S_2O_8$ solution; b) $2 \cdot 10^{-3}$ N $K_2S_2O_8$ + 1 N Na_2SO_4 solution. Rotation speed (rev/sec): 1) 0.25; 2) 1.85; 3) 4.50; 4) 9.50; 5) 2.30; 6) 4.20; 7) 9.0.

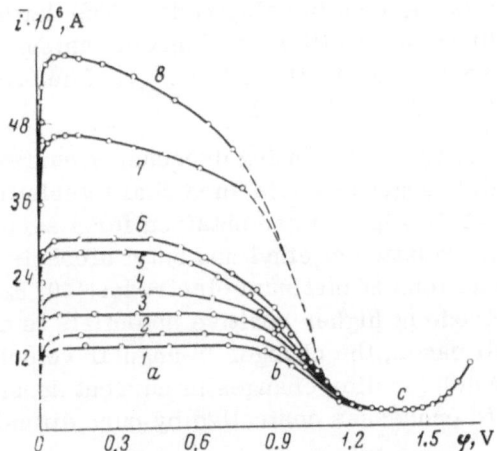

Fig. 3.9. Passivation of a bright platinum electrode during ionization by hydrogen in 1 N H_2SO_4 [47]. Rotation speed (rev/sec): 1) 1.5; 2) 2.0; 3) 3.7; 4) 5.1; 5) 8.7; 6) 11.6; 7) 23.4; 8) 47.7.

b). Here the rates of the diffusional and chemical steps become commensurate (mixed kinetics). Upon further increase of overpotential, passivation beomes more pronounced, the rate constant of the chemical step decreases, and H_2 ionization proceeds only at a very slow rate. In region c, the current is independent of the rate of stirring (chmical kinetic control). The current increase above 1.6 V is connected with a new reaction (oxygen evolution).

The diffusion rate increases with increasing rate of stirring rate, resulting in a narrowing of the potential range over which the chemical kinetic process is faster than the diffusional one (cf. Fig. 3.9). Increase of the angular velocity of the electrode allows kinetic control to be maintained over the entire overpotential region. Therefore, at a sufficiently high stirring rate, the current remains independent of the angular velocity at all potential values. The polarization curve is then the normal direct characteristic of the dependence of the rate constant of the heterogeneous step on potential.

The limiting value of the angular velocity at which transition from diffusional to chemical kinetic control occurs depends on the surface activity of the rotating electrode. The latter can be decreased by addition of specifically adsorbing anions to the electrolyte, which decrease the $Pt - H$ bond energy. The effect increases in the series $SO_4^{2-} < Cl^- < Br^- < I^-$.

Polarization curves exhibiting decreasing current with increasing potential (negative resistance characteristics) similar to the plots shown in Fig. 3.9 are obtained for a series of anodic processes, e.g., oxidation of ethyl and butyl alcohols [48, 49] and of some inorganic ions at platinum electrodes [50, 52]. Passivation of the electrode at higher positive potentials is due to surface oxidation. In all cases, the changes in angular velocity are accompanied by much smaller changes in current density than those corresponding to processes controlled by pure diffusion.

The current decrease observed in polarization curves for anodic oxidation of iodide at Pt in K_2SO_4 solution is due to the formation of a liquid-iodine phase at the anode [53], and in some cases apparently to the formation of iodine crystals [54] which block the surface.

Anodic oxidation of divalent iron and ferrocyanide ions at a rotating gold disc electrode is also inhibited at potentials above 1 V owing to oxygen adsorption. Nevertheless gold is less susceptible to passivation with adsorbed oxygen than platinum.† Therefore, a rotating gold disc electrode is more suitable for analytical determinations at higher anodic potentials [55].

Passivation is encountered also in studies of reduction processes. Thus, polarograms or oxygen reduction at active platinum electrodes have regions of limiting diffusion currents (whose density is proportional to $\sqrt{\omega}$). Potentiostated electrodes passivate in time, and the ionization current decreases and becomes independent of the stirring rate [56]. A similar phenomenon was observed in the reduction of tetraphenylborate ions. During the time of recording of the polarization curve, the platinum surface oxidizes (especially quickly in alkaline solutions), and in consecutive polarograms the wave of reduction for $B(C_6H_5)_4^-$ disappears. Passivation of the electrode was also observed during reduction of p-nitrobenzene and p-nitrotoluene at rotating disc electrodes of gold, platinum, and silver [37].

§ 3.7. Mixed Kinetics of Heterogeneous Reactions at the Surface of a Nonuniform Disc

In the analysis of the effect of the finite rate of the heterogeneous step on the kinetics of the process in § 3.2, it was assumed that the whole disc area was uniformly active. This assumption resulted in the same density of the diffusional flux (3.17) over the whole disc surface. Thus, uniform accessibility of the rotating disc was maintained.

We have also seen in § 2.8 that deviations from uniform activity (e.g., by "coating" the central part of the disc) immediately result in nonuniform accessibility. As was shown in § 2.8, the limiting current density j_d on the "noncoated" part of the disc

---

†Here the authors probably mean that surface oxidation of Au occurs at much more anodic potentials (1.37 V vs 0.8 V for Pt) than at Pt. — Editor.

depends on the distance r from the rotation axis, as described by
Eq. (2.73). The total diffusion flux $\bar{j}_{\mathscr{R}d}$ to the rotating disc with
a "coated" central part is given by

$$\bar{j}_{\mathscr{R}d} = \bar{j}_d \left[1 - \left(\frac{r_{10}}{r_{10} + \Delta r} \right)^3 \right]^{2/3}. \tag{3.34}$$

Here $\bar{j}_d = (Dc^*/\delta_d)\, \pi \, (r_{10} + \Delta r)^2$ is the total limiting diffusion flux
to the disc of radius $r_{10} + \Delta r$; r_{10} is the radius of the "coated"
part.

If the overall rate of the process is limited by the specific
rate of the heterogeneous step and the reaction occurs in the
kinetic region, the flux to the "noncoated" (active) part of the disc
is determined by the area of this part of the surface:

$$\bar{j}_{\mathscr{R}k} = \bar{j}_k \left[1 - \left(\frac{r_{10}}{r_{10} + \Delta r} \right)^2 \right], \tag{3.35}$$

where $\bar{j}_k = kc^* \pi \, (r_{10} + \Delta r)^2$ is the flux to the rotating disc in the
kinetic region under conditions where the whole surface is active.

Relations (3.34) and (3.35) characterize the current on the
"noncoated" active part of the rotating disc in the diffusion and
kinetic regions, respectively.

In the intermediate case (mixed kinetics), a simple analytical
expression describing the flux to a partly "coated" disc cannot be
obtained. Calculations of the current density at the active part
result in an integral equation of the Volterra type. Its numerical
solution was given by Rosner [58], and the results of his computations are shown in Fig. 3.10. The abscissa expresses the relative
area of the "coated" part. The ordinate corresponds to the ratio
of the flux at a partly "coated" disc to the flux at a uniformly active
disc. The family of curves in Fig. 3.10 has been constructed for
various values of the parameter $\eta_0 = (1 + k\delta_d / D)^{-1}$. The two outer
curves of this family obtained for $\eta_0 = 0$ and for $\eta_0 = 1$ correspond
to purely diffusional (1) and purely kinetic (2) conditions of the
process. They are analytically represented by expressions (3.34)
and (3.35).

It can be seen from Fig. 3.10 that the uniform accessibility
of a partly "coated" disc is maintained only in the kinetic region.

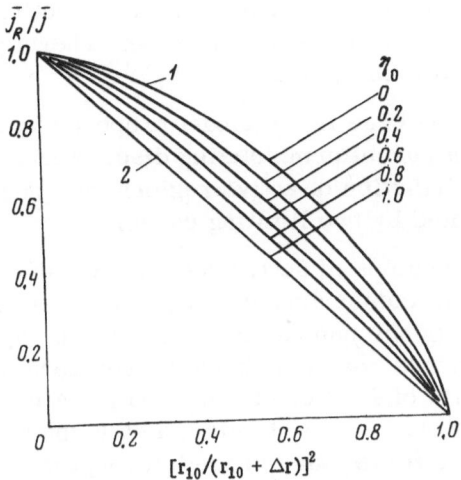

Fig. 3.10. The rate of a heterogeneous reaction at a partly "coated" disc surface as a function of the nonactive surface fraction [58]. The values of the parameter $\eta_0 = (1 + k\delta_d /D)^{-1}$ are indicated on the curves.

This property disappears under condition of mixed kinetic control and diffusion control.

It has been noted by Rosner that in the case of narrow active rings, when $\Delta r \ll r_{10} + \Delta r$, $\bar{j}_R/\bar{j}$ can be represented to a high accuracy by

$$\bar{j}_R/\bar{j} \approx \frac{1 - \left(\dfrac{r_{10}}{r_{10} + \Delta r}\right)^2}{\eta_0} \approx \frac{2\Delta r}{\eta_0 r_{10}} , \qquad (3.36)$$

where Δr is the width of the ring. Obviously, Eq. (3.36) becomes invalid in the diffusion region when $\eta_0 = 0$.

The above case of a nonuniform disc surface with an active ring coaxial with the rotation axis is the simplest example of non-uniformity. This case can be treated by a strictly theoretical analysis.

A much more frequently encountered case is that of a non-uniform surface with active centers distributed in the form of small sections over the disc surface. A theoretical description of mixed kinetics is difficult in this case. Therefore, models are usually utilized for this purpose.

An interesting case is the model proposed by Landsberg and Thiele [59] for a nonuniform rotating disc. According to these authors, the distribution of active regions on a nonuniform surface can be represented by two limiting cases.

1. The dimensions of active sections and of distance between them are small in comparison with the thickness of the diffusion layer δ_d. The surface can be considered uniform, and the rate of the heterogeneous reaction should be replaced by an effective rate. The results of § 3.2 can be utilized for a mathematical description of this case. In particular, a straight i^{-1} vs $\omega^{-1/2}$ line is obtained with a finite value of the intercept on the ordinate.

2. The dimensions of active centers and of distance between them are much larger than δ_d. Qualitatively, this case can be reduced to that of a partly "coated" disc (cf. § 2.8 and the beginning of this section). For example, the limiting diffusion current is still proportional to the square root of the angular velocity. However, the slope of the straight i vs $\sqrt{\omega}$ line is less than that calculated according to Eq. (2.37).

In order to calculate the diffusion limiting currents for intermediate cases, Landsberg and Thiele assumed that circular active sections of radius r' are uniformly distributed over the disc surface. The average distance between them is $2r''$. Using the Nernst model for convective diffusion, the authors calculated the value

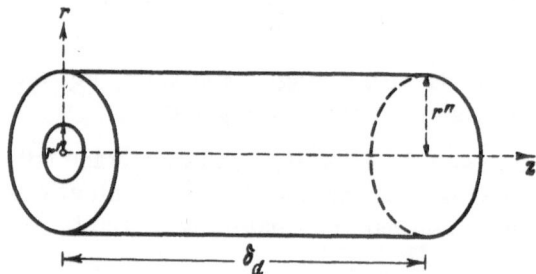

Fig. 3.11. The model of Landsberg and Thiele [59] for calculations of diffusion to a nonuniform disc.

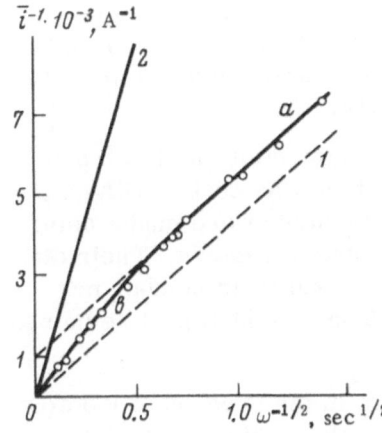

Fig. 3.12. Reduction of $K_3Fe(CN)_6$ at a model platinum rotating disc electrode [63]. 1, 2) Theoretical lines calculated from Eq. (3.37) for a half-active (1) or fully active (2) disc surface area; points are experimental data.

of current per active region collected from a cylindrical volume of solution of height δ_d and radius r'' (Fig. 3.11). Following Nagy et al. [60], they replaced diffusional calculations for this model by calculations of conductivity of a conductor of the same geometry, arriving at the expression

$$\frac{1}{i_d} = \frac{1.61v^{1/6}}{nFD^{2/3}c^*}\frac{1}{\sqrt{\omega}} + \frac{\left|\sum_n A_n \tanh\left(\frac{x_n\delta_d}{r''}\right)\right|}{nFDc^*}, \qquad (3.37)$$

where A_n and X_n are constants calculated in [61].

At low angular velocities ($r'' \ll \delta_d$), the hyperbolic tangents in the second term may be replaced by unity, and Eq. (3.37) becomes

$$\frac{1}{i_d} = \frac{1.61v^{1/6}}{nFD^{2/3}c^*}\frac{1}{\sqrt{\omega}} + \frac{\left|\sum_n A_n\right|}{nFDc^*}. \qquad (3.38)$$

This dependence is shown in Fig. 3.12 by the straight line a. †

---

†A similar dependence of i_d^{-1} on $\omega^{-1/2}$ is also obtained for a uniform disc surface if the electrode reaction is preceded by a chemical step in the bulk (cf. Chap. 6). Criteria allowing a distinction to be made between the two cases are discussed in [62].

At higher angular velocities ($r'' \gg \delta_d$), tanh ($x_n \delta_d / r''$) can be replaced by ($x_n \delta_d / r''$). Then both terms in Eq. (3.37) are proportional to $\omega^{-1/2}$, and the i^{-1} vs $\omega^{-1/2}$ plot is a straight line passing through the origin (section b in Fig. 3.12).

The above theory was verified experimentally [63] at a special model platinum electrode. Its surface was coated with an insulating polymer layer in which circular holes were made (using the photoresistance method), imitating active centers. Their diameter varied from $6 \cdot 10^{-4}$ to $5 \cdot 10^{-2}$ cm, and their number per square centimeter was 100–160,000. The insulated part remained inactive.

This electrode served as a cathode in the process of $K_3Fe(CN)_6$ reduction in 1 N KCl. It can be seen in Fig. 3.12 that the i^{-1} vs $\omega^{-1/2}$ plot consists of two intersecting straight lines, one of which passes through the origin. Consequently, both limiting cases of Eq. (3.37) can be experimentally realized using a sufficiently wide range of angular velocities. The slope of section a depends on concentration and on the diffusion coefficient of the discharging species. Transition from region a to b occurs at a definite value of the r'/δ_d ratio ($r'' > \delta_d$) [64].

In spite of the agreement between theory and experiment, the highly approximate character of the Landsberg–Thiele theory must be stressed, especially in the case of high angular velocities. Indeed, when $r'' > \delta_d$ the concept of a constant diffusion layer thickness across the entire active and inactive surface turns out to be invalid. Diffusion to each active center is essentially nonuniform, and a "diffusional cylinder" similar to that in Fig. 3.11 satisfactorily representing the real diffusion situation cannot be constructed. Therefore, the conclusions of Landsberg and Thiele pertaining to the $r'' > \delta_d$ case cannot be considered valid.

Nevertheless, the above theory has undoubtedly played an important role in the development of studies of the macroscopic nonuniformity of the electrode surface† using the rotating disc method. It explained qualitatively the kinetic properties of a series of electrode reactions, such as reduction of H_2O_2, O_2, and ClO^- and oxidation of MnO_4^{2-} and $Fe(CN)_6^{4-}$ at carbon and graphite electrodes

---

†As opposed to nonuniformity at the atomic level, a concept widely used in the analysis of the kinetics of electrode processes and catalytic reactions.

in terms of nonuniformity (or rather partial blockage) of their surface. This type of explanation is supported by the very method of preparation of these electrodes in which conducting particles (graphite) are pressed with an insulating binder (diphenyl, azobenzene), or porous carbon or graphite is saturated with paraffin or polyfluoroethylene resins.

The concept of macrononuniformity (of a domain type) was used by Povarov and Lukovtsev et al. [65-67] in their studies of the decrease with time of the diffusion limiting current of a series of redox reactions at rotating disc electrodes of platinum and other noble metals. This phenomenon obviously plays an important role in hydrogen ionization on platinum [60] and in lead dissolution in chromate solutions [68].

References

1. A. N. Frumkin, V. S. Bagotskii, Z. A. Iofa, and B. N. Kabanov, Kinetics of Electrode Processes, Izd. MGU (1952).
2. K. Vetter, Electrochemischa Kinetik, Springer-Verlag, Berlin (1961).
3. D. A. Frank-Kamenetskii, Diffusion and Heat Transfer in Chemical Kinetics, Nauka, Moscow (1967).
4. D. Olander, Ind. Eng. Chem. Fundamen., 6:178 (1967).
5. A. N. Frumkin and G. A. Tedoradze, Dokl. Akad. Nauk SSSR, 118:530 (1958); Z. Elektrochem., 62:251 (1958).
6. A. N. Frumkin and E. A. Aikazyan, Dokl. Akad. Nauk SSSR, 100:315 (1955); Izv. Akad. Nauk SSSR, Otd. Khim. Nauk, 1959:202.
7. A. C. Riddiford, Advances in Electrochemistry and Electroenemical Engineering, Vol. 4, P. Delahay, ed., Interscience Publishers, New York (1966), p. 47; W. A. Ledger and A. C. Riddiford, Nature, 194:1233 (1962).
8. M. Enyo and T. Yokoyama, Electrochim. Acta, 15:183 (1970).
9. T. Dickinson, R. Greef, and Lord Wynne-Jones, Electrochim. Acta, 14:467 (1969).
10. V. V. Sobol, A. A. Dmitrieva, and A. N. Frumkin, Élektrokhimiya, 3:1040 (1967); J. Electroanalyt. Chem., 13:179 (1967); M. P. Makowski, E. Heitz, and E. Yeager, J. Electrochem. Soc., 113:204 (1966).
11. L. Müller, Z. Phys. Chem., 241:185 (1969); Electrochim. Acta, 14:293 (1969); L. Myuller (Müller), P. Yanets, and R. Landsberg, Élektrokhimiya, 5:1001 (1969).
12. J. D. E. McIntyre, J. Phys. Chem., 71:1196 (1967); 73:4102 (1969); E. M. Podgaetskii and V. Yu. Filinovskii, Élektrokhimiya, 6:1178 (1970).
13. L. N. Nekrasov and T. K. Zolotova, Élektrokhimiya, 4:864 (1968).
14. M. R. Tarasevich and F. Z. Sabirov, Élektrokhimiya, 5:643 (1969).

15. M. C. Giordano, J. C. Bazan, and A. J. Arvia, Electrochim. Acta, 11:1553 (1966); 14:389 (1969); Extended Abstracts, 19th CITCE Meeting, Detroit (1968), p. 149; I. E. Barbasheva, Yu. M. Povarov, and P. D. Lukovtsev, Élektrokhimiya, 3:1149, 1202 (1967).

16. J. D. Newson and A. C. Riddiford, J. Electrochem. Soc., 108:699 (1961).

17. Yu. V. Pleskov, Zh. Fiz. Khim., 35:2540 (1961).

18. R. Landsberg, R. Nitzsche, and W. Geissler, Electrochim. Acta, 11:495 (1966).

19. S. Azim and A. C. Riddiford, J. Polarogr. Soc., 12:20 (1966).

20. L. Myuller (Müller), Élektrokhimiya, 4:199 (1968).

21. O. Schwarzer and R. Landsberg, Electrochim. Acta, 19:391 (1968).

22. R. Landsberg and R. Thiele, Chem. Techn., 15:627 (1963).

23. Z. Galus and R. N. Adams, J. Phys. Chem., 67:862 (1963).

24. T. Mussini, Chim. Ind., 45:1075 (1963); G. Faita, G. Fiori, and T. Mussini, Electrochim. Acta, 13:1765 (1968).

25. B. N. Rybakov, N. V. Nikolaeva-Fedorovich, and G. V. Zhutaeva, Zh. Fiz. Khim., 38:500 (1964); I. V. Nikolaeva-Fedorovich, B. N. Rybakov, and K. A. Radyushkina, Élektrokhimiya, 3:1086 (1967).

26. R. E. Meyer, J. Electrochem. Soc., 112:684 (1965); R. E. Meyer and S. M. Zettl, J. Electrochem. Soc., 112:1092 (1965).

27. Z. Zembura, W. Ziolkowska, and H. Kolny, Bull. Acad. Polon. Sci., Ser. Chim., 13:487 (1965).

28. R. Landsberg, H. Fürtig, and L. Müller, Z. Phys. Chem., 216:199, 212 (1961).

29. A. A. Ravdel and V. N. Kupriyanov, Zh. Prikl. Khim., 40:1734 (1967).

30. S. P. Antonov and D. P. Zosimovich, Ukr. Khim. Zh., 29:1111 (1963).

31. G. Faita, G. Fiori, and J. W. Augustynski, J. Electrochem. Soc., 116:928 (1969).

32. L. Müller, J. Electroanalyt. Chem., 16:67 (1968); Electrochim. Acta, 13:2005 (1968).

33. B. P. Nesterov and N. V. Korivin, Élektrokhimiya, 2:1296 (1966).

34. D. Jahn and W. Vielstich, J. Electrochem. Soc., 109:849 (1962); D. Jahn, Dissertation, Bonn (1961).

35. J. E. B. Randles and K. W. Somerton, Trans. Faraday Soc., 48:951 (1952).

36. Z. Galus and R. N. Adams, J. Phys. Chem., 67:866 (1963); R. N. Adams, Rev. Polarogr., 11:71 (1963).

37. L. Holleck, B. Kastening, and H. Vogt, Electrochim. Acta, 8:255 (1963).

38. V. A. Macagno, M. C. Giordano, and A. J. Arvia, Electrochim. Acta, 14:335 (1969).

39. T. Iwasita and M. C. Giordano, Electrochim. Acta, 14:1045 (1969).

40. M. R. Tarasevich, N. A. Shumilova, and R. Kh. Burshgein, Izv. Akad. Nauk SSSR, Ser. Khim., 1966:32.

41. A. S. Afanasev and V. V. Shevchenko, Ukr. Khim. Zh., 38:1061 (1962).

42. F. Lohmann and W. Mehl, Electrochim. Acta, 13:1469 (1968).

43. J. Kuta and E. Yeager, Extended Abstracts, 19th CITCE Meeting, Detroit (1968), p. 116.

44. M. E. Martins, A. J. Calandra, and A. J. Arvia, Electrochim. Acta, 15:111 (1970).

45.   S. V. Gorbachev and V. A. Belyaeva, Zh. Fiz. Khim., 35:2157 (1961);
       36:229, 1794 (1962); 37:197 (1963); 39:2576 (1965); S. V. Gorbachev and
       I. I. Aryamova, Tr. MKhTI im. D. I. Mendeleeva, 32:5 (1961); S. V.
       Gorbachev and L. P. Kholpanov, Zh. Fiz. Khim., 36:855, 859, 1074 (1962);
       38:3020 (1964); L. N. Ivanovskaya and S. V. Gorbachev, Tr. MKhITI im.
       D. I. Mendeleeva, 44:50, 59 (1963); S. V. Gorbachev and Yu. A. Korostelin,
       Zh. Fiz.Khim., 39:1469 (1965).

46.   G. M. Florianovich and A. N. Frumkin, Zh. Fiz. Khim., 29:1827 (1955).

47.   E. A. Aikazyan and A. I. Fedorova, Dokl. Akad. Nauk SSSR, 86:1137 (1952);
       E. A. Aykazyan, Zh. Fiz. Khim., 33:1016 (1959); L. T. Shanina, Dokl. Akad.
       Nauk SSSR, 134:141 (1960); Izv. Akad. Nauk Kaz. SSR, Ser. Khim., 1960:94.

48.   E. A. Aykazyan and Yu. V. Pleskov, Zh. Fiz. Khim., 31:205 (1957).

49.   G. A. Bogdanovskii and A. I. Shlygin, Zh. Fiz.Khim., 31:1732 (1957).

50.   K. I. Rozental and V. I. Veselovskii, Zh. Fiz. Khim., 34:57 (1960).

51.   E. A. Aikazyan and R. A. Arkelyan, Izv. Akad. Nauk Arm. SSR, Khim.
       Nauk, 13:225 (1960).

52.   G. A. Tedoradze, Nauchn. Dokl. Vyssh. Shkoly, Khim. i Khim. Tekhnol.,
       No. 2, p. 250 (1958).

53.   A. Ya. Gokhshtein, Élektrokhimiya, 1:285 (1965).

54.   A. M. Averbukh, M. A. Novitskii, L. A. Sokolov, and P. D. Lukovtsev,
       Élektrokhimiya, 1:251 (1965).

55.   O. L. Kabanova, Zh. Fiz. Khim., 35:2465 (1961); Zh. Anal. Khim., 16:135
       (1961).

56.   A. I. Oshe, V. I. Tikhomirova, and V. S. Bagotskii, Élektrokhimiya, 1:688 (1965).

57.   W. R. Turner and P. J. Elving, J. Phys. Chem., 69:1067 (1965).

58.   D. E. Rosner, J. Electrochem. Soc., 113:624 (1966).

59.   R. Landsberg and R. Thiele, Electrochim. Acta, 11:1243 (1966); R. Landsberg,
       S. Müller, and R. Thiele, Acta Chim. Acad. Sci. Hung., 51:85 (1967);
       S. Müller and R. Landsberg, Ber. Bunsenges., 70:586 (1966); R. Landsberg,
       S. Müller, and R. Thiele, Z. Phys. Chem., 236:261 (1967); R. Landsberg,
       S. Müller, and F. Scheller, Abhand. Sachsischen Akad. Wiss. zu Leipzig,
       Mathem.-Naturwiss. Kl., B49, p. 125 (1968).

60.   F. Nagy, G. Horanyi, and G. Vertes, Acta Chim. Acad. Sci. Hung., 34:35
       (1962); Magyar Kemiai Folyoirat, 68:198, 202 (1962).

61.   W. R. Smythe, J. Appl. Phys., 24:70 (1953).

62.   F. Scheller, R. Landsberg, and H. Wolf, Z. Phys. Chem., 243:345 (1970).

63.   F. Scheller, S. Müller, R. Landsberg, and H. J. Spitzer, J. Electroanalyt.
       Chem., 19:187 (1968); 20:375 (1969).

64.   F. Scheller, R. Landsberg, and H. Wolf, Electrochim. Acta, 15:525 (1970).

65.   Yu. M. Povarov, L. V. Eroshkina, and P. D. Lukovtsev, Élektrokhimiya, 4:464
       (1968).

66.   I. E. Barbasheva, Yu. M. Povarov, and P. D. Lukovtsev, Élektrokhimiya, 6:92,
       175, 306 (1970).

67.   Yu. M. Povarov, A. M. Trukhan, and P. D. Lukovtsev, Élektrokhimiya, 6:425,
       602 (1970).

68.   H. Rickert and G. Holzäpfel, Ber. Bunsenges., 70:171 (1966).

Electric Current in a Cell with a Rotating Disc Electrode

§ 4.1. General Equations

Electrochemical processes at the electrode −solution interface represent a special type of heterogeneous reaction. All the stages of the transformation listed in § 2.1 can also be separated in the case of electrochemical reactions.

However, electrode reactions have a special property; the heterogeneous process itself is accompanied by charge transfer. Therefore, at least one of the reaction components is electrically charged. Transfer of this component to (or from, in the case of ions formed as a reaction product) the phase boundary results in an electric current. Electrical phenomena which accompany electrode reactions at phase boundaries result in a series of special features.

It will be useful first to consider the transport of reactants to the electrode surface. A galvanostatic process, i.e., a constant current i, flowing through an electrochemical cell will be discussed.

It is well known [1, 2] that under constant current conditions the distribution of field and of bulk charges can be described by the following equations:

$$\text{div}\, i = 0, \tag{4.1}$$

$$\text{div}\, E = 4\pi\rho/\varepsilon_0, \tag{4.2}$$

where $\mathbf{E}$ is the intensity of the electric field, ρ is the charge density, and ε_0 is the electric permeability of the medium. Under stationary conditions, the electric field can be expressed by the relation

$$\mathbf{E} = -\operatorname{grad} \varphi \qquad (4.3)$$

involving the potential φ.

Equations (4.1) and (4.2) must be supplemented by relations describing the mechanism of the current flow in the cell and the corresponding boundary conditions.

Current in an electrochemical cell results from the flow of ions of various kinds. The net current $\mathbf{i}$ is an algebraic sum of the currents due to various ions:

$$\mathbf{i} = \sum_j \mathbf{i}_j, \qquad (4.4)$$

where $\mathbf{i}_j$ is the current due to the j-th kind of ion and the summation is carried out over all kinds of ions present in the cell.

Apart from Eq. (4.4), the continuity equation should obtain for each ion kind separately, i.e.,

$$\operatorname{div} \mathbf{i}_j = 0. \qquad (4.1a)$$

It has been mentioned already in § 2.1 that the movement of ions in the solution is due to ionic diffusion, to migration in the electric field, and to convective transport of ions and other components of the solution resulting from gradients of temperature, pressure, etc.

The analysis here will be restricted to isothermal cells with low ionic concentration in solution. Neglecting the interactions between the moving ions and the possible effect of the current on the solution motion, the following expression for current due to the j-th kind of ions can be written:

$$\mathbf{i}_j = n_j F\,(c_j \mathbf{v} - D_j \operatorname{grad} c_j + \gamma_j n_j c_j \mathbf{E})\,, \qquad (4.5)$$

where c_j is the concentration of the j-th kind of ion per cubic centimeter, n_j is the ionic charge, γ_j is the ionic mobility, F is the Faraday number, and **v** is the velocity of flow of the solution.

According to conventions accepted in electrodynamics [1], a positive current is that due to positive charges. Also, a positive field is that which causes motion of positive charges.

Substituting Eq. (4.5) into (4.1a) and using the incompressibility condition (div **v** = 0), we can easily obtain the following expression:

$$\mathbf{v}\,\mathrm{grad}\,c_j = \mathrm{div}\,(D_j\,\mathrm{grad}\,c_j - \gamma_j n_j c_j \mathbf{E}). \tag{4.6}$$

The bulk charge can be expressed in terms of ionic concentrations:

$$\rho = \sum_j n_j c_j F. \tag{4.7}$$

Substitution of (4.7) into Eq. (4.2) gives

$$\mathrm{div}\,\mathbf{E} = \frac{4\pi F}{\varepsilon_0}\sum_j n_j c_j. \tag{4.8}$$

Equations (4.6) and (4.8) † form a complete system describing constant current flow in an electrochemical cell.

Equation (4.6) was derived assuming the absence of effects of charge transport on the character of motion of the solution. In other words, the flow of the electric current i through the cell was assumed not to result in a movement of the solution as a whole. This assumption is obviously valid at low ionic concentrations.

The above simplification allows the velocity distribution (**v**) in the solution to be considered known. Therefore the function **v** in Eq. (4.6) is assumed to be known from the solution of the hydrodynamic problem. For example, in the case of a cell with a rotating disc electrode the hydrodynamic problem was analyzed in detail in Chap. 1.

---

†Calculations of electric-field and concentration distribution in electrochemical cells usually utilize the electroneutrality condition $\sum_j n_j c_j = 0$ instead of Eq. (4.8).

The current distribution in a given cell is determined by the character of the convective electrolyte transport and by processes occurring at the cell walls and at the electrodes. Therefore, the general equations (4.6) and (4.8) must be supplemented by a suitable system of boundary conditions.

In particular, in the case of a simple first-order redox process, the current may be related, as shown in § 3.1, to the concentrations of the oxidized and reduced species and to the electrode polarization by expression (3.11).

Calculations of electric-field and ionic-concentration distributions in electrochemical cells are usually very complex. The theoretical analysis becomes considerably simplified, however, in the case of cells with a rotating disc electrode.

The properties of a rotating disc [already described in the discussion of convective diffusion of electrically neutral species (§ 2.3)] allow a considerable simplification to be made of the general equations describing the electric-field and concentration distributions, (4.6) and (4.8).

§ 4.2. Electric Current Flow in the Cell

The physical processes caused by charge transport in electrochemical cells have been discussed in detail in a number of monographs (cf., e.g., [3, 4]). Here the current flow in cells containing one rotating disc electrode will be discussed in detail.

It will be supposed that the solution contains three kinds of ions: two kinds of cations, denoted below by subscripts 1 and 2, and one kind of anion, denoted by subscript 3.

The rotating disc will be taken as a cathode, and it will be assumed that only cations of type 1 undergo reduction. The reaction product is assumed to be electrically neutral.† Cations of type 2 and anions do not take part in the cathodic reaction and are electrochemically inert. They are a part of the inert, or supporting electrolyte, which to a first approximation does not affect the current flow in the cell.

---

†The electrode process may occur in the opposite direction as well; oxidation of an electrically neutral species results in cations 1. In this case, electrical current obviously flows in the opposite direction.

The anode is assumed to be sufficiently distant from the cathode and to have a sufficiently large area not to affect the rotating cathode. It can be assumed then to have no effect on motion of charges in the vicinity of the rotating disc. The effect of the anode on the general electrical properties of the cell can, however, be easily taken into account.

Before the concentration and electric-field distributions in the cell are quantitatively evaluated, the phenomena caused by passage of the electric current will be qualitatively considered. It is well known [3] that a double layer exists at the electrode – solution interface. Let the electrode be charged positively, and let the thin solution layer at the electrode contain an excess of anions.† The ionic part of the double layer extends to the distance of the Debye radius λ_D and occupies an extremely small part of the volume of the solution in the cell. Thus, at not too low concentrations, $\lambda_D \sim 10^{-7}$ to 10^{-6} cm. Then the electrolyte can be divided in two regions: 1) the double-layer region where some deviations from electroneutrality arise, and 2) the remainder of the solution where there is no net charge and which is thus electroneutral.

The formation of a double layer at the electrode – solution interface affects the distribution of the electric field in a cell (a potential difference arises, of course, between the electrode and solution).

The charge in the ionic part of the double layer, the effective thickness of the latter, and the electric-field distribution depend on the intensity of the current passing. A strict analysis of charge transport in the double-layer region requires, generally speaking, a quantum-mechanical approach. However, as will be seen later, a number of useful conclusions can also be drawn on the basis of phenomenological analysis.

The bulk solution (outside the double-layer region) in a cell with stirred electrolyte can be divided into the following regions: a) the diffusion layer at phase boundary, thickness δ_d; b) the hydrodynamic boundary layer, δ_0; and c) the intensively stirred bulk solution region outside of the hydrodynamic boundary layer (Fig.

---

†Below, effects connected with specific adsorption of ions at the electrode are neglected.

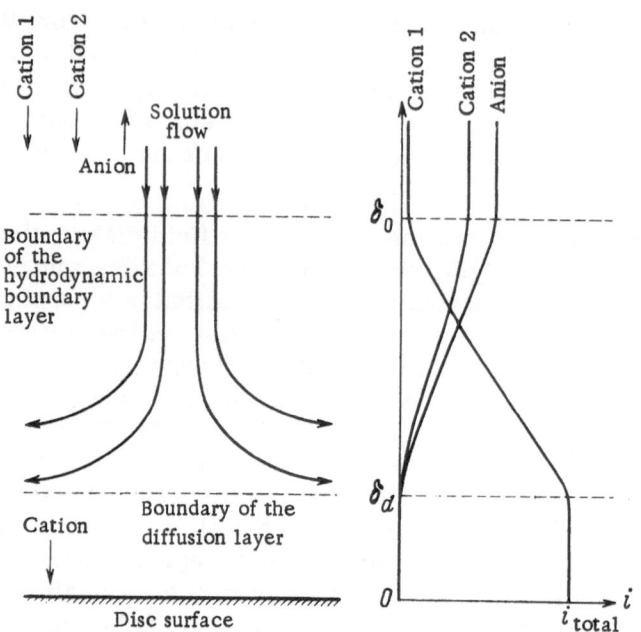

Fig. 4.1. Distribution of the electric current among separate ion kinds in the vicinity of a rotating disc electrode.

4.1). The passage of current will be considered in each of these regions. Equation (4.5) describing the various modes of charge transport will be utilized.

Discussion of the convective transport pertains mainly to the analysis of the axial component of the ionic flux, since the radial and azimuthal flux components do not affect the net current passing through the cell.

It can be assumed for a uniform rotating disc electrode that close to its surface $u \approx 0$, $v \approx r\omega$, and $w \approx 0$, i.e., the liquid inside the diffusion layer is assumed to rotate with the same velocity as the disc (Nernst model, § 2.8). The following expressions can be written for the radial, azimuthal, and normal components of the current density:

$$
\begin{aligned}
&i_{jr} \sim 0, \\
&i_{j\varphi} \sim n_j F c_j r\omega, \\
&i_{jz} \sim n_j F\left(-D_j \frac{dc_j}{dz} + \gamma_j n_j c_j E_z\right).
\end{aligned}
\tag{4.9}
$$

It has been assumed here that the concentrations and electric field vary only in the direction perpendicular to the rotating disc. This assumption is valid for a uniform rotating disc electrode of infinite radius.

Substituting Eq. (4.9) into Eq. (4.1a) which describes the continuity of fluxes gives $di_{jz}/dz = 0$, i.e., $i_{jz} = $ const.

Since the electric current at the disc surface is determined by reduction of cations of type 1, the following relations obtain inside the diffusion layer δ_d:

$$i_{1z} = n_1 F \left(- D_1 \frac{dc_1}{dz} + \gamma_1 n_1 c_1 E_z \right) = i,$$
$$i_{2z} = n_2 F \left(- D_2 \frac{dc_2}{dz} + \gamma_2 n_2 c_2 E_z \right) = 0, \qquad (4.10)$$
$$i_{3z} = - n_3 F \left(- D_3 \frac{dc_3}{dz} - \gamma_3 n_3 c_3 E_z \right) = 0.$$

It can be easily established that the azimuthal component does not result in the appearance of a current. Indeed, owing to the electroneutrality condition, $\sum_j i_{j\varphi} = 0$.

It follows from Eqs. (4.10) that inside the diffusion layer δ_d ionic motions causing the passage of electric current i are due to the simultaneous action of concentration and electric-field gradients. The distribution of ion concentrations and of the electric field is such that it ensures the supply to the electrode of cations 1 in amounts necessary to support current i. Cations 2 and anions do not contribute to current passage, and they rotate only together with the solution around the disc axis.

Outside the diffusion layer δ_d all ions are uniformly distributed; their concentrations remain constant and equal to the respective bulk concentrations.

The components of current density in the region between the external (facing the solution) boundary of the diffusion layer and the external boundary of the hydrodynamic layer (i.e., in the range $\delta_d < z < \delta_0$) can be described by

$$i_{jr} = n_j F c_j^* u,$$
$$i_{j\varphi} = n_j F c_j^* v, \qquad (4.11)$$
$$i_{jz} = n_j F c_j^* w + \gamma_j n_j^2 F c_j^* E_z.$$

With increasing distance from the disc surface, the velocity components, u, v, and w vary in a rather complex manner (cf. Fig. 1.2). In particular, the radial and normal fluxes increase.

Substituting Eqs. (4.11) into the equation of continuity of current (4.1a) we obtain

$$\text{div } i_j = n_j F \dot{c}_j \text{ div } v + \gamma_j n_j^2 c_j^* F \frac{dE_z}{dz} = 0.$$

Taking into account the condition of the liquid's incompressibility (1.1), we can easily show that the above equation is equivalent to the condition

$$\frac{dE_z}{dz} = 0. \tag{4.12}$$

The physical meaning of Eq. (4.12) is that the electric field E_z is constant beyond the hydrodynamic boundary layer, $z > \delta_0$. Then the migration flux of each ion is also constant. The convective flux varies with the corresponding velocity component which changes with distance from the electrode surface. Thus, outside the diffusion layer boundary the values of separate components of ionic fluxes are variable. The direction of the ionic motion changes.

It can be concluded from Eq. (4.11) that all three ion kinds take part in the current passage in the $\delta_d < z < \delta_0$ region. Indeed, summation of the last equation of (4.11) over all ion kinds, carried out with consideration of the electroneutrality condition, gives

$$\sum_j i_{jz} = E_z \sum_j \gamma_j n_j^2 c_j^* F = i. \tag{4.13}$$

The relative contribution of each kind of ion to the resulting current i depends on the individual concentrations c_j^* and mobilities γ_j. The quantity $\sum_j \gamma_j n_j^2 c_j^* F$ characterizes the conductivity of the solution.

The above qualitative analysis shows that redistribution of current among the various ion kinds occurs when the current passes from the diffusion to the hydrodynamic boundary layer. Within the diffusion layer, the total current is carried by cations 1 only, whereas outside this layer all three kinds of ions contribute to the total current. Obviously, the part played by each kind depends on their relative concentrations.

In the presence of an excess of the inert electrolyte, when $c_2^* \gg c_1^*$, the current within the diffusion layer is carried mainly by ions of the supporting electrolyte. The current fraction due to cations 1 is quite small in this region ($i_{1z} \ll i_{2z}$). The effect of the electric field on the flow of cations 1 also turns out to be negligible inside the diffusion layer δ_d. The transport of cations 1 to the electrode surface occurs mainly by diffusion.

Thus, in a solution containing an excess of supporting electrolyte, the distribution of electric field, ohmic resistance of the cell, and the structure of the double layer are determined by the supporting electrolyte. However, the electrode kinetics and the magnitude of the current are controlled by diffusional limitations connected with the transport of cations 1 within the diffusion layer δ_d.

Outside the hydrodynamic boundary layer δ_0, the liquid flows toward the electrode surface. The axial velocity component w turns out to be constant in this region and is equal to $w_\infty = -0.89 (\nu \omega)^{1/2}$. The current-density components for $z > \delta_0$ can be described by

$$i_{jr} = 0,$$
$$i_{j\varphi} = 0,$$
$$i_{jz} = n_j F c_j^* w_\infty + \gamma_j n_j^2 F c_j^* E_z. \qquad (4.14)$$

It can be easily shown that the electric field E_z should remain constant beyond the limits of the hydrodynamic boundary layer ($z > \delta_0$). The resulting electric current is given by Eq. (4.13). It should be noted that, as opposed to the conditions in the region of the hydrodynamic boundary layer ($z < \delta_0$), the normal components of current in the solution bulk due to separate ion kinds become again constant, i.e., i_{jz} = const. However, for $z > \delta_0$, i_{jz} values are different from those in Eq. (4.10), obtained for the region inside the diffusion layer δ_d. The current distribution in the cell is shown in Fig. 4.1.

The potential drop in the cell with a rotating disc electrode consists of the following parts: potential drop at the electrode—

solution interface, potential drop in the layer of varying concentration (inside δ_d), and ohmic drop inside the hydrodynamic boundary layer and in the solution bulk. In general, an analogous potential drop close to the anode surface should also be taken into account.

The above qualitative picture of the current passage through a cell with a rotating disc electrode remains valid for all cells with stirred electrolyte.

The exact calculation of ionic concentrations and of the electric field in the vicinity of a rotating disc cathode will now be considered. First, the current passage inside the hydrodynamic boundary layer δ_0 will be calculated. The distribution of ionic concentrations in the diffusion layer and in the electric double layer will be discussed in detail.

Calculations of the ohmic resistance of the electrolyte outside the hydrodynamic boundary layer δ_0 will be made separately since a knowledge of the geometry of the cell and of the electrodes is required.

§ 4.3. Distribution of the Electric Field and of Ionic Concentrations in the Vicinity of the Electrode

The general equations (4.6) and (4.8), characterizing distribution of ionic concentrations and of the electric field in the cell with a stirred electrolyte, can be rewritten for a rotating disc electrode as follows:

$$w\frac{dc_1}{dz} = D_1\left[\frac{d^2c_1}{dz^2} + \frac{n_1F}{RT}\frac{d}{dz}(c_1E_z)\right],$$

$$w\frac{dc_2}{dz} = D_2\left[\frac{d^2c_2}{dz^2} + \frac{n_2F}{RT}\frac{d}{dz}(c_2E_z)\right],$$

$$w\frac{dc_3}{dz} = D_3\left[\frac{d^2c_3}{dz^2} - \frac{n_3F}{RT}\frac{d}{dz}(c_3E_z)\right],\qquad (4.15)$$

$$\frac{|dE_z}{dz} = -\frac{4\pi F}{\varepsilon_0}(n_1c_1 + n_2c_2 - n_3c_3).$$

Here $w = w(z)$ is the axial velocity component; $E_z = E_z(z)$ is the axial component of the electric field intensity. The concentrations c_1, c_2, and c_3 and the field intensity E_z were assumed to depend on the distance from the cathode only. The ionic mobility γ_j is related to the diffusion coefficient by Einstein's equation $\gamma_j = D_jF/RT$.

In the assumed arrangement (cathode at $z = 0$, anode $z \to \infty$), the cathodic current resulting from motions of cations 1 in the direction opposite to that of the z axis should be considered negative. Simultaneously, the electric field intensity is considered negative.† Correspondingly, electrodynamics [1] requires the cations to move in the direction of the lower electric potential. Assuming additionally that far from the cathode $\varphi \to 0$ (for $z \to \infty$), the disc potential acquires a negative sign.

Equations (4.15) must be supplemented by a system of boundary conditions. Let the electrolyte far from the cathode be electro-neutral. The concentrations in the solution bulk are then constant and satisfy the relation

$$n_1 \overset{\bullet}{c_1} + n_2 \overset{\bullet}{c_2} - n_3 \overset{\bullet}{c_3} = 0. \tag{4.16}$$

Therefore, as $z \to \infty$,

$$c_1 \to \overset{\bullet}{c_1}, \quad c_2 \to \overset{\bullet}{c_2}, \quad \text{and} \quad c_3 \to \overset{\bullet}{c_3}. \tag{4.17}$$

The present case concerns reduction of cations 1 only. Therefore, at $z = 0$,

$$i = n_1 F D_1 \left(\frac{dc_1}{dz} + \frac{n_1 F}{RT} c_1 E_z \right),$$
$$0 = n_2 F D_2 \left(\frac{dc_2}{dz} + \frac{n_2 F}{RT} c_2 E_z \right),$$
$$0 = - n_3 F D_3 \left(\frac{dc_3}{dz} - \frac{n_3 F}{RT} c_3 E_z \right). \tag{4.18}$$

Apart from the boundary conditions (4.18), another condition must be formulated for the vicinity of the electrode, which characterizes the rate of the electrochemical reaction at the cathode. With this condition the mathematical problem would be fully described. However, owing to reasons made clear below, this condition will be formulated later.

The discussion will be restricted to a detailed consideration of the simplest case — a system of three kinds of equally charged ions ($n_1 = n_2 = |n_3| = n$), with equal diffusion coefficients ($D_1 = D_2 = $

----

†Denoting by 1_z the unit vector directed in the positive direction of the z axis, one can obtain Eqs. (4.15) and boundary conditions (4.18) from Eqs. (4.6), (4.8), and (4.5) by the substitutiong $\mathbf{i} = -i1_z$ and $\mathbf{E} = -E_z 1_z$.

$D_3 = D$). This simplification has no substantial effect on the physical picture of the process, and it makes the mathematical treatment much easier. It is convenient to introduce dimensionless variables:

$$\xi = z/\delta_d; \quad C_1 = c_1/c_1^*, \quad C_2 = c_2/c_1^*, \quad C_3 = c_3/c_1^*;$$

$$\dot{I} = \frac{i\delta_d}{nFDc_1^*} = i/i_{d_1}; \quad \mathscr{E} = \frac{nF\delta_d E_z}{RT}. \tag{4.19}$$

Thus, concentrations are measured in units of concentration of electrochemically active cations, c_1^*, and distances are measured in units of the thickness of the diffusion layer δ_d, equal for all ions and determined by formula (2.20). Current is measured in units of the limiting diffusion current of cations 1.

Using the above dimensionless quantities and retaining the first term only of the expansion (2.18a) for the axial velocity component w, we can write the system (4.15) as follows:

$$\frac{d^2C_1}{d\xi^2} + \frac{d}{d\xi}(C_1\mathscr{E}) + a\xi^2 \frac{dC_1}{d\xi} = 0,$$

$$\frac{d^2C_2}{d\xi^2} + \frac{d}{d\xi}(C_2\mathscr{E}) + a\xi^2 \frac{dC_2}{d\xi} = 0,$$

$$\frac{d^2C_3}{d\xi^2} - \frac{d}{d\xi}(C_3\mathscr{E}) + a\xi^2 \frac{dC_3}{d\xi} = 0, \tag{4.20}$$

$$\varepsilon_e^2 \frac{d\mathscr{E}}{d\xi} = -(C_1 + C_2 - C_3),$$

where $a = 2.13$ is a numerical coefficient and

$$\varepsilon_e^2 = RT\varepsilon_0/4\pi n^2 F^2 c_1^* \delta_d^2. \tag{4.21}$$

The parameter ε_e has the same order of magnitude as the ratio of the Debye radius to the thickness of the diffusion layer: λ_D/δ_d. It has been mentioned previously that ε_e is small in comparison with unity.

The boundary conditions (4.18) can be rewritten using dimensionless variables as follows:

for $\xi \to \infty$

$$C_1 \to 1; \quad C_2 \to \frac{c_2^*}{c_1^*} = \gamma; \quad C_3 \to \frac{c_3^*}{c_1^*} = 1 + \gamma; \tag{4.22a}$$

for $\xi = 0$

$$\dot{I} = \frac{dC_1}{d\xi} + C_1 \mathscr{E},$$

$$0 = \frac{dC_2}{d\xi} + C_2 \mathscr{E},$$

$$0 = \frac{dC_3}{d\xi} - C_3 \mathscr{E}. \qquad (4.22b)$$

The parameter $\gamma = c_2^* / c_1^*$ characterizes the amount of the supporting electrolyte in the solution.

When $\gamma = 0$, the cell contains a binary electrolyte consisting of cations 1 and anions only. When $\gamma \gg 1$, the cell contains an excess of supporting electrolyte.

Introducing two new functions

$$C = C_1 + C_2 + C_3$$

$$\qquad (4.23)$$

$$\rho = C_1 + C_2 - C_3$$

characterizing the total concentration and density of electric charges, respectively, we can replace Eqs. (4.20), (4.22a), and (4.22b) by

$$\frac{d^2 C}{d\xi^2} + \frac{d}{d\xi}(\rho \mathscr{E}) + a\xi^2 \frac{dC}{d\xi} = 0,$$

$$\frac{d^2 \rho}{d\xi^2} + \frac{d}{d\xi}(C \mathscr{E}) + a\xi^2 \frac{d\rho}{d\xi} = 0,$$

$$\frac{d^2 C_1}{d\xi^2} + \frac{d}{d\xi}(C_1 \mathscr{E}) + a\xi^2 \frac{dC_1}{d\xi} = 0, \qquad (4.24)$$

$$\varepsilon_e^2 \frac{d\mathscr{E}}{d\xi} = -\rho.$$

When $\xi \to \infty$,

$$C \to 2(1 + \gamma), \quad \rho \to 0, \quad C_1 \to 1. \qquad (4.25a)$$

When $\xi = 0$,

$$\dot{I} = \frac{dC}{d\xi} + \mathscr{E}\rho,$$

$$\dot{I} = \frac{d\rho}{d\xi} + \mathscr{E}C,$$

$$\dot{I} = \frac{dC_1}{d\xi} + \mathscr{E}C_1. \qquad (4.25b)$$

It should be remembered that the full solution of the problem requires that the boundary conditions (4.25b) be supplemented by another criterion describing the kinetics of the cathodic electrode reaction.

The presence of the small parameter ε_e in Eqs. (4.20) allows a special method of approximation to be used to solve the problem. Application of this method to electrochemical problems was suggested by Chernenko [5] and Newman [6]. Since the method is rather general and can be used in solution of other problems, it will be explained here in detail.

The approximation is as follows: The small parameter ε_e^2 in front of the derivative $d\mathscr{E}/d\xi$ on the left side of the last equation of system (4.20) indicates the fully determined functional character of the quantity ρ, i.e., there exists a region of relatively small variation of the electric field $\mathscr{E}$, where net electrical charge is absent ($\rho \approx 0$), and a region of larger $\mathscr{E}$ variations where the electric charge differs from zero (region of the double layer). Correspondingly, separate solutions can be made for the bulk solution (when $\rho \approx 0$) and for the double-layer region.

The solution of the system (4.20) for bulk solution (outside the double layer) can be presented in the form of series expansions in powers of the parameter ε_e^2:

$$\bar{C} = \bar{C}^{(0)} + \varepsilon_e^2 \bar{C}^{(1)} + \varepsilon_e^4 \bar{C}^{(2)} + \ldots,$$

$$\bar{\rho} = \bar{\rho}^{(0)} + \varepsilon_e^2 \bar{\rho}^{(1)} + \varepsilon_e^4 \bar{\rho}^{(2)} + \ldots,$$

$$\bar{\mathscr{E}} = \bar{\mathscr{E}}^{(0)} + \varepsilon_e^2 \bar{\mathscr{E}}^{(1)} + \varepsilon_e^4 \bar{\mathscr{E}}^{(2)} + \ldots, \qquad (4.26)$$

$$\bar{C}_1 = \bar{C}_1^{(0)} + \varepsilon_e^2 \bar{C}_1^{(1)} + \varepsilon_e^4 \bar{C}_1^{(2)} + \ldots.$$

Substituting these expressions into (4.20) and equating terms containing equal powers of the parameter ε_e^2, we obtain equations which determine the terms of the zero, first, and following orders. Thus, in the zeroth approximation

$$\frac{d^2\bar{C}^{(0)}}{d\xi^2} + \frac{d}{d\xi}(\bar{\rho}^{(0)}\bar{\mathscr{E}}^{(0)}) + a\xi^2\frac{d\bar{C}^{(0)}}{d\xi} = 0,$$

$$\frac{d^2\bar{\rho}^{(0)}}{d\xi^2} + \frac{d}{d\xi}(\bar{C}^{(0)}\bar{\mathscr{E}}^{(0)}) + a\xi^2\frac{d\bar{\rho}^{(0)}}{d\xi} = 0,$$

$$\frac{d^2\bar{C}_1^{(0)}}{d\xi^2} + \frac{d}{d\xi}(\bar{C}_1^{(0)}\bar{\mathscr{E}}^{(0)}) + a\xi^2\frac{d\bar{C}_1^{(0)}}{d\xi} = 0, \qquad (4.27)$$

$$\bar{\rho}^{(0)} = 0.$$

A similar substitution of Eqs. (4.26) into the boundary conditions (4.25a) and (4.25b) results in boundary conditions for any desired approximation with respect to ε_e^2.

Before solutions are sought for the functions $\bar{C}^{(0)}, \bar{C}_1^{(0)}, \bar{\mathscr{E}}^{(0)} \dots$, it will be noted that the expansion of the function $\bar{\rho}$ starts with a quantity of the order of ε_e^2. In fact, as follows from the last equation of the system (4.27), $\bar{\rho}^{(0)} = 0$. This property of the function $\bar{\rho}$ reflects the electroneutrality of the solution (with an accuracy of the order of ε_e^2) outside of the double-layer region.

Therefore, ionic concentrations and the electric field intensity close to the rotating disc electrode can be described by the following equations:

$$\frac{d^2\bar{C}^{(0)}}{d\xi^2} + a\xi^2\frac{d\bar{C}^{(0)}}{d\xi} = 0,$$

$$\frac{d}{d\xi}(\bar{C}^{(0)}\bar{\mathscr{E}}^{(0)}) = 0,$$

$$\frac{d^2\bar{C}_1^{(0)}}{d\xi^2} + \frac{d}{d\xi}(\bar{C}_1^{(0)}\bar{\mathscr{E}}^{(0)}) + a\xi^2\frac{d\bar{C}_1^{(0)}}{d\xi} = 0, \qquad (4.28)$$

$$\bar{\rho}^{(0)} = 0,$$

with the following boundary conditions:

for $\xi \to \infty$

$$\bar{C}^{(0)} \to 2(1+\gamma), \quad \bar{\rho}^{(0)} \to 0, \quad \bar{C}_1^{(0)} \to 1; \qquad (4.29a)$$

for $\xi = 0$

$$I = \frac{d\bar{C}^{(0)}}{d\xi}, \quad I = \bar{\mathscr{E}}^{(0)}\bar{C}^{(0)},$$

$$I = \frac{d\bar{C}_1^{(0)}}{d\xi} + \bar{\mathscr{E}}^{(0)}\bar{C}_1^{(0)}. \qquad (4.29b)$$

Solutions for $\bar{C}^{(0)}$ and $\bar{\mathscr{E}}^{(0)}$ are of the form

$$\bar{C}^{(0)} = 2\,(1 + \gamma) - i\int\limits_{\xi}^{\infty} \exp\,(-\,at^3/3)\,dt = 2\,(1 + \gamma)\,F\,(\xi), \qquad (4.30)$$

$$\bar{\mathscr{E}}^{(0)} = \frac{i}{\bar{C}^{(0)}} = \frac{i}{2\,(1 + \gamma)\,F\,(\xi)} = \frac{i}{2\,(1 + \gamma) - i\int\limits_{\xi}^{\infty} \exp\,(-\,at^3/3)\,dt}, \qquad (4.31)$$

where

$$F\,(\xi) = 1 - \frac{i}{2\,(1 + \gamma)}\int\limits_{\xi}^{\infty} \exp\,(-\,at^3/3)\,dt. \qquad (4.32)$$

Equations (4.30) and (4.31) express the distribution of the total ionic concentraion and of the electric field intensity in the diffusion-layer region.

Substitution of Eq. (4.31) into the equation describing variations of the concentration of the discharging ion gives

$$\frac{d^2\bar{C}_1^{(0)}}{d\xi^2} + \frac{d}{d\xi}\left[\frac{i\bar{C}_1^{(0)}}{2\,(1 + \gamma)\,F\,(\xi)}\right] + a\xi^2\,\frac{d\bar{C}_1^{(0)}}{d\xi} = 0. \qquad (4.33)$$

The solution of this equation which satisfies the corresponding boundary conditions is of the form

$$\bar{C}_1^{(0)} = F\,(\xi)\left\{1 + \frac{2\gamma\,[2\,(1 + \gamma) - i]}{2\Gamma\,[2\,(1 + \gamma) - i] - 2\,(1 + \gamma)}\int\limits_{\xi}^{\infty} \frac{F'}{F^2}\exp\left(-\int\limits_{0}^{t} \frac{i\,dz}{2\,(1 + \gamma)\,F}\right)dt\right\},$$

$$(4.34)$$

where

$$\Gamma = \int\limits_{0}^{\infty} \frac{F'}{F^2}\exp\left(-\int\limits_{0}^{t} \frac{i\,dz}{2\,(1 + \gamma)\,F}\right)dt. \qquad (4.35)$$

Expressions (4.30), (4.31), and (4.34), together with the electroneutrality condition, allow the distribution in the diffusion layer of each ion and of the electric field to be found.

Unfortunately, expression (4.34) which describes the distribution of the discharging cation is rather cumbersome. This renders it difficult to make a direct analysis of the results obtained. A convenient way to obtain simpler expressions consists in using the qualitative Nernst model. As previously (§ 2.8) assumed, within the diffusion-layer region of thickness δ_d the solution will rotate together with the disc, and there will be no motion in the axial direction. The electrolyte outside the diffusion layer is thoroughly stirred.† Since application of the Nernst model to mass-transport phenomena in stirred media has been strongly criticized [8], it should be stressed that the results below have only a qualitative value. Nevertheless, the character of the distribution of ions at the disc and at large distance from the latter should be considered valid within the framework of the Nernst model. Close to the external limit of the diffusion layer δ_d, the actual distribution of concentrations and of the electric field differs considerably from that obtained using the Nernst model.

Using the Nernst model, we can express the function $F(\xi)$ which appears in Eqs. (4.30) and (4.31) in the form

$$F(\xi) = \begin{cases} 1 - \dfrac{I}{2(1+\gamma)}(1-\xi) & 0 \leqslant \xi \leqslant 1, \\ 1 & \xi \geqslant 1. \end{cases} \qquad (4.36)$$

Equations (4.30), (4.31), and (4.34) are also considerably simplified.

Equation (4.30) describing the total concentration $\bar{C}^{(0)}(\xi)$ now becomes

$$\bar{C}^{(0)}(\xi) = \begin{cases} 2(1+\gamma) - I(1-\xi) & 0 \leqslant \xi \leqslant 1, \\ 2(1+\gamma) & \xi \geqslant 1. \end{cases} \qquad (4.37)$$

---

†It can easily be shown that the Nernst model is equivalent to an electrochemical system which has been discussed by a number of authors [3, 7]. It consists of a capillary of a finite length l connecting the electrode with a large volume of a stirred solution. The results of the work quoted coincide with those presented in Chap. 4 if δ_d is replaced by l.

The concentration of the discharging cations $\bar{C}_1^{(0)}(\xi)$ is now described by

$$\bar{C}_1^{(0)}(\xi) = \begin{cases} (1+\gamma)\left[1 - \dfrac{I(1-\xi)}{2(1+\gamma)}\right] - \dfrac{\gamma}{\left[1 - \dfrac{I(1-\xi)}{2(1+\gamma)}\right]} & 0 \leqslant \xi \leqslant 1, \\ 1 & \xi > 1 \end{cases} \quad (4.38)$$

[instead of Eq. (4.34)], and Eq. (4.31) for the distribution of the electric field is replaced by

$$\bar{\mathscr{E}}^{(0)}(\xi) = \begin{cases} \dfrac{I}{2(1+\gamma) - I(1-\xi)} & 0 \leqslant \xi \leqslant 1, \\ \dfrac{I}{2(1+\gamma)} & \xi > 1. \end{cases} \quad (4.39)$$

The character of variations of ionic concentrations and of the electric field described by Eqs. (4.37), (4.38), and (4.39) is shown in Fig. 4.2. The values of surface concentrations $\bar{C}_S^{(0)}$ and $\bar{C}_{1S}^{(0)}$ and of the electric field at the disc surface $\bar{\mathscr{E}}_S^{(0)}$ can be easily

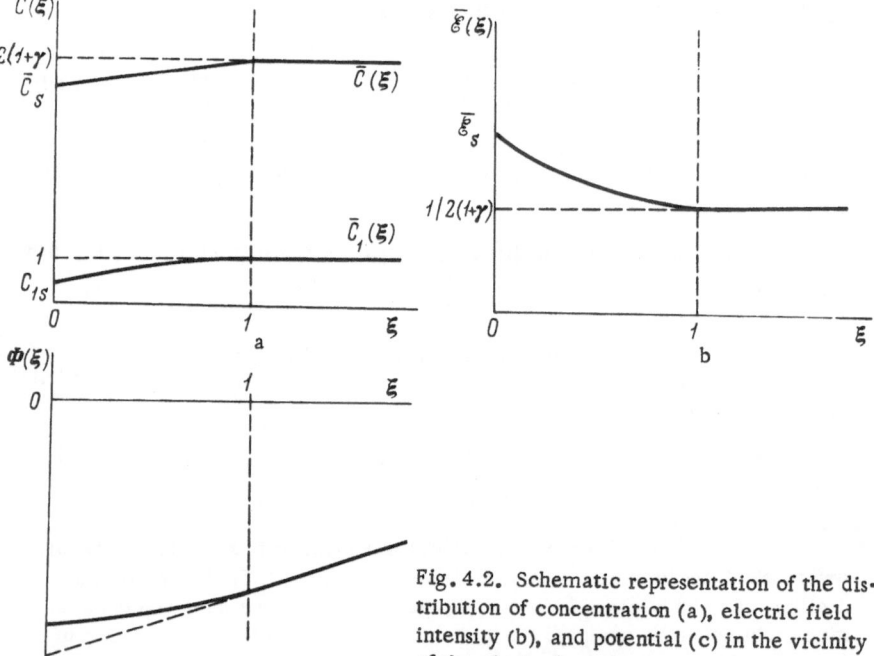

Fig. 4.2. Schematic representation of the distribution of concentration (a), electric field intensity (b), and potential (c) in the vicinity of the electrode surface.

found by substituting $\xi = 0$ in expressions (4.37)–(4.39). The results thus obtained will be explicitly presented later.

It can be seen from Fig. 4.2a that the concentration of electrochemically active cations, $\bar{C}_1^{(0)}(\xi)$, decreases in the diffusion layer. The change of the total concentration, $\bar{C}^{(0)}(\xi)$, virtually coincides with that of $\bar{C}_1^{(0)}(\xi)$. It should be mentioned, however, that the relative change of $\bar{C}^{(0)}(\xi)$ turns out to be different. In a solution of a binary electrolyte, i.e., in the absence of cations of type 2 ($\gamma = 0$), the variation of $\bar{C}^{(0)}(\xi)$ in the diffusion layer is quite significant [change of anion concentration coincides with that of $\bar{C}_1^{(0)}(\xi)$]. Conversely, in the presence of the supporting electrolyte ($\gamma \gg 1$), $\bar{C}^{(0)}(\xi)$ hardly changes in the diffusion layer, and the concentrations of ions of the supporting electrolyte are practically uniform over the entire volume of the cell.

The expressions derived above describing the distribution of ionic concentrations and of the electric field close to the rotating disc electrode can be used to find contributions of each kind of ion to the electric current. This can best be done using Eqs. (4.37)–(4.39). Within the diffusion layer ($\xi \leq 1$), the whole current is carried by cations of type 1. It is easily verified, by a direct substitution of Eqs. (4.38) and (4.39), that for $\xi \leq 1$,

$$\frac{dC_1}{d\xi} + C_1 \mathscr{E} = I,$$

$$\frac{dC_2}{d\xi} + C_2 \mathscr{E} = 0,$$

$$\frac{dC_3}{d\xi} - C_3 \mathscr{E} = 0.$$

Outside the diffusion layer ($\xi > 1$), the partial currents for the separate ions are

$$\dot{I}_1 = C_1 \mathscr{E} = \frac{i}{2(1+\gamma)},$$

$$\dot{I}_2 = C_2 \mathscr{E} = \frac{\gamma i}{2(1+\gamma)},$$

$$\dot{I}_3 = C_3 \mathscr{E} = \frac{(1+\gamma) i}{2(1+\gamma)}.$$

Thus, at large distance from the electrode the fraction of current carried by the given ion is proportional to its concentration,

in agreement with the qualitative considerations of the former
section. Redistribution of the current among the different kinds
of ions occurs close to the diffusion-layer boundary δ_d.

The distribution of the electric field in the vicinity of the rotating disc cathode will now be found. According to Eq. (4.3), the
electric field potential φ is described in the present case by
$E_z = d\varphi/dz$, which can be transformed by introducing a dimensionless
potential

$$\Phi = \frac{nF\varphi}{RT}, \tag{4.40}$$

into

$$\frac{d\Phi}{d\xi} = \mathscr{E}. \tag{4.41}$$

The change of the electric potential between two arbitrary
points ξ_1 and ξ_2 is given by

$$\Phi(\xi_1) - \Phi(\xi_2) = \int_{\xi_2}^{\xi_1} \mathscr{E}dy. \tag{4.42}$$

Thus, the potential difference between a point ξ near the surface
of the disc electrode and a point placed at infinity can be found by
substitution of Eq. (4.31) in Eq. (4.42):

$$\Delta\Phi = \Phi(\xi) - \Phi_\infty = -\int_{\xi}^{\infty} \frac{\bar{i}dy}{2(1+\gamma) - \bar{i}\int_{y}^{\infty} \exp(-at^3/3)\,dt}. \tag{4.43}$$

Here Φ_∞ is the potential at infinity, assumed to be zero.

Following previously accepted conventions, the electric potential decreases with decreasing distance from the cathode
$[\Phi(\xi) < \Phi_\infty]$, and consequently $\Phi(\xi)$ must be negative.

The above picture of the passage of current is valid for a real
electrochemical cell at a large but finite distance from the cathode,
$l = L\delta_d$ (naturally, $L \gg 1$). At greater distances, factors which
were not hitherto taken into account begin to play a significant role,
e.g., the presence of the anode, the finite dimensions of the elec-

trodes and cell, etc. Therefore, a stricter treatment requires replacement of Eq. (4.43) by

$$\Delta\Phi = \Phi(\xi) - \Phi(L) = -\int\limits_{\xi}^{L} \frac{I\,dy}{2(1+\gamma) - I\int\limits_{y}^{\infty} \exp(-at^3/3)\,dt},$$

where $\Phi(L)$ is the potential at a point far from the cathode $[\Phi(L) = 0]$.

Using the model of convective diffusion described above and substituting the approximate relation (4.39) into Eq. (4.42), we obtain

$$\Delta\Phi = \Phi(\xi) = -\frac{I(L-\xi)}{2(1+\gamma)} \qquad (4.44a)$$

if the point ξ is located outside the diffusion layer $(1 \leq \xi \leq L)$, and

$$\Delta\Phi = \Phi(\xi) = -\ln\frac{2(1+\gamma)}{2(1+\gamma) - I(1-\xi)} - \frac{I(L-1)}{2(1+\gamma)} \qquad (4.44b)$$

if the point ξ is located inside the diffusion layer (in the region of varying concentration) $(0 \leq \xi \leq 1)$.

It can easily be shown that the quantity $1/2(1+\gamma)$ is the dimensionless specific resistivity of the solution. The quantity

$$\Delta\Phi_{ohm} = -\frac{I(L-1)}{2(1+\gamma)} \qquad (4.45)$$

in Eq. (4.44b) represents the ohmic potential drop in the electrolyte.

The total potential drop $\Delta\Phi$ between a point ξ inside the diffusion layer and a point far from the cathode may be expressed as

$$\Delta\Phi = \Delta\Phi_{dif} + \Delta\Phi_{ohm}.$$

The first term

$$\Delta\Phi_{dif} = -\ln\frac{2(1+\gamma)}{2(1+\gamma) - \tilde{I}(1-\xi)} \qquad (4.46)$$

describes the potential drop within the diffusion layer. In particular, this quantity reflects the effect of concentration gradients in the vicinity of the cathode. It should not, however, be identified with concentration polarization, which will be discussed later.

The potential distribution at a rotating disc cathode described by Eqs. (4.44a) and (4.44b) is shown in Fig. 4.2c.

§ 4.4. Distribution of the Electric Field and of Ionic Concentrations in the Double-Layer Region

It has already been mentioned that the distributions of electric field and concentration derived in the former section are valid only in the region of electroneutrality of the solution, where the variation of field with distance is relatively small. The functional dependence on distance of field and concentration is totally different in the electric double layer at the interface.

The ionic part of the double layer can be divided into the two well-known regions: one, called the "compact" or Helmholtz layer, and the other, the "diffuse" or Gouy layer. The compact layer consists of ions held by coulombic or specific adsorption forces in immediate contact with the surface. Descriptions of the compact layer usually rely on model (most often electrostatic) considerations [9]. The diffuse layer consists of ions also attracted to the electrode by electric forces but with thermal motions resulting, however, in diffuseness of the charge distribution. The structure of the diffuse layer in the absence of current flow is similar to the Debye ionic atomsphere surrounding a given ion in electrolyte solutions. The thickness of the diffuse layer is characterized by the Debye radius λ_D.

Passage of current destroys the equilibrium distributions of electric field and ionic concentration, which now depend on the magnitude of current i.

The theory of the double layer under nonequilibrium conditions was first discussed by Levich [10]. A detailed analysis of the structure of a nonequilibrium double layer in a binary electrolyte was made by Grafov and Chernenko [5, 7].

The distribution of the electric field and ionic concentrations in the double layer will now be evaluated. A constant current i is assumed to be passing through the solution containing, as previously, three kinds of ion. Equations (4.20) can be rewritten in the form

$$C = \bar{C} + \tilde{C},$$
$$\rho = \bar{\rho} + \tilde{\rho},$$
$$\mathscr{E} = \bar{\mathscr{E}} + \frac{1}{\varepsilon_e}\tilde{\mathscr{E}},$$
$$C_1 = \bar{C}_1 + \tilde{C}_1. \tag{4.47}$$

Here $\bar{C}$, $\bar{\rho}$, $\bar{\mathscr{E}}$, and $\bar{C}_1$ represent the solutions of Eqs. (4.20) found in the former section. The effect of the double layer, i.e., of the nonelectroneutral region, is described by the functions $\tilde{C}$, $\tilde{\rho}$, $\tilde{\mathscr{E}}$,† and $\tilde{C}_1$ which rapidly vary at distances of the order of λ_D from the electrode. Analysis of such a "micromodel" requires some changes in the scale of distance measurements. The distance will be "stretched" by designating

$$\xi = \varepsilon_e t. \tag{4.48}$$

Thus, all distances will be measured in units of the Debye radius λ_D.

Substituting Eqs. (4.47) and (4.48) into Eqs. (4.20) and using expressions (4.28) for the functions $\bar{C}$, $\bar{\rho}$, $\bar{\mathscr{E}}$, and $\bar{C}_1$, we obtain

$$\tilde{C}'' + (\tilde{\rho}\bar{\mathscr{E}} + \varepsilon_e \tilde{\rho}\bar{\mathscr{E}})' + a\varepsilon_e^3 t^2 \tilde{C}' = 0,$$
$$\tilde{\rho}'' + (\tilde{C}\bar{\mathscr{E}} + \bar{C}\tilde{\mathscr{E}} + \varepsilon_e \tilde{C}\bar{\mathscr{E}})' + a\varepsilon_e^3 t^2 \tilde{\rho}' = 0,$$
$$\tilde{C}_1'' + (\tilde{C}_1\bar{\mathscr{E}} + \bar{C}_1\tilde{\mathscr{E}} + \varepsilon_e \tilde{C}_1\bar{\mathscr{E}})' + a\varepsilon_e^3 t^2 \tilde{C}_1' = 0,$$
$$(\tilde{\mathscr{E}} + \varepsilon_e \bar{\mathscr{E}})' = -\tilde{\rho}. \tag{4.49}$$

The prime designates differentiation with respect to the variable t.

The boundary conditions (4.25a) and (4.25b) may be similarly transformed. The new system of boundary conditions is then as follows:

for $t \to \infty$

$$\tilde{C} \to 0, \quad \tilde{\rho} \to 0, \quad \tilde{C}_1 \to 0, \tag{4.50a}$$

---

†The factor $1/\varepsilon_e$ in front of $\tilde{\mathscr{E}}$ allows all functions within the double layer to be considered of the same (small) order of magnitude. This will be made clear in the subsequent text.

for $t = 0$

$$C_1' = -(\tilde{\mathscr{E}} + \varepsilon_e \bar{\mathscr{E}})\tilde{\rho},$$
$$\tilde{\rho}' = -(\tilde{\mathscr{E}}C + \tilde{\mathscr{E}}C + \varepsilon_e \bar{\mathscr{E}}C), \qquad (4.50b)$$
$$\tilde{C}_1' = -(\tilde{\mathscr{E}}C_1 + \tilde{\mathscr{E}}C_1 + \varepsilon_e \bar{\mathscr{E}}C_1).$$

The solution of Eqs. (4.49) with the boundary conditions (4.50a) can be sought in the form of a series expansion in powers of the small parameter ε_e:

$$C = C^{(0)} + \varepsilon_e C^{(1)} + \dots,$$
$$\tilde{\rho} = \tilde{\rho}^{(0)} + \varepsilon_e \tilde{\rho}^{(1)} + \dots,$$
$$\tilde{\mathscr{E}} = \tilde{\mathscr{E}}^{(0)} + \varepsilon_e \tilde{\mathscr{E}}^{(1)} + \dots, \qquad (4.51)$$
$$C_1 = C_1^{(0)} + \varepsilon_e C_1^{(1)} + \dots.$$

Collecting quantities in the same power of ε_e, we obtain the equations for determining the corresponding expansions terms. Thus, the functions of zeroth order are easily shown to be

$$C^{(0)''} + (\tilde{\rho}^{(0)}\tilde{\mathscr{E}}^{(0)})' = 0,$$
$$\tilde{\rho}^{(0)''} + (C^{(0)}\tilde{\mathscr{E}}^{(0)} + \bar{\sigma}_s \tilde{\mathscr{E}}^{(0)})' = 0,$$
$$C_1^{(0)''} + (C_1^{(0)}\tilde{\mathscr{E}}^{(0)} + \bar{\sigma}_{1s}\tilde{\mathscr{E}}^{(0)})' = 0, \qquad (4.52)$$
$$\tilde{\mathscr{E}}^{(0)'} = -\tilde{\rho}^{(0)}.$$

It should be remembered in the derivation of Eq. (4.52) that the transformation of variables [from ξ to t, Eq. (4.48)] must be made also in the functions $\bar{C}$, $\bar{\mathscr{E}}$, and $\bar{C}_1$. Upon such a transformation of variables, the latter functions become ε_e dependent. Expansion in a power series, analogous to (4.51), must also be made in the functions $\bar{C}$, $\bar{\mathscr{E}}$, and $\bar{C}_1$. It can be easily shown, however, that the terms in zero powers of ε_e in the latter functions are identical with the values of these functions at $\xi = 0$. Therefore,

$$\bar{\sigma}_s = 2(1 + \gamma) - 1,$$
$$\bar{\mathscr{E}}_s = \frac{1}{2(1 + \gamma) - 1}, \qquad (4.53)$$
$$\bar{\sigma}_{1s} = \frac{(1 + \gamma)[1 - 1/2(1 + \gamma)]^2 - \gamma}{1 - 1/2(1 + \gamma)}.$$

The boundary conditions (4.50b) for functions of zeroth order with respect to the parameter ε_e are then for $t = 0$

$$C^{(0)\prime} = - \mathcal{E}^{(0)} \widetilde{\rho^{(0)}},$$

$$\widetilde{\rho}^{(0)\prime} = - \mathcal{E}^{(0)} (C^{(0)} + \bar{C}_S),$$

$$C_1^{(0)\prime} = - \mathcal{E}^{(0)} (C_1^{(0)} + \bar{C}_{1S}).$$

(4.54)

Further considerations are restricted to zeroth-order terms (with respect to ε_e) only. Therefore, for reasons of simplicity, the superscript is omitted in the subsequent text.

It can be easily checked by direct substitution that Eqs. (4.55a)-(4.55) are solutions of the system (4.52) with the boundary conditions (4.50a) and (4.54):

$$C = \frac{\bar{C}_S}{2} \left[\tanh^2 \frac{\sqrt{\bar{C}_S}}{2} (t + A) + \coth^2 \frac{\sqrt{\bar{C}_S}}{2} (t + A) - 2 \right], \quad (4.55a)$$

$$\widetilde{\rho} = \frac{\bar{C}_S}{2} \left[\coth^2 \frac{\sqrt{\bar{C}_S}}{2} (t + A) - \tanh^2 \frac{\sqrt{\bar{C}_S}}{2} (t + A) \right], \quad (4.55b)$$

$$\mathcal{E} = \frac{2 \sqrt{\bar{C}_S}}{\sinh \sqrt{\bar{C}_S} (t + A)}, \quad (4.55c)$$

$$C_1 = \bar{C}_{1S} \left[\tanh^2 \frac{\sqrt{\bar{C}_S}}{2} (t + A) - 1 \right]. \quad (4.55d)$$

Here A is a constant, the value of which is dependent on the kinetics of the electrochemical reaction of cations of type 1 and will be determined later.

By substituting Eqs. (4.55a)-(4.55d) into Eq. (4.47), we obtain the following expressions which describe the concentration and field distribution in the double layer during the passage of a constant current $\dot{I}$:

$$C = \bar{C}(\xi) + \frac{\bar{C}_S}{2} \left[\tanh^2 \frac{\sqrt{\bar{C}_S}}{2\varepsilon_e} (\xi + A) + \coth^2 \frac{\sqrt{\bar{C}_S}}{2\varepsilon_e} (\xi + A) - 2 \right], \quad (4.56a)$$

$$\rho = \frac{\bar{C}_S}{2} \left[\coth^2 \frac{\sqrt{\bar{C}_S}}{2\varepsilon_e} (\xi + A) - \tanh^2 \frac{\sqrt{\bar{C}_S}}{2\varepsilon_e} (\xi + A) \right], \quad (4.56b)$$

$$C_1 = \bar{C}_1(\xi) + \bar{C}_{1S} \left[\tanh^2 \frac{\sqrt{\bar{C}_S}}{2\varepsilon_e} (\xi + A) - 1 \right], \quad (4.56c)$$

$$\mathscr{E} = \frac{1}{\overline{C}(\xi)} + \frac{2\sqrt{\overline{C}_S}}{\varepsilon_e \sinh \frac{\sqrt{\overline{C}_S}}{\varepsilon_e}(\xi + A)}. \qquad (4.56d)$$

The potential of the electric field Φ inside the ionic part of the double layer may be found by direct integration of Eq. (4.56d) using relation (4.42). The potential difference between a point ξ located inside the double layer and a point at infinity is

$$\Phi(\xi) = \Delta\Phi + \ln \tanh^2 \frac{\sqrt{\overline{C}_S}}{2\varepsilon_e}(\xi + A), \qquad (4.57)$$

where $\Delta\Phi$ is given by Eq. (4.43).

It is easy now to establish on the basis of Eqs. (4.56) and (4.57) the effect of the ionic part of the double layer on the distributions of ionic concentration, bulk charge density, and electric field in the cell. First, it should be noted that according to Eq. (4.56b) the bulk charge region extends over a distance $\xi_D \approx \varepsilon_e \times (\overline{C}_S)^{1/2}$. The thickness of the double layer can easily be obtained by substitution of the equation for $\overline{C}_S$ from (4.53) into the latter expression for ξ_D:

$$z_D \approx \lambda_D \sqrt{\frac{2(1+\gamma)}{2(1+\gamma)-1}}, \qquad (4.58)$$

where the Debye radius is given by

$$\lambda_D = [RT\varepsilon_0/4\pi n^2 F2c_1^*(1+\gamma)]^{1/2}. \qquad (4.59)$$

Equation (4.58) characterizes the thickness of the nonequilibrium double layer subject to the passage of constant current. In the absence of current ($\overset{*}{I} = 0$), the thickness of the double layer is identical with the Debye radius ($z_D = \lambda_D$), and expressions (4.56a)-(4.56d) and (4.57) are transformed into the corresponding equilibrium relations [6]. It should also be noted that a similar relation is obeyed in other cases as well. If, for example, the current passing through the cell is small compared with the limiting diffusion current for cation reduction, $\overset{*}{I} \ll 1$, then $z_D \approx \lambda_D$, so that the passage of current practically does not effect the equilibrium charge distribution in the cell.

If the cell solution contains an excess of inert electrolyte, the quantity γ characterizing the concentration of the supporting electrolyte is large ($\gamma \gg 1$). In this case, the ionic part of the double layer is formed almost exclusively from ions of the inert electrolyte. The flow of cations discharged at the cathode occurs in a field formed by the background electrolyte and does not significantly affect the structure of the double layer. Therefore, $z_D \approx \lambda_D$.

It can be seen from Eq. (4.58) that in the case of cation reduction passage of current causes an increase of the double-layer thickness. This increase is due to a decrease of cation concentration at the cathode to a value lower than that of the bulk. This results in a certain increase of the Debye radius in comparison with the λ_D value corresponding to the bulk concentration.

It must be mentioned that the relations derived above become meaningless when the current passing through a cell with a binary electrolyte ($\gamma = 0$) approaches the limiting value ($\dot{I} \rightarrow 2$). This case requires a special analysis (cf. [6]).

If the electrochemical process is anodic, i.e., it results in formation of cations, all the preceding equations remain valid after replacing $\dot{I}$ by $-\dot{I}$. This reflects the opposite direction of the current.

The thickness of the double layer is now expressed by

$$z_D \approx \lambda_D \sqrt{\frac{2(1+\gamma)}{2(1+\gamma)+|\dot{I}|}} \, . \tag{4.60}$$

It can be seen that anodic processes result in a decrease of the double-layer thickness. This is connected with an increase of cation concentration at the electrode above the bulk value, c_1^*.

§ 4.5. Voltammetric Curves

The voltammetric curve for an electrochemical cell can be obtained by supplementing the equations describing the transport of ions by kinetic equations for their reactions at the electrode surface. The form of the latter equations depends on the kinetics of the given electrochemical process, and this kinetic equation is the omitted boundary equation mentioned in § 4.3.

Let the cations of type 1 become reduced at the cathode to neutral particles. A reaction the products of which are not subject to mass transport in solution, e.g., reduction of metal ions, will be considered, viz.

$$\text{Me}^{n+} + ne^- \underset{k_b}{\overset{k_f}{\rightleftharpoons}} \text{Me}. \tag{4.61}$$

It is well known [3, 11] that the rates of electrochemical reactions depend on the concentration of discharging ions and on the potential difference at the electrode−solution interface. Assuming that the charge transfer involves ions located at the distance of closest approach to the electrode,† the reaction rate is described by an expression of the form

$$-i = nF \{k_f^0 c_{1S} \exp\left[-\alpha nF(\varphi_M - \psi_1)/RT\right] - k_b^0 \exp\left[(1-\alpha) nF(\varphi_M - \psi_1)/RT\right]\}. \tag{4.62}$$

Rate constants of the forward and back reactions (k_f^0 and k_b^0) are independent of the potential difference ($\varphi_M - \psi_1$) at the interface; c_{1S} is the concentration of the discharging ions in the plane of closest approach to the electrode surface, and the minus sign on the left side of Eq. (4.62) indicates that, in agreement with accepted conventions, the reduction current is considered negative. The quantity $\varphi_M - \psi_1$ is the potential difference between the electrode (φ_M) and the plane of closest approach of the ions (ψ_1).

The concentration c_{1S} and potential ψ_1 of this plane can be obtained from Eqs. (4.56c) and (4.57) by putting $\xi = 0$. However, it is extremely difficult to describe the voltammetric curve, i.e., the dependence of current i on electrode potential (φ_M), in the general case. Therefore, the discussion will be restricted to simpler cases.

By the introduction of dimensionless quantities [Eqs. (4.19) and (4.40)], we transform Eq. (4.62) into

$$-I = \frac{\delta_d nFk_f^0 c_1^*}{nFDc_1^*} \left\{ C_{1S} \exp\left[-\alpha(V_M - \Psi_1)\right] - \frac{k_b^0}{k_f^0 c_1^*} \exp\left[(1-\alpha)(V_M - \Psi_1)\right] \right\}, \tag{4.63}$$

where $V_M = nF\varphi_M/RT$ and $\Psi_1 = nF\psi_1/RT$.

---

†This is the simplest model. A more general case is discussed in the modern quantum-mechanical theory of electrode reactions [12-15].

Replacing V_M and Ψ_1 in Eq. (4.63) by the respective equilibrium vaues V_e and Ψ_{1e}, and putting $\dot{I} \equiv 0$, we obtain

$$k_f^0 c_1^* C_{1Se}/k_b^0 = \exp(V_e - \Psi_{1e}). \qquad (4.64)$$

The concentration of cations of type 1, C_{1Se}, in the $\xi = 0$ plane and the Ψ_{1e} potential are determined by the double-layer equations (cf. § 4.4). In particular, it can be easily shown using Eqs. (4.56c) and (4.57) in the absence of current that the equilibrium concentration of cations of type 1 obeys the Boltzmann distribution:

$$C_{1Se} = \exp(- \Psi_{1e}) \qquad (4.65)$$

The equilibrium potential of the system (4.61) is expressed by the Nernst equation obtained by combining Eqs. (4.64) and (4.65):

$$V_e = \ln \frac{k_f^0 c_1^*}{k_b^0}. \qquad (4.66)$$

Combining Eqs. (4.66) and (4.63), we obtain the standard equation describing the rate of electrochemical reactions (4.61):

$$- I = I_0 \{C_{1S} \exp[-\alpha(\eta - \Psi_1)] - \exp[(1 - \alpha)(\eta - \Psi_1)]\}, \qquad (4.67)$$

where the exchange current density

$$I_0 = nF(k_f^0 c_1^*)^{(1-\alpha)} k_b^{0\alpha} \delta_d/nFDc_1^* \qquad (4.68)$$

is written in dimensionless units. The quantity $\eta = V_M - V_e$ represents the overpotential (i.e., the excess electrode−solution potential difference as compared with the equilibrium value), which is necessary to maintain the passage of the constant current $\dot{I}$ in the cell. We recall that Ψ_1 is the electric potential in the plane of closest approach of ions, different from that in the solution bulk.

For simplicity, it will be assumed that the whole potential drop at the electrode−solution interface occurs within the compact part (Helmholtz layer) † of the double layer.

---

†This approach corresponds to the model concepts of the kinetics of electrochemical processes introduced by Volmer and Erdey-Gruz [16].

To obtain the values of the cation concentration $\overline{C}_{1S}$ and of the potential Ψ_1 in this case, it is sufficient to utilize Eqs. (4.38) and (4.44b) derived for the diffusion layer. Putting $\xi = 0$ in the latter expressions and substituting values of C_{1S} and $\Psi_1 = \Phi(0) = \Delta\Phi_S$, we obtain

$$-\dot{I} = \dot{I}_0 \{\overline{C}_{1S} \exp[-\alpha(\dot{\eta} - \Delta\Phi_S)] - \exp[(1-\alpha)(\eta - \Delta\Phi_S)]\}. \quad (4.69)$$

Then, substituting the explicit expressions for $\overline{C}_{1S}$ and $\Delta\Phi_S$, and after some further transformations, we obtain the voltammetric characteristics of the cell

$$-\dot{I} = \frac{\dot{I}_0}{2[2(1+\gamma) - \dot{I}]^{(1-\alpha)}[2(1+\gamma)]^{\alpha}} \{[(2(1+\gamma) - \dot{I})^2 - \\ - 4\gamma(1+\gamma)]\exp[-\alpha(\eta - \Delta\Phi_{ohm})] - \\ - 4(1+\gamma)\exp[(1-\alpha)\eta - \Delta\Phi_{ohm})]\}, \quad (4.70)$$

where $\Delta\Phi_{ohm}$ is the ohmic potential drop in the bulk solution. Its value is determined by Eq. (4.45).

The polarization curve (4.70) allows an analysis to be made of the dependence of the net rate of an electrochemical reaction on the following factors which affect the passage of current.

A. If the transport of discharging ions to the electrode surface is sufficiently fast and the electrochemical reaction itself has a low rate constant, the concentrations of the reactions at the electrode are practically unaffected by current so that concentrations in the cell are everywhere equal to their values. Assuming that the current in Eq. (4.70) is much lower than the limiting diffusion current (i.e., $\dot{I} \ll 1$), the voltammetric characteristics can be easily transformed into

$$-\dot{I} = \dot{I}_0 \{\exp[-\alpha(\eta - \Delta\Phi_{ohm})] - \exp[(1-\alpha)(\eta - \Delta\Phi_{ohm})]\}. \quad (4.71)$$

Under these conditions, the polarization of an electrolytic cell is purely electrochemical in nature. The total potential drop in the cell during passage of current $\dot{I}$, consists of the ohmic drop $(-\Delta\Phi_{ohm})$ and the interfacial potential difference $(V_M - \Delta\Phi_{ohm})$. The rate of the electrochemical reaction depends on the magnitude of the overpotential $(V_M - V_e - \Delta\Phi_{ohm})$. The potential distribution in the cell under conditions of activation (electrochemical) polariza-

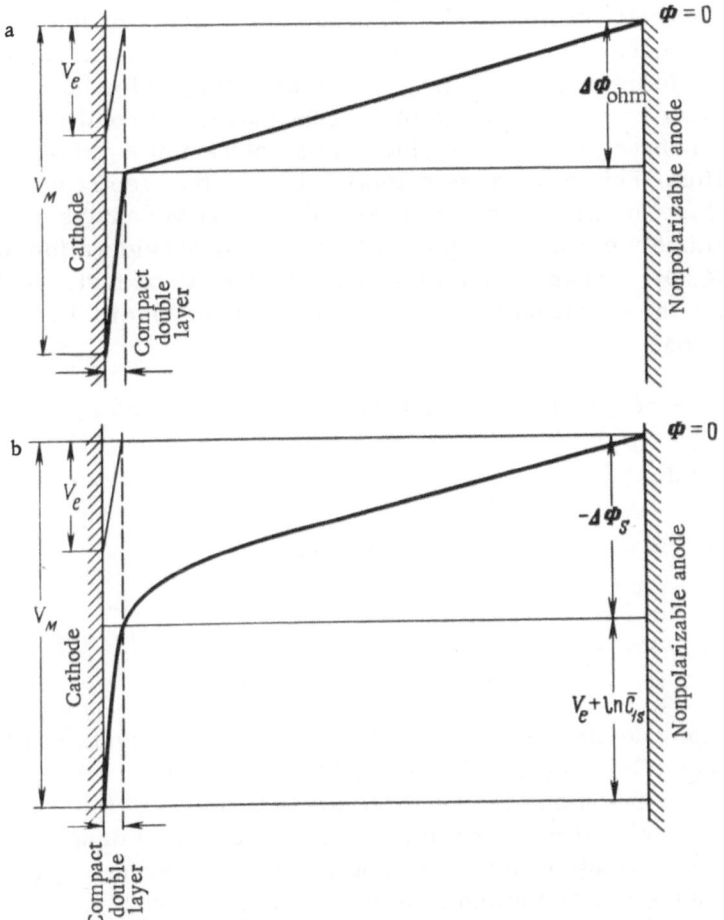

Fig. 4.3. Potential distribution in a cell with a rotating disc cathode under equilibrium and polarization conditions. a) Fast transport of reactants and slow electrode reaction; b) slow transport of reactants and fast electrode reaction.

tion is shown in Fig. 4.3. The equilibrium distribution is also shown for comparison.

B. The opposite limiting case is that where the rate is conrolled by transport of reactants to the reaction site, with the electrochemical reaction itself being fast. In this case it may be assumed that the equilibrium of the electrode—electrolyte interface is practically undisturbed. However, the nonequilibrium distribution of reactant concentrations (i.e., when concentrations at the electrode interface are higher, or lower, than in the bulk solution) resulting from the passage of current makes the equilibrium electrode behave as if it were placed in a solution of a different composition. This causes the potential drop at the electrode−solution interface to change. It is convenient to start the analysis for the voltammetric characteristic in this case by consideration of Eq. (4.69). Assuming the electrode reaction to be fast, and inserting the condition $I_0 \gg 1$ into Eq. (4.69), we obtain the following condition:

$$\bar{C}_{1S} \exp\left[-\alpha(\eta - \Delta\Phi_S)\right] = \exp\left[(1-\alpha)(\eta - \Delta\Phi_S)\right],$$

from which

$$\eta = \Delta\Phi_S + \ln \bar{C}_{1S} \qquad (4.72)$$

or

$$V_M = V_e + \ln \bar{C}_{1S} + \Delta\Phi_S. \qquad (4.73)$$

Here the question of concentration polarization arises, and the total potential drop consists of the potential drop in the solution, $\Delta\Phi_S$ (i.e., in and outside the diffusion layer), and of the additional (as compared with the equilibrium value) potential drop† at the electrode−solution interface required for net passage of current. The potential distribution in this case is shown in Fig. 4.3b.

Substitution of the value of $\overline{C}_{1S}$ [Eq. (4.56b)] into Eq. (4.72) gives

$$\eta = \Delta\Phi_{ohm} + \ln \frac{[2(1+\gamma) - I]^2 - 4\gamma(1+\gamma)}{4(1+\gamma)}, \qquad (4.72a)$$

where $\Delta\Phi_{ohm}$ is the ohmic drop in the cell volume (outside the diffusion layer). Equation (4.72a) allows the concentration polarization to be calculated for a cell with a rotating disc electrode containing

---

†This potential drop can be conveniently treated as the change of the equilibrium potential of the system.

electrochemically active cations and inert electrolyte in any concentration ratio.

Frumkin [3, 17] demonstrated that the effect of the diffuse double layer on electrode kinetics can be neglected [as was done in deriving Eq. (4.70)] only in those cases where the total concentration is sufficiently high. In dilute solutions charge transport occurs within the diffuse part of the double layer. Therefore, the concentration of discharging ions, as well as the effective electrode−solution potential difference controlling the rate of the electrode reaction, differs from the corresponding values in the bulk solution.[†]

In order to obtain the new concentration C_{1S} and the Ψ_1 potential values at the electrode, it is now necessary to use Eqs. (4.56a) and (4.57). Unfortunately, an exact mathematical treatment of the nonequilibrium double-layer equations is extremely difficult and has not yet been accomplished.

Consider, however, one of the conclusions of § 4.4. In the analysis of the structure of the nonequilibrium double layer, it was shown that the nonequilibrium character of the double layer affects the process only at currents commensurate with the limiting diffusion current. At low currents the equilibrium distribution in the diffuse part of the double layer remains practically undisturbed. This indicates that the distribution of discharging ions in the double layer can still be described by the Boltzmann distribution (4.65). Only the changes of concentration outside the double layer have to be taken into account. Instead of Eq. (4.65), the following should be written[‡]:

$$C_{1S} = \bar{C}_{1S} \exp\left[-(\Psi_1 - \Delta\Phi_S)\right], \tag{4.74}$$

where $\bar{C}_{1S}$ is the concentration of cations of type 1 in the vicinity of the electrode (but outside the diffuse part of the double layer) during passage of cell current $\dot{I}$. The term $(\Psi_1 - \Delta\Phi_S)$ represents the potential drop in the diffuse layer.

[†]For simplicity, the effect of specific adsorption is, as previously, neglected.
[‡]Expression (4.74) can be easily obtained directly from Eqs. (4.56c) and (4.57), assuming $\bar{C}_S$ to be independent of the cell current.

Substituting Eq. (4.74) into (4.67), we obtain

$$-I = I_0 \{\bar{C}_{1S} \exp[-(\Psi_1 - \Delta\Phi_S)] \exp[-\alpha(\eta - \Psi_1)] - \exp[(1-\alpha)(\eta - \Psi_1)]\}.$$

It is convenient to transform the above relation into

$$-I = I_0 \exp[-(1-\alpha)(\Psi_1 - \Delta\Phi_S)] \{\bar{C}_{1S} \exp[-\alpha(\eta - \Delta\Phi_S)] - \exp[(1-\alpha)(\eta - \Delta\Phi_S)]\}. \qquad (4.75)$$

It can then easily be seen by comparing Eq. (4.75) with Eq. (4.69) that the presence of the electric double layer results in a change of the exchange current density $\dot{I}_0$. In fact, introducing the factor $\exp[-(1-\alpha)(\Psi_1 - \Delta\Phi_S)]$ into Eq. (4.68) which describes the exchange current density, we find that the rate constant k_f of the forward reaction should be replaced in the double layer by an apparent rate constant $k_{f\ eff} = k_f \exp[-(\Psi_1 - \Delta\Phi_S)]$. A similar change of the rate of the forward reaction must be taken into account in the interpretation of the experimental data. In particular, the double-layer structure affects the value of the experimentally determined transfer coefficient α [3, 18].

Using Eqs. (4.38) and (4.44b), we can easily obtain the voltammetric characteristics of the cell in the form of Eq. (4.70). However, the exchange current $\dot{I}_0$ should be replaced in this case by its effective value (i.e., as related to k_{eff} above).

§ 4.6.   The Limiting Current

At high cathodic polarizations ($|\eta| \gg 1$), the reduction rate becomes so high that practically all ions supplied to the electrode are immediately discharged. In this case, the observed rate of the electrochemical transformation is determined by the rate of supply of cations of type 1 to the electrode surface. The concentration $\bar{C}_{1S}$ of cations of type 1 at the electrode becomes equal to zero. Equating the expression for $\bar{C}_S$ [Eq. (4.53)] to zero, we obtain the equation

$$[2(1+\gamma) - I_{\lim}]^2 - 4\gamma(1+\gamma) = 0, \qquad (4.76)$$

from which the maximum cathodic current $\dot{I}_{\lim}$ in the cell can be determined.

Solving Eq. (4.76) for $\dot{I}_{lim}$ gives

$$I_{lim} = 2\left[(1 + \gamma) - \sqrt{\gamma(1 + \gamma)}\right].$$ (4.77)

Designating the ratio of the bulk concentration of the anion and cation of type 1, c_3^*/c_1^*, by M (i.e., $M = c_3^*/c_1^* = 1 + \gamma$), we can transform Eq. (4.77) into

$$\dot{I}_{lim} = 2\left[M - \sqrt{M(M - 1)}\right].$$ (4.77a)

Equation (4.77a) describing the limiting current in a solution containing three equally charged ions was first obtained by Eucken [19] on the basis of the Nernst model of the diffusion layer.

Equation (4.77) can be used for the analysis of the dependence of the limiting current I_{lim} on the concentration of the indifferent electrolyte. In an excess of the latter ($\gamma \gg 1$), it follows from Eq. (4.77) that

$$I_{lim} = 1,$$ (4.78)

or, in dimensional units,

$$i_{lim} = i_{d1} = nFDc_1^*/\delta_d.$$ (4.78a)

Equation (4.78a) is identical with the limiting diffusion current at a rotating disc electrode (2.27).

For a binary electrolyte ($\gamma = 0$), the limiting current is

$$I_{lim} = 2,$$ (4.79)

or, in dimensional units,

$$i_{lim} = 2i_{d1} = 2nFDc_1^*/\delta_d.$$ (4.79a)

Thus, the simultaneous action of convective transport and of ion migration in the electric field results in a doubled value (as compared to the diffusion-limited current) of the limiting current.

The predictions of Eqs. (4.78a) and (4.79a) are in good agreement with experimental data [20, 21]. Values of limiting currents for intermediate cases (arbitrary γ) can be found from Eq. (4.77) or (4.77a).

All the above results were derived assuming that the diffusion coefficients and ionic charges of all ions in solution are equal. In practice, however, this situation is rarely encountered. Attempts to make corrections which would take into account differences in the diffusion coefficients and in ionic charges result in serious mathematical complications, although the physical picture remains unchanged.

Only in the case of a binary electrolyte has the problem been fully solved [6, 8]. The expression for the limiting current at a rotating disc electrode in a solution of binary electrolyte is as follows [8]:

$$i_{\lim} = n_1 F D_1 \left(1 + \frac{n_1}{n_2}\right) \frac{c_1^*}{\delta_{d\,eff}}, \qquad (4.80)$$

where the effective thickness of the diffusion boundary layer,

$$\delta_{d\,eff} = 1.61 \, (D_{eff}\,/\nu)^{1/3} \, (\nu/\omega)^{1/2}, \qquad (4.81)$$

is usually expressed by the effective diffusion coefficient

$$D_{eff} = \frac{D_1 D_2 \, (n_1 + n_2)}{n_1 D_1 + n_2 D_2}. \qquad (4.82)$$

When $n_1 = n_2 = n$ and $D_1 = D_2 = D$, Eqs. (4.80)-(4.82) reduce to Eq. (4.79a).

Satisfactory theoretical expressions describing the limiting current at a disc electrode rotating in a solution of three different ions have not hitherto been evaluated. The first attempts in this direction were made by Zembura [22, 23]. He presents a solution of three different ion kinds by a combination of two binary electrolytes, one of which is under nonequilibrium conditions. Distributions of concentrations and potential in the latter electrolyte were described by the usual expressions for a binary electrolyte [8]. An equilibrium distribution of electric charges and field was assumed for the second electrolyte (at equilibrium) [24]. However, this approach results in one kind of ion being partly in equilibrium and partly under nonequilibrium conditions. A similar am-

biguity appears with respect to the potential distribution. Therefore, the attempt at calculations of the limiting current described
above must be considered unsatisfactory.

Albery's calculations [25] for two simple cases (identical
diffusion coefficient for all three types of ions and excess of indifferent electrolyte) are based on physically correct but mathe-
matically not fully valid assumptions. Although this author did
not seek directly the dependence of the limiting current on the
solution composition, his expressions can be used to establish
this dependence for the two cases mentioned above.

Gordon, Newman, and Tobias [26] made approximate calculations of the effect of an inert electrolyte on the magnitude of the
limiting current. They introduced a certain effective electric
field of magnitude derived on the basis of intuitive physical concepts. The intensity of the electric field is initially assumed to
be constant over the whole solution volume, including the diffusion
layer, and equal to the value which maintains passage of current
in the bulk solution. Obviously, such an approximation can be valid
only at high concentrations of the inert electrolyte.

In order to account for the change of the effective field in the
diffusion-layer region, Gordon et al. [26] introduced additional
corrections, still assuming, however, a constancy of the field in
this region. This approach cannot be considerd rigorous, so that
the results obtained [26] can serve only as an illustration rather
than a quantitative description.

Later, Newman [27] made exact calculations for a series of
actual electrochemical systems. The solution of the system of
nonlinear differential equations describing the distribution of ionic
concentrations and of electric field was found by computer-
assisted numerical integration. The dependence of the limiting
current on the concentration of the inert electrolyte obtained for
the system $K_3Fe(CN)_6 + K_4Fe(CN)_6 + KOH$ is shown in Fig. 4.4.
At low concentrations of the inert electrolyte (KOH), the limiting
current differs considerably from the diffusion-limited current.
The difference is positive for the anodic process [oxidation of
$Fe(CN)_6^{4-}$] and negative for the cathodic one [reduction of $Fe(CN)_6^{3-}$].
In the first case, the migration flux adds to the diffusional one,
and in the second its value is subtracted.

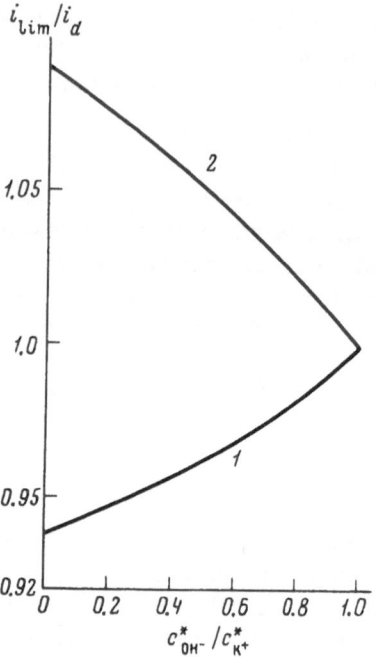

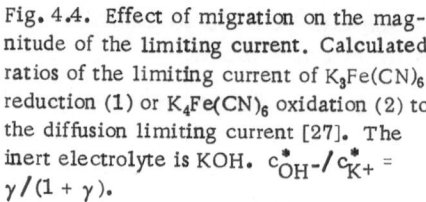

Fig. 4.4. Effect of migration on the magnitude of the limiting current. Calculated ratios of the limiting current of $K_3Fe(CN)_6$ reduction (1) or $K_4Fe(CN)_6$ oxidation (2) to the diffusion limiting current [27]. The inert electrolyte is KOH. $c_{OH^-}^* / c_{K^+}^* = \gamma / (1 + \gamma)$.

The dependence of the limiting current at a rotating disc electrode on the concentration of inert electrolyte was calculated by Malev and Durdin [28]. Analyzing the case of a solution containing three kinds of ions having equal charges, these authors used a transformation which considerably simplified the mathematical problem. In fact, in solving the system (4.15) with the condition $n_1 = n_2 = n_3$ in the region of electroneutral solution $(c_1 + c_2 = c_3)$, it is possible to introduce an auxiliary (yet unknown) function $\varphi(z) = (dc_2/dz)/(dc_3/dz)$. The function $\varphi(z)$ enabled the authors to express the concentration distributions of all three ions in a relatively simple form. Nevertheless, the equation from which the function $\varphi(z)$ can itself be determined is extremely complex. The authors restricted their treatment to a system of ions having the same mobility. Later, Malev [29] calculated the magnitude of the limiting current for the case of small differences in mobilities of two ions.

Experimental verification [30] of the calculations of Malev and Durdin was carried out for the case of reduction of Cu^{2+} ions at an amalgamated platinum disc and of H^+ ions at a platinized-

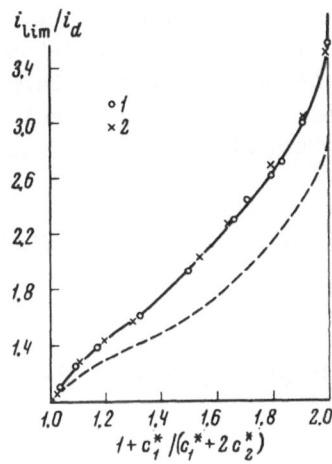

Fig. 4.5. Effect of migration on the magnitude of the limiting current [31]. The dashed line is calculated [30]. 1) Experimental data for HCl + KCl solutions; 2) experimental data for $HNO_3 + KNO_3$ solutions.

platinum rotating disc; $CdSO_4$ and KCl (as well as NaCl and LiCl) were used as inert electrolytes, respectively. The concentration of the inert electrolyte was varied from 10-12 times excess (with respect to the concentration of discharging ions) down to a very low value (subsequently Durdin and Svetasheva [31] measured the limiting currents also in the absence of inert electrolyte, i.e., in a binary electrolyte). The dependence of the limiting current on the solution composition is in qualitative agreement with the calculated behavior, as exemplified by the results in Fig. 4.5. The latter figure shows the dependence of the limiting current for H^+ ion reduction on the ratio of ionic concentrations in the solution: $1 + c_1^*/(c_1^* + 2c_2^*)$, where c_1^* is the concentration of the discharging ion (H^+) and c_2^* is the concentration of the inert electrolyte. The observed deviations between the calculated and experimental curves were explained by the authors in terms of changes of the diffusion coefficients of the discharging ion with changing concentration.

The above brief review shows that the behavior of systems containing three kinds of ions has hitherto only been satisfactorily described for a few special cases.

§ 4.7. The Rotating Disc Electrode
in Solutions of a Binary Electrolyte

The expressions describing the distribution of ionic concentrations (4.30) and (4.34), and of the electric field, (4.31), become considerably simplified in the case of a binary electrolyte. In order to apply these expressions to this practically important case, it is sufficient to equate to zero the concentration of the nondischarging cation, i.e., $c_2^* = 0$. Assuming $\gamma = c_2^*/c_1^* = 0$, the equation for the electric field in the vicinity of a rotating disc electrode, (4.31), is transformed to

$$\overline{\mathscr{E}}^{(0)} = \frac{I}{2 - I \int\limits_{\xi}^{\infty} \exp\left(-at^3/3\right) dt}. \tag{4.83}$$

The concentration distribution of ions discharging at the cathode is described [cf. Eq. (4.34)] by

$$\bar{C}_1^{(0)} = 1 - \frac{I}{2} \int\limits_{\xi}^{\infty} \exp\left(-at^3/3\right) dt. \tag{4.84}$$

Also the potential distribution in the vicinity of a rotating cathode can easily be derived from Eq. (4.43). At an arbitrary point ξ, the potential is given by

$$\Phi(\xi) = -\int\limits_{\xi}^{\infty} \frac{I \, dy}{2 - I \int\limits_{y}^{\infty} \exp\left(-at^3/3\right) dt}. \tag{4.85}$$

It should be mentioned again here (cf. § 4.3) that the finite value of the electric potential in the vicinity of the cathode is obtained by assuming the anode to be located at a sufficiently large but finite distance from the cathode. Therefore, the upper integration limit in Eq. (4.85) should be replaced by a certain high value.

Expressions describing concentration and field distributions within the nonequilibrium double layer can be easily obtained from Eqs. (4.56a)–(4.56d) and (4.57) by putting $\gamma = 0$. It can be easily checked that the equations thus obtained are identical with the corresponding expressions derived by Chernenko [5].

The voltammetric characteristics of a cell with a rotating disc electrode in a solution of a binary electrolyte can be obtained from Eq. (4.70) by putting $\gamma = 0$.

The treatment here will be restricted to the case of a fast electrochemical reaction where the current passed through the cell, I, is small in comparison with the exchange current I_0. In this case, the polarization curve for the cell exhibits pure concentration polarization behavior, and its form can be obtained from Eq. (4.72a) by putting $\gamma = 0$; thus

$$\eta = \Delta\Phi_{ohm} + 2\ln\frac{2-I}{2}.$$

Taking into account that the dimensionless limiting current for a symmetrical binary electrolyte is $\dot{I}_{lim} = 2$, we can write the above expression in the form

$$\eta = \Delta\Phi_{ohm} + 2\ln(1 - I/I_{lim}). \tag{4.86}$$

The voltammetric cell characteristics are often found in the literature in another form:

$$\dot{I} = \dot{I}_{lim}\left\{1 - \exp\left[\frac{1}{2}(\eta - \Delta\Phi_{ohm})\right]\right\}. \tag{4.87}$$

In Fig. 4.6 an example of a polarization curve is shown, calculated by Newman [32]; points correspond to experimental data obtained for copper deposition from 0.1 M $CuSO_4$ (2−2) electrolyte. The plot gives an idea of the relative contribution of various components of overvoltage to the polarization curve; concentration and

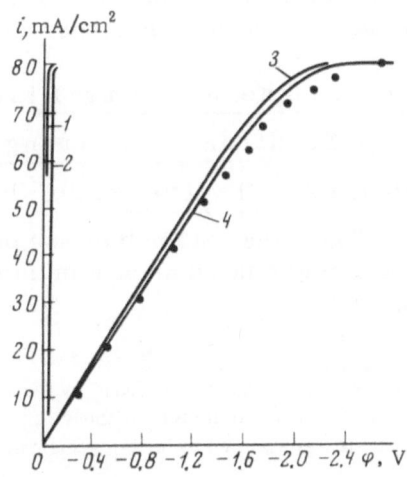

Fig. 4.6. Voltammetric characteristics of a cell with a rotating disc electrode in a binary solution. The solid line is calculated; experimental points are obtained for cathodic copper deposition from 0.1 M $CuSO_4$. Calculated values of concentration (1) and activation (2) polarization, of ohmic drop (3), and of total overvoltage (4) are drawn separately.

activation polarization turn out to be much less than the ohmic potential drop in the bulk solution.

The expressions obtained at the beginning of this section were derived for a symmetric binary electrolyte ($n_1 = n_2$) assuming the same diffusion coefficient for cations and anions ($D_1 = D_2$). A general case of a binary electrolyte ($n_1 = n_2$, $D_1 \neq D_2$) was discussed by Levich [8] and Newman [6]. The papers contain expressions for distribution of ions, electric field, and potential in the vicinity of the rotating disc electrode. In particular, the voltammetric characteristics of a cell with a rotating disc cathode may have, in the general case, a form close to that of Eq. (4.87):

$$ i = i_{\lim}\left\{1 - \exp\left[\frac{Fn_1n_2}{RT(n_1 + n_2)}(\Delta\varphi_M - \Delta\varphi_{\text{ohm}})\right]\right\}, \qquad (4.88) $$

where $\Delta\varphi_M$ and $\Delta\Phi_{\text{ohm}}$ are the potential drops at the electrodes and the ohmic drop in the solution, respectively; $i_{\lim}$ is the limiting current density at a rotating disc electrode described by relations (4.80)-(4.82).

The calculations of the potential distribution and of the voltammetric cell characteristics made by Kholpanov [33, 34] for the case discussed are repetitions of calculations previously made by Levich [8]. Correcting errors committed in the former paper [33],[†] Kholpanov arrives, in conclusion, at Eq. (4.88). However, the form of his expression (potential drop in the cell is given in terms of the diffusion-layer resistance, which changes during passage of current) is inconvenient for practical use.

§ 4.8.   The Rotating Disc Electrode in a Solution Containing an Excess of Inert Electrolyte

The situation for the case of current passing in a solution containing a large amount of an inert electrolyte was discussed in § 4.2.

[†] The potential distribution found by the author contained the field intensity at the electrode as a parameter. Although this quantity depends on the magnitude of current, it was assumed to be an experimental constant.

The field and concentration distribution may be easily obtained for this case from the general relations of § 4.3 by putting everywhere $\gamma \gg 1$ (i.e., $c_2^* \gg c_1^*$) and remembering that $\bar{I} \leq 1$.

According to Eq. (4.30), the total ion concentration is uniform throughout the solution and equals the concentration of the indifferent electrolyte:

$$\bar{C}^{(0)} \approx 2\gamma. \tag{4.89}$$

Concentration changes connected with the depletion of cations of type 1 at the electrode turn out to be small and proportional to $\dot{I}/2\gamma$.

The distribution of the electric field in the vicinity of the rotating disc is, in this case [cf. Eq. (4.31)], given by

$$\bar{\mathscr{E}}^{(0)} \approx I/2\gamma. \tag{4.90}$$

i.e., as has been mentioned already in § 4.2, the field is constant.[†] The quantity $1/2\gamma$ represents the dimensionless specific electric conductivity of the solution. Therefore, the conductivity of solutions containing excess of inert electrolyte is purely ohmic in nature. Expressions for the potential drop (4.44a) and (4.44b) become, in this case,

$$\Delta\Phi \approx \Delta\Phi_{ohm}; \tag{4.91}$$

i.e., the potential drop is determined exclusively by the ohmic potential drop in the solution. Corrections to the ohmic losses connected with concentration changes of cations of type 1 are proportional to $\dot{I}/2\gamma$ and are extremely small.

The expressions for concentrations of electrochemically active ions, $\bar{C}_1^{(0)}(\xi)$, which previously were of the form of Eqs. (4.34) and (4.35), are transformed, when $\gamma \gg 1$, into the following relation:

$$\bar{C}_1^{(0)}(\xi) = 1 - I \int_{\xi}^{\infty} \exp(-at^3/3)\, dt. \tag{4.92}$$

[†]In §4.10 it will be useful to return to the question of calculations of the electric field distribution in the vicinity of a rotating disc of finite dimensions, i.e., to the case where the problem is no longer one-dimensional.

Comparing Eq. (4.92) with (2.48) for the concentration distribution of the inert substance in the vicinity of a rotating disc, it is easy to establish the identity of both expressions. This identity signifies that in the presence of excess inert electrolyte, discharging ions are supplied to the disc surface by convective diffusion. Ionic migration plays only an insignificant role.

This can be established also by a direct analysis of Eq. (4.33) describing the transport of cations of type 1 to the rotating disc surface. In the presence of excess inert electrolyte ($\gamma \gg 1$), the second term on the left side of this equation (which represents the contribution of ionic migration) is small and can be neglected. Thereafter, the equation describing the transport of cations of type 1 can be transformed into the usual equation for convective diffusion (2.45).

The voltammetric curve for a cell with a disc electrode (in the absence of double-layer effects) rotating in a solution containing excess inert electrolyte can be obtained from Eq. (4.70) by putting $\gamma \gg 1$; thus,

$$-\dot{I} = \dot{I}_0 \left\{ (1 - \dot{I}) \exp\left[-\alpha \left(\eta - \Delta\Phi_{ohm}\right)\right] - \exp\left[(1 - \alpha)\left(\eta - \Delta\Phi_{ohm}\right)\right] \right\}.$$

Upon further transformation, the polarization curve is obtained in the form

$$\dot{I} = \frac{1 - \exp\left(\eta - \Delta\Phi_{ohm}\right)}{1 - \exp\left[\alpha\left(\eta - \Delta\Phi_{ohm}\right)\right]/\dot{I}_0}. \tag{4.93}$$

Since in the presence of excess inert electrolyte the diffusion-limited current I_{lim} is equal to 1, Eq. (4.93) can be rewritten as follows:

$$\dot{I} = \dot{I}_{lim} \frac{1 - \exp\left(\eta - \Delta\Phi_{ohm}\right)}{1 - \exp\left[\alpha\left(\eta - \Delta\Phi_{ohm}\right)\right]/\dot{I}_0}; \tag{4.94}$$

Eq. (4.94) has the usual form for an electrochemical reaction occuring under diffusion control of one of its components [35].

For fast electrochemical reactions, when $\dot{I}_0 \gg 1$, the above relation simplifies to

$$\dot{I} = \dot{I}_{lim} \left[1 - \exp\left(\eta - \Delta\Phi_{ohm}\right)\right], \tag{4.95}$$

thus acquiring the form characteristic of concentration polarization in the cell.

In the case of a completely irreversible electrode process, we have

$$I = I_{\lim} \frac{I_0 \exp\left[-\alpha\left(\eta - \Delta\Phi_{ohm}\right)\right]}{I_0 \exp\left[-\alpha\left(\eta - \Delta\Phi_{ohm}\right)\right] - 1}. \tag{4.96}$$

Using dimensional quantitites, Eqs. (4.95) and (4.96) are transformed into

$$i = i_d\left\{1 - \exp\left[\frac{nF}{RT}\left(\Delta\varphi_M - \Delta\varphi_{ohm}\right)\right]\right\} \tag{4.95a}$$

and

$$i = i_d \frac{\exp\left[-\frac{\alpha nF}{RT}\left(\Delta\varphi_M - \Delta\varphi_{ohm}\right)\right]}{\exp\left[-\frac{\alpha nF}{RT}\left(\Delta\varphi_M - \Delta\varphi_{ohm}\right)\right] - i_d/i_0}, \tag{4.96a}$$

respectively. Here $\Delta\varphi_M$ is the cell voltage, $\Delta\Phi_{ohm}$ is the ohmic potential drop in the solution, and i_d and i_0 are the diffusion-limited and exchange current densitites, respectively.

§ 4.9. Voltammetric Characteristics of the Cell in the Case of a Redox Reaction

It has been shown in § 4.8 that in the presence of excess inert electrolyte, the concentration changes of the depolarizer at a rotating disc electrode are solely determined by convective diffusion. Effects of migration are negligible. This simplifies the analysis of reactions of electrochemically active species in processes other than those discussed in § 4.8.

As an example, consider again (cf. § 3.1) a simple redox reaction:

$$Ox + ne^- \underset{k_b}{\overset{k_f}{\rightleftarrows}} Red. \tag{4.97}$$

In order to obtain the polarization curve of the cell with the above reaction proceeding at the rotating disc electrode, let us express, by analogy to § 4.5, the reaction rate in the form

$$-i = nF\left\{k_f^0 c_{OS} \exp\left[-\frac{\alpha nF}{RT}(\varphi_M - \psi_1)\right] - \right.$$
$$\left. - k_b^0 c_{RS} \exp\left[\frac{(1-\alpha)nF}{RT}(\varphi_M - \psi_1)\right]\right\}, \qquad (4.98)$$

where c_{OS} and c_{RS} are the respective concentrations of the oxidized and reduced species at the disc surface, φ_M is the electrode potential, and Ψ_1 is the potential at the plane of closest approach of discharging ions. It has been shown in § 4.8 that in the presence of the excess inert electrolyte the Ψ_1 potential is equal to the ohmic drop in the solution:

$$\psi_1 = \Delta\varphi_{ohm}. \qquad (4.99)$$

Introducing the reversible potential of the system

$$\varphi_{Me} = \frac{RT}{nF} \ln \frac{k_f^0 c_O^*}{k_b^0 c_R^*} \qquad (4.100)$$

and the exchange current density

$$i_0 = nF(k_f^0 c_O^*)^{(1-\alpha)}(k_b^0 c_R^*)^\alpha, \qquad (4.101)$$

we can transform Eq. (49) into

$$-i = i_0\left\{\frac{c_{OS}}{c_O^*}\exp\left[-\frac{\alpha nF}{RT}(\Delta\varphi_M - \Delta\varphi_{ohm})\right] - \right.$$
$$\left. - \frac{c_{RS}}{c_R^*}\exp\left[\frac{(1-\alpha)nF}{RT}(\Delta\varphi_M - \Delta\varphi_{ohm})\right]\right\}, \qquad (4.102)$$

where c_O^* and c_R^* are the concentrations of the respective species in the solution bulk, and $\Delta\varphi_M = \varphi_M - \varphi_{Me}$ is the cell voltage.†

Ionic concentrations at the rotating disc electrode are, in this case, related to the bulk concentrations by

$$c_{OS}/c_O^* = 1 - i/i_{dO}, \qquad (4.103a)$$

———————

†As indicated previously, processes occurring at the second electrode are neglected.

$$c_{RS}/c_R^{\bullet} = 1 + i/i_{dR}, \tag{4.103b}$$

where i_{dO} and i_{dR} are the respective limiting diffusion currents (cf. § 2.4). The opposite direction of flux of the reduced particles (with respect to that of the oxidized form) has been accounted for in Eq. (4.103b).

Substitution of Eqs. (4.103a) and (4.103b) into Eq. (4.102) gives

$$i = i_{dO} \frac{1 - \exp\left[\dfrac{nF}{RT}(\Delta\varphi_M - \Delta\varphi_{ohm})\right]}{1 + \dfrac{i_{dO}}{i_{dR}}\exp\left[\dfrac{nF}{RT}(\Delta\varphi_M - \Delta\varphi_{ohm})\right] - \dfrac{i_{dO}}{i_0}\exp\left[\dfrac{\alpha nF}{RT}(\Delta\varphi_M - \Delta\varphi_{ohm})\right]} \cdot$$

$$\tag{4.104}$$

Equation (4.104) describes the voltammetric characteristics of a cell with a rotating disc electrode in the case of simple redox processes (4.97).

When the reaction product (reduced form) is quickly removed from the surface (particularly, if the Red-species is the electrode material and deposits on the disc), $i_{dO} \ll i_{dR}$ and expression (4.104) is transformed into the voltammetric characteristic (4.94) derived in § 4.8.

At high cathodic polarization, i.e., when $\Delta\varphi_M$ is large and negative, $i = i_{dO}$ and the limiting cathodic current is determined by the rate of the diffusional transport of the oxidized form. Similarly, at high anodic polarizations, i.e., when $\Delta\varphi_M$ is large and positive, $i = -i_{dR}$, i.e., the limiting current is equal to the rate of supply of the reduced form.

§ 4.10. The Ohmic Potential

Drop in the Cell

The voltammetric cell characteristics derived in the former sections contain the ohmic potential drop in the cell, $\Delta\varphi_{ohm}$. The latter depends on the magnitude of current, solution resistivity, cell geometry, etc.

It has been assumed in § 4.3 that the disc electrode is an infinite rotating plane. This assumption enabled the problem of electric field distribution in the cell to be treated as a one-dimensional one; i.e., the field distribution was represented as a function only of the distance from the disc surface. Solution of the respective equations resulted in a linear dependence of the potential outside the diffusion layer on distance as described by Eq. (4.44a).

However, practical cells must contain a rotating disc of finite dimensions. Therefore, a more realistic model corresponds to an electrode of radius r_0 located in the center of an infinite rotating insulator plane. The potential distribution in such a cell has an essentially different character from that corresponding to Eq. (4.44a), and it is much smaller than the limiting current i_d. In the latter case, the solution concentration in the entire cell (including the region at the electrode) is practically uniform and concentration polarization can be neglected. The potential drop in the cell is then† of a purely ohmic character. Therefore, the potential distribution in the cell is given by the Laplace equation

$$\Delta \varphi = 0, \qquad (4.105)$$

and the cell current is

$$i = - \varkappa \, \mathrm{grad} \, \varphi, \qquad (4.106)$$

i.e., it is determined by the solution conductivity $\varkappa = \frac{F^2}{RT} \sum_j n_j^2 D_j c_j^*$.

The boundary conditions which determine the potential distribution in the cell depend on the kinetics of the electrode processes.

The case of a rotating disc cathode will be considered. The anode is, as previously, considered to be sufficiently large and far removed from the cathode as to have no influence on the behavior of the disc. The anode potential will be chosen as the reference zero.

Neglecting the changes of solution concentration, the cathodic current density can be related to the Ψ_1 potential ($\Psi_1 = \Delta \varphi_{\mathrm{ohm}}$) at the electrode by

$$i = - \varkappa \left(\frac{\partial \varphi}{\partial n} \right)_S = f \, [\Delta \varphi_M - \psi_1], \qquad (4.107)$$

where $(\partial \varphi / \partial n)_S$ is the normal (to the cathode surface) component

---

†That is, when both electrodes of the cell are of identical materials. – Editor.

of the electric field and $\Delta\varphi_M = \varphi_M - \varphi_{Me}$ is the departure from the equilibrium emf. The form of the function $f[\Delta\varphi_M - \Psi_1]$ is determined by the type of the given disc electrode reaction and has been expressed for some special cases by Eqs. (4.71), (4.96a), and (4.104).

In some cases indicated below the general boundary condition (4.107) can be simplified.

1. At nonpolarizable electrodes, electrochemical reaction is reversible, activation overpotential is practically absent, and interfacial potential differences remain upon passage of current equal to their equilibrium values. In this case, the changes of the cell voltage $\Delta\varphi_M$ equal (see footnote on p.160) the ohmic potential drop in the solution $\Delta\varphi_{ohm}$ since all other sources of potential difference are absent, i.e.,

$$\psi_1 = \Delta\varphi_M. \tag{4.108}$$

Equation (4.108) expresses the fact that the potential in the vicinity of the electrode is uniform over the whole electrode surface.

2. In the case of electrodes at which the polarization is small, the functional dependence (4.107) can be replaced by the linear relation

$$-\varkappa\left(\frac{\partial\varphi}{\partial n}\right)_s = K\,(\Delta\varphi_M - \psi_1), \tag{4.109}$$

where K is a constant.

3. Finally, in the case of considerable overpotential, the main potential drop can be assumed to occur at the electrode—solution interface. The ohmic drop can be relatively small in this case and can be neglected, to a first approximation, on the right side of Eq. (4.107). The boundary condition at the electrode surface is then of the form

$$-\varkappa\left(\frac{\partial\varphi}{\partial n}\right)_s = i_K, \tag{4.110}$$

where $i_K = f[\Delta\varphi_M]$. Equation (4.110) shows that in this case (as expected) the current density i_K is uniform over the entire electrode surface.

Condition (4.110) was formulated by Levich and Frumkin [36] in their discussion of the dissolution of local metal cathodes accompanied by hydrogen evolution. Their condition of low ohmic potential drop in the cell has the following form:

$$\frac{F|\psi_1|}{2RT} \ll 1. \qquad (4.111)$$

The fulfillment of this condition can be checked after solving the problem of the ohmic potential drop.

No current passes through the surface of the insulator (surrounding the disc) beyond the disc cathode boundary. Therefore,

$$-\varkappa\left(\frac{\partial\varphi}{\partial n}\right)_S = 0. \qquad (4.112)$$

The ohmic drop in the vicinity of a disc cathode with a radius r_0 will now be calculated. For this purpose, a cylindrical coordinate system will be introduced with the origin at the disc center. The disc surface lies in the plane $z = 0$. The electrolytic cell occupies the space $z > 0$. Potential changes are described by the Laplace equation (4.105) which in cylindrical coordinates has the following form:

$$\frac{\partial^2\varphi}{\partial r^2} + \frac{1}{r}\frac{\partial\varphi}{\partial r} + \frac{\partial^2\varphi}{\partial z^2} = 0. \qquad (4.113)$$

Far from the cathode, the potential is constant and equal to zero:

$$\varphi \to 0 \quad \text{for } z \to \infty,\ 0 < r < \infty. \qquad (4.114)$$

At the surface of the surrounding insulator, according to Eq. (4.112),

$$\left(\frac{\partial\varphi}{\partial z}\right)_S = 0 \quad \text{for } z = 0,\ r > r_0. \qquad (4.115)$$

The boundary condition at the cathode surface is determined by the kinetics of the electrode process. For the three special cases discussed above, the boundary condition has the following respective forms:

$$\psi_1 = \Delta\varphi_M{}^\dagger \quad \text{for } z = 0, \quad 0 < r < r_0; \tag{4.116a}$$

$$\left(\frac{\partial\varphi}{\partial z}\right)_S = \frac{K}{\varkappa}(\psi_1 - \Delta\varphi_M) \quad \text{for } z = 0, \quad 0 < r < r_0; \tag{4.116b}$$

$$\left(\frac{\partial\varphi}{\partial z}\right)_S = -\frac{i_K}{\varkappa} \quad \text{for } z = 0, \quad 0 < r < r_0. \tag{4.116c}$$

The solution of Eq. (4.113) will be sought in the form of the following integral:

$$\varphi(r, z) = \int_0^\infty \Phi(\rho, z) J_0(\rho r) \rho d\rho, \tag{4.117}$$

where $J_0(\rho r)$ is a cylindrical function of zeroth order. The function $\Phi(\rho, z)$ is as yet unknown.

Substituting Eq. (4.117) into Eq. (4.113) and utilizing the properties of cylindrical functions, an equation can be obtained which should be satisfied by a function $\Phi(\rho, z)$ given by

$$\frac{d^2\Phi}{dt^2} - \rho^2\Phi = 0. \tag{4.118}$$

It can easily be shown that a solution of the latter equation which satisfies condition (4.114) must be of the form

$$\Phi(\rho, z) = A(\rho) \exp(-\rho z). \tag{4.119}$$

Introducing Eq. (4.119) into Eq. (4.117) gives

$$\varphi(r, z) = \int_0^\infty A(\rho) \exp(-\rho z) J_0(\rho r) \rho d\rho. \tag{4.120}$$

---

†This case was discussed by Newman [37]. It should be mentioned that the problem solved by Newman is known in mathematical physics as "Weber's problem." Its solution is described in several textbooks concerning the theory of Bessel functions (cf., e.g., [38-40]).

Substitution of Eq. (4.120) into the system of boundary conditions at the z = 0 plane results in a system of two integral equations which allow the function $A(\rho)$ to be found. For the case described by conditions (4.116a) and (4.115), the system of "paired" integral equations is as follows:

$$\int_0^\infty A(\rho) J_0(\rho r) \rho d\rho = \Delta \varphi_M \qquad \text{for } 0 \leqslant r \leqslant r_0, \qquad (4.121a)$$

$$\int_0^\infty A(\rho) J_0(\rho r) \rho^2 d\rho = 0 \qquad \text{for } r > r_0. \qquad (4.121b)$$

The solution, $A(\rho)$, of the system of "paired" integral equations (4.121a) and (4.121b) is of the form [38-40]

$$A(\rho) = \frac{2\Delta\varphi_M}{\pi} \frac{\sin r_0\rho}{\rho^2}. \qquad (4.122)$$

The potential distribution in the cell is described by

$$\varphi(r, z) = \frac{2\Delta\varphi_M}{\pi} \int_0^\infty \frac{\exp(-\rho z) J_0(\rho r) \sin \rho r_0}{\rho} d\rho$$

$$= \frac{2\Delta\varphi_M}{\pi} \arcsin \frac{2r_0}{[(r - r_0)^2 + z^2]^{1/2} + [(r + r_0)^2 + z^2]^{1/2}}. \qquad (4.123)$$

Newman [37] solved this problem in elliptical coordinates, related to the usual cylindrical coordinates by

$$r = r_0 \sqrt{(1 + \xi^2)(1 - \eta^2)}, \qquad z = r_0 \xi\eta.$$

Upon transformation of coordinates, Newman's solution becomes identical with (4.123).

Figure 4.7 (reprinted from [37]) represents the family of equipotential planes in the vicinity of a rotating disc. It can be seen that the main part of the potential drop connected with ohmic losses in the solution occurs in a thin layer of electrolyte adjacent to the disc electrode.

Properties due to the finite dimensions of the disc become insignificant sufficiently far from the disc surface. In fact, the following expression can easily be obtained from Eq. (4.123):

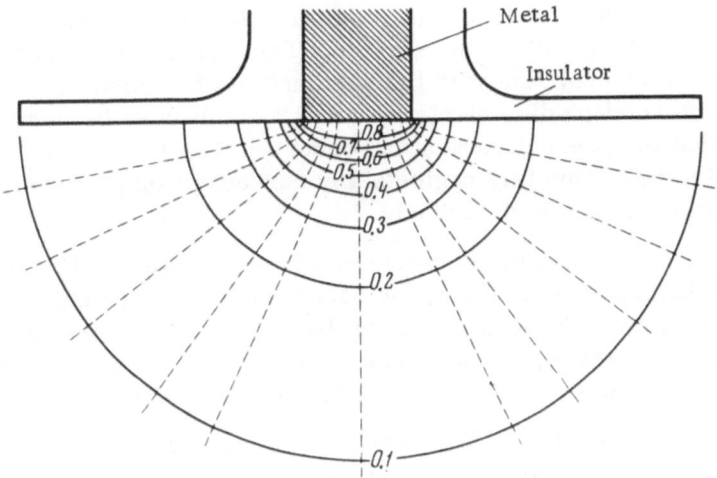

Fig. 4.7. Potential distribution in the solution in the vicinity of a rotating disc electrode of finite dimensions surrounded by an insulating sheath [37]. The dashed lines are current lines; the solid lines are cross sections of equipotential surfaces. Potential values are referred to the disc potential, assumed to be unity; the potential of the solution at infinity is assumed to be equal to zero.

$$\varphi(r, z) \sim \frac{2\Delta\varphi_M}{\pi} \frac{r_0}{(r^2 + z^2)^{1/2}} \qquad \text{for } r^2 + z^2 \gg r_0^2 \qquad (4.124)$$

for potential distribution in the region far removed from the electrode. Taking into account that $(r^2 + z^2)^{1/2}$ is the distance of the point (r, z) from the origin of coordinates, we find that Eq. (4.124) coincides with the formula for potential distribution close to a point source.

Using the potential distribution (4.123), it is possible now to find the current density distribution over the disc surface. The normal (to the cathode surface) current-density component is

$$i_z = -\varkappa\left(\frac{\partial\varphi}{\partial z}\right)_S = \frac{2\Delta\varphi_M}{\pi}\int_0^\infty J_0(\rho r)\sin\rho r_0 d\rho =$$

$$= \begin{cases} \dfrac{2\Delta\varphi_M\varkappa}{\pi\sqrt{r_0^2 - r^2}} & \text{for } 0 < r < r_0 \text{ (on the disc),} \\ 0 & \text{for } r > r_0 \text{ (beyond the disc boundary).} \end{cases} \qquad (4.125b)$$

It can be seen from Eq. (4.125a) that the current density differs at various points of the surface. This result should not be considered inconsistent with the assumption of uniform accessibility of the rotating disc electrode (cf. § 2.3). It must be remembered that the present section is concerned with the case of low currents and completely neglects any concentration gradients in the vicinity of the disc electrode.

Experimentally, the current density distribution at the disc surface has been studied [41] for copper deposition (using currents much lower than the limiting value) by measurements of the thickness of the metal deposit at various points over the disc surface. The experimental results confirm the calculations [37].

The total current at the rotating disc electrode can easily be obtained by integration of Eq. (4.125a):

$$i = 2\pi \int_0^\infty i_z r\,dr = 4\varkappa\Delta\varphi_M \int_0^{r_0} \frac{r\,dr}{\sqrt{r_0^2 - r^2}} = 4\varkappa\Delta\varphi_M r_0, \qquad (4.126)$$

and the total ohmic resistance of the solution is given by

$$R_{\text{OM}} = i/\Delta\varphi_M = 4\varkappa r_0. \qquad (4.127)$$

Calculated values of the effective ohmic solution resistance between the disc electrode and a point (r, z) are given in Table 4.1. It can be seen from the table that the ohmic resistance of the solution cannot be neglected even at a distance as small as 0.05 cm

TABLE 4.1. The Effective Solution Resistance R_{eff} between the Disc Electrode and a Point (r, z) in Solution [37]†

r, cm	z, cm	R_{eff}, ohm	r, cm	z, cm	R_{eff}, ohm
0	0.05	14.48	2.5	0	107.39
0	0.1	27.78	2.7	0	107.93
0	2.5	107.43	∞	∞	14.7

†Disc radius 0.25 cm; specific conductivity of solution 0.00872 ohm^{-1} · cm^{-1}.

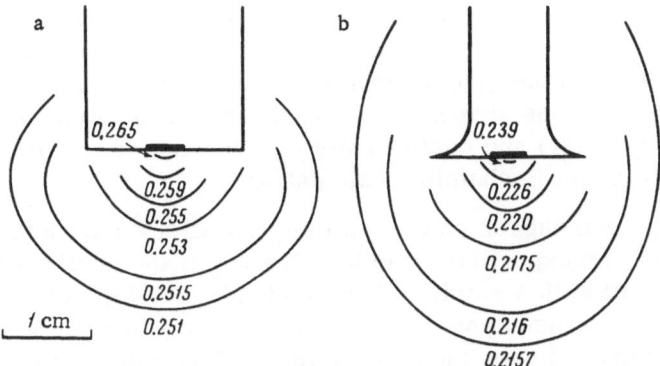

Fig. 4.8. Equipotentials in the vicinity of a disc electrode [43]. Platinum disc (cathode) 0.205 cm in radius in 0.1 M KI + 3.5 · 10^{-2} M I_2.
Current 0.347 mA. Potential values (V) measured against a saturated
calomel electrode are shown on the equipotentials. a and b are electrodes with various shapes of the insulating sheath.

from the disc surface. Therefore, the ohmic potential drop is always difficult to eliminate in poorly conducting solutions, even if
the tip of a Luggin capillary is located close to the disc surface.
Far from the electrode, the ohmic drop varies little with distance.

In spite of the relative simplicity of the above considerations,
the results obtained have a rather qualitative character. In fact,
using Eq. (4.116a) as a boundary condition at the disc surface, we
completely neglect activation overpotential (as mentioned earlier)
and restrict the discussion to polarizable electrodes.

The potential distribution in the vicinity of a polarizable disc
electrode was analyzed by Hochstein. In particular, he demonstrated [42] that the disc electrode is nonequipotential at high
polarizations. The ratio of the potential at the disc center $\varphi(0, 0)$
to the potential at the edges of the disc $\varphi(r_0, 0)$ is given by

$$\varphi(0, 0)/\varphi(r_0, 0) = \pi/2.$$

Experimentally, the potential distribution in the vicinity of a
disc electrode was studied by Angell, Dickinson, and Greef [43].
By accurate cathetometric measurements of the location of the tip

of a movable Luggin cappillary, they arrived at the distribution shown in Fig. 4.8. The pattern of equipotential surfaces in the solution close to the disc surface is virtually the same for electrodes with various shapes of the insulating sheath (truncated cone and cylinder) and qualitatively corresponds to the predictions of Newman's and Hochstein's calculations.

It has been suggested that an interrupted current method can be used for the experimental determination of the ohmic potential drop in a cell with a rotating disc electrode [44, 45]. Obviously, in the case of nonequipotential surfaces, the method can yield only a value, averaged over the disc surface, of the ohmic drop.

References

1. I. E. Tamm, Theoretical Principles of Electricity, Nauka, Moscow (1966).
2. V. G. Levich, Theoretical Physics, Vol. 1, Fizmatgiz, Moscow (1962).
3. A. N. Frumkin, V. S. Bagotskii, Z. A. Iofa, and B. N. Kabanov, Kinetics of Electrode Processes, Izd. MGU (1952).
4. V. S. Vorovkov, B. M. Grafov, A. A. Novikov, M. A. Novitskii, and L. A. Sokolov, Electrochemical Information Transformers, Nauka, Moscow (1966).
5. A. A. Chernenko, Dokl. Akad. Nauk SSSR, 153:1129 (1963).
6. J. Newman, Trans. Faraday Soc., 61:2229 (1965); W. H. Smyrl and J. Newman, Trans. Faraday Soc., 63:207 (1967).
7. E. M. Grafov and A. A. Chernenko, Dokl. Akad. Nauk SSSR, 146:135 (1962).
8. V. G. Levich, Physicochemical Hydrodynamics, Fizmatgiz, Moscow (1959).
9. V. S. Krylov, in: Principles of Modern Theoretical Electrochemistry, Mir, Moscow (1965), p. 222.
10. V. G. Levich, Dokl. Akad. Nauk SSSR, 67:309 (1949); 124:869 (1959).
11. K. Vetter, Elektrochemische Kinetik, Springer Verlag, Berlin (1961).
12. B. G. Levich, Advances in Electrochemistry and Electrochemical Engineering, Vol. 3, Interscience Publishers, New York (1966), p. 249.
13. R. R. Dogonadze and A. M. Kuznetsov, Progress in Science. Electrochemistry, 1967, VINITI, Moscow (1969), p. 5.
14. R. A. Marcus, in: Principles of Modern Theoretical Electrochemistry, Mir, Moscow (1965), p. 11.
15. H. Gerischer, Z. Phys. Chem., N. F., 26:223, 325 (1960); 27:48 (1961).
16. T. Erdey-Gruz and M. Folmer, Z. Phys. Chem. 150A:203 (1930).
17. A. Frumkin, Z. Phys. Chem., 164A:121 (1933); A. N. Frumkin, Zh. Fiz. Khim, 24:244 (1950).
18. B. B. Damaskin, Principles of Modern Methods of Investigation of Electrochemical Reactions, Izd. MGU (1965).
19. A. Eucken, Z. Phys. Chem., 59:72 (1907).

20.   V. D. Yukhtanova, Dokl. Akad. Nauk SSSR, 124:377 (1959); Zh. Fiz. Khim.,
       35:2778 (1961).

21.   B. N. Rybakov, N. V. Nikolaeva-Fedorovich, and G. B. Zhutaeva, Zh. Fiz.
       Khim., 38:782 (1964).

22.   Z. Zembura, Roczn. Chem., 34:1509 (1960); Bull. Acad. Polon. Sci., Ser.
       Sci. Chim., 9:531 (1961).

23.   Z. Zembura, A. Fulinski, and M. Bierowski, Z. Elektrochem., 65:887 (1961).

24.   A. N. Frumkin, Dokl. Akad. Nauk SSSR, 117:102 (1957).

25.   W. J. Albery, Trans. Faraday Soc., 61:2063 (1965).

26.   S. L. Gordon, J. S. Newman, and C. W. Tobias, Ber. Bunsenges., 70:414 (1966).

27.   J. Newman, Ind. Eng. Chem. Fundam., 5:525 (1966).

28.   V. V. Malev and Ya. V. Durbin, Élektrokhimiya, 2:1354 (1966).

29.   V. V. Malev and R. V. Balukov, Élektrokhimiya, 4:348 (1968).

30.   Ya. V. Durbin and V. V. Malev, Élektrokhimiya, 3:1056, 1412 (1967).

31.   Ya. V. Durbin and E. S. Svetasheva, Élektrokhimiya, 4:354, 1217 (1968);
       5:590 (1969); 6:480 (1970).

32.   J. Newman, J. Electrochem. Soc., 113:1235 (1966); 114:239 (1967); Intern.
       J. Heat Mass Transfer, 10:983 (1967).

33.   L. P. Kholpanov, Zh. Fiz. Khim., 35:1539 (1961); Tr. MKhTI im. D. I.
       Mendeleeva, 32:27, 47 (1961).

34.   L. P. Kholpanov, Zh. Fiz. Khim., 35:2759 (1961); 36:214 (1962); 37:1576
       (1963).

35.   J. O'M. Bockris, in: Modern Aspects of Electrochemistry, IL, Moscow (1958),
       p. 219.

36.   V. G. Levich and A. N. Frumkin, Zh. Fiz. Khim., 15:748 (1941).

37.   J. Newman, J. Electrochem. Soc., 113:501 (1966).

38.   E. Titchmarsh, Introduction to the Theory of Fourier Integrals, Oxford Univer-
       sity, Press (1937).

39.   A. Gray and G. B. Matthews, Treatise on Bessel Functions and Their Applica-
       tions to Physics, IL, Moscow (1953).

40.   I. Sneddon, Fourier Transforms, IL, Moscow (1955).

41.   V. Marathe and J. Newman, J. Electrochem. Soc., 116:1704 (1969).

42.   A. Yu. Gokhshtein and A. K. Frumkin, Dokl. Akad. Nauk SSSR, 144:821 (1962).

43.   D. H. Angell, T. Dickinson, and R. Greef, Electrochim. Acta, 13:120 (1968).

44.   J. Newman, J. Electrochem. Soc., 117:507 (1970).

45.   J. D. E. McIntyre and W. F. Peck, J. Electrochem. Soc., 117:747 (1970).

Nonstationary Processes at the Rotating Disc Electrode

Various processes of convective diffusion were till now examined under stationary conditions. A nonstationary regime of diffusion, however, is often encountered. Nonstationary methods have been widely used in recent years to study electrode kinetics. Sometimes, particularly in the study of slow electrode reactions, the interpretation of experimental results and the calculations of concentration polarization are made difficult by effects of natural convection. In order to suppress the latter, attempts were made [1, 2] to carry out nonstationary measurements in cells with stirred electrolyte. Rotating disc electrodes with or without a ring, polarized with triangular and trapezoidal pulses, have been used in studies of the kinetics of oxygen reduction at various metals [3].

It should be mentioned that the application of nonstationary methods to cells with stirred electrolytes often allows simultaneous determination of several kinetic parameters of an electrode reaction to be made. Here a rotating disc electrode has clear advantages over other systems.

§ 5.1. General Comments on Nonstationary Diffusion

Convective transport processes can proceed under nonstationary conditions as a result of the change of the liquid flux in time, as a result of changing mass- or heat-transfer conditions (when boundary conditions are not constant; in the presence of periodically changing mass or heat sources, etc.), or as a result of both causes.

Various regimes of nonstationary liquid flow at a rotating disc differ considerably among themselves. A detailed description of them is beyond the scope of this book. Therefore the discussion will be restricted to a single, practically important, case of a sudden change of the angular velocity of the rotating disc.

The establishment of a stationary regime in the course of a sudden switch-on and speeding up of an infinite disc to an angular velocity (or a sudden halt of a uniformly rotating disc) is determined by the character of viscous stresses appearing in the liquid.

Upon a sudden acceleration of the disc, only a thin layer of liquid immediately adjacent to the disc surface is initially entrained in the rotating motion. Thereafter, farther, more distant layers are progressively entrained, and the character of motion changes: at distances sufficiently far from the disc, the flow of the liquid occurs in a direction perpendicular to the disc surface. The time τ_0 necessary to establish a stationary regime described in § 1.2 can be evaluated from the condition $\tau_0 \sim \delta_0^2/\nu$, where δ_0 is the thickness of the hydrodynamic boundary layer at a disc rotating with angular velocity ω (cf. § 1.2) and ν is the kinematic viscosity of the liquid. It follows from the latter formula and Eq. (1.17) that $\tau_0 \sim 1/\omega$.

The establishment of the hydrodynamic boundary layer at a disc brought into rotating motion and of damping of the motion of the liquid upon bringing the disc to a standstill has been discussed by many authors. The essential results of their studies can be found in [4].

As opposed to hydrodynamic processes, the relaxation time of diffusional processes connected with mass transport near the surface of a rotating disc is given by $\tau_d \sim \delta_d^2/D$, where δ_d is the thickness of the boundary diffusion layer. The latter formula is derived in § 5.3.

In order to compare the magnitudes of τ_0 and τ_d, Eq. (2.22) derived in § 2.3 for high Schmidt numbers Sc is substituted into the expression for τ_d. This gives

$$\frac{\tau_0}{\tau_d} \sim \left(\frac{D}{\nu}\right)^{1/3} = Sc^{-1/3}.$$

The above expression shows that at high Sc numbers, the hydro-dynamic flow regime is established faster than the relaxation of the concentration gradient. At low Sc numbers (e.g., diffusion in the gas phase) relaxation times of hydrodynamic and concentra-tion perturbations are practically the same.

It will be assumed in the following discussion of nonstationary convective diffusion that the hydrodynamic liquid flow near the rotating disc is fully established.

§ 5.2. Equations for Nonstationary Convective Diffusion

It has been previously shown (§ 2.1) that the equation of convec-tive mass transport in a moving incompressible liquid has the following general form:

$$\frac{\partial c}{\partial t} + (v\,\text{grad})\,c = D\Delta c, \tag{5.1}$$

where v is the velocity of motion of the liquid.

In the vicinity of a uniformly rotating disc, Eq. (5.1) can be transformed into

$$\frac{\partial c}{\partial t} + u\frac{\partial c}{\partial r} + w\frac{\partial c}{\partial z} = D\frac{\partial^2 c}{\partial z^2}, \tag{5.2}$$

where u and w are the radial and axial components of the velocity of the liquid, respectively.

If the disc surface is uniformly accessible to diffusion and if the disc dimensions are sufficiently large for edge effects to be neglected, Eq. (5.2) can be simplified to

$$\frac{\partial c}{\partial t} + w\frac{\partial c}{\partial z} = D\frac{\partial^2 c}{\partial z^2}. \tag{5.3}$$

Equation (5.3) must be supplemented by boundary (in the present case, nonstationary) conditions.

For simplicity, only systems with high Schmidt numbers, $Sc = \nu/D \gg 1$, will be discussed. In this case $w = -0.51\omega^{3/2}\nu^{-1/2}z^2$, and δ_d is given by Eq. (2.20).

Introducing the dimensionless variables

$$\eta = Dt/\delta_a{}^2 \quad \text{and} \quad \xi = z/\delta_d \tag{5.4}$$

and the dimensionless concentration

$$C(\xi, \eta) = \frac{c^* - c(z, t)}{c^*}, \tag{5.5}$$

we can transform Eq. (5.3):

$$\frac{\partial C}{\partial \eta} = \frac{\partial^2 C}{\partial \xi^2} + a\xi^2 \frac{\partial C}{\partial \xi}, \tag{5.6}$$

where c* is the bulk concentration and $a = 2.13$.

The nonstationary processes arising as a result of changing surface properties of the disc will now be discussed. This can be effected in electrochemical systems by a change of polarizing voltage or of the cell current. The boundary conditions at the disc surface are then defined by one of the following equations:

$$c(0, t) = f(t) \qquad \text{for } z = 0, \quad t > 0; \tag{5.7a}$$

$$D\left(\frac{\partial c}{\partial z}\right)_S = \varphi(t) \qquad \text{for } z = 0, \quad t > 0; \tag{5.7b}$$

$$D\left(\frac{\partial c}{\partial z}\right)_S = \psi\left[t, c_S, \left(\frac{\partial c}{\partial z}\right)_S\right] \qquad \text{for } z = 0, \quad t > 0. \tag{5.7c}$$

The above boundary conditions can be rewritten for the dimensionless concentration $C(\xi, \eta)$ as follows:

$$C(0, \eta) = F(\eta) \qquad \text{for } \xi = 0, \quad \eta > 0; \tag{5.8a}$$

$$\left(\frac{\partial C}{\partial \xi}\right)_S = \Phi(\eta) \qquad \text{for } \xi = 0, \quad \eta > 0; \tag{5.8b}$$

$$\left(\frac{\partial C}{\partial \xi}\right)_S = \Psi\left[\eta, C_S, \left(\frac{\partial C}{\partial \xi}\right)_S\right] \qquad \text{for } \xi = 0, \quad \eta > 0. \tag{5.8c}$$

Far from the rotating disc the concentration approaches its bulk value c*. Therefore,

$$C \to 0 \quad \text{as } \xi \to \infty, \ \eta \geqslant 0. \tag{5.9}$$

Initially, the concentration of the transferred species can be presented in the form of a certain distribution:

$$C\,(\xi,\ 0) = C_0\,(\xi)\quad \text{for } \eta = 0,\ \xi > 0. \tag{5.10}$$

The present discussion will be restricted to the case of uniform concentration, i.e., before electrolysis is commenced† :

$$C_0\,(\xi) = 0. \tag{5.11}$$

Among various nonstationary problems, that connected with the establishment of a stationary convective diffusion regime is of special importance. Examples are processes accompanying an instantaneous change of surface properties from one stationary state to another.

If, for example, the surface concentration at the rotating disc has a stationary value c_{1S} before the change, the relaxation processes arising upon an instantaneous change of the surface concentration to c_{2S} can be considered. Utilizing Eq. (5.8a), we can present the problem in the form

$$C\,(0, \eta) = F\,(\eta) = \begin{cases} C_{1S} & \text{for } \xi = 0, \quad \eta < 0, \\ C_{2S} & \text{for } \xi = 0, \quad \eta > 0, \end{cases} \tag{5.12}$$

where $C_S = (c^* - c_S)/c^*$.

An investigation of the establishment of a stationary state is an important aspect of the study of kinetics of diffusion processes and kinetics of reactions occurring at the electrode surface. The methods used by various authors for calculations on nonstationary convective diffusion can be divided in two groups: 1) attempts at a direct solution of Eq. (5.3), and 2) construction of a suitable model describing the processes in question.

Levich [5] solved the problem of establishment of a stationary regime of convective diffusion at the rotating disc using the method of successive approximations. The effect of convective stirring was assumed to be insignificant in the initial moment, when the diffusion front is located in the vicinity of the electrode surface. The diffusion process was regarded as occurring in the

---

†The present method of calculation can also be used for the case of a nonuniform distribution of the diffusing species at t = 0.

same way as at a stationary electrode. Therefore, the last term on the right side of Eq. (5.6), which describes convection, can be omitted as a first approximation. The effect of stirring is taken into account in the next approximation. Obviously, the method is only valid for the initial stages of the process and does not completely describe the establishment of a stationary regime.

Hale [6] used a numerical method to calculate the time necessary to establish a stationary state. The differential equation (5.6) was replaced in the usual way by equations with finite differences which were then solved numerically. In that way, relaxation of the diffusion boundary layer was studied at given constant diffusional flux density at the surface and, in the case of electrode reactions, in systems with kinetic complications. A numerical solution of the problem of establishment of a stationary regime at the rotating disc electrode under galvanostatic conditions was presented also by Nanis and Klein [7].

The greatest success in this field was achieved by Krylov [8] who succeeded in obtaining an exact analytical solution of Eq. (5.6). Krylov assumed the solution of Eq. (5.6) to be in the form of a power series of parabolic cylindrical functions and found explicit expressions for the coefficients of the series in two cases: 1) for stationary concentration of the transferred species at the disc surface, and 2) for constant current at the rotating disc electrode. The series obtained converge sufficiently fast and allow a relatively easy calculation to be made of concentration changes near the rotating disc electrode.

Calculations of nonstationary heat transfer to the rotating disc surface belong in the same group. It should be stressed, however, that theoretical analysis of nonstationary heat transfer is a more complex problem. In fact, as was previously mentioned, the heat number Pr varies over a wide range. Correspondingly, the ratio of the thickness of heat and hydrodynamic layers strongly varies. Therefore, it is not possible to solve Eq. (5.3) by assuming the simple expression for the axial velocity component w, as was done in deriving Eq. (5.6).

The first attempt at calculation of nonstationary heat transfer to a rotating disc surface was made by Cess and Sparrow [9]. They sought the solution of the nonstationary problem by interpolation of expressions which they obtained for short and long times.

The resulting solution is very cumbersome and valid only in the intermediate range of Pr numbers (1 < Pr < 100). The theory was later extended by Cess [10] to small Pr numbers. The nonstationary regime of heat and mass transfer in the system "rotating disc—rotating fluid" was studied also by Olander [11] who numerically inegrated the equations of heat and mass transfer for the case of incompressible liquids. Nonstationary heat transfer from a disc rotating in a compressible fluid, whose viscosity depends linearly on temperature, was described by Andrews and Riley [12].

The second group of methods concerning calculations of nonstationary mass transfer in the vicinity of the rotating disc is based on model considerations. The model, proposed first by Siver [13] and subsequently used by other authors, is conceptually close to that of Nernst for the diffusion layer. According to the latter (cf. § 2.8), the liquid volume is divided into two regions: 1) a layer immediately adjacent to the electrode of thickness δ_d, and 2) the remaining space. The electrolyte is assumed to be motionless in the first region. Mass transfer occurs here by diffusion only. In the second region, stirring is assumed to be so intensive that the concentration of the diffusing species is close to its bulk value. All nonstationary processes are thus assumed to occur within the δ_d layer.

The above model was used to calculate the time of establishment of a stationary regime [1, 13] and to calculate currents flowing upon polarization of the electrode to potentials linearly changing with time [14, 15] (cf. § 5.7). It should be mentioned, however, that the method based on the Nernst model results in very cumbersome expressions. This hinders considerably its use in studies of nonstationary processes.† Furthermore, the Nernst model does not allow δ_d to be determined. The latter parameter is, however, considered to be known from the solution of the stationary problem. The applicability of the Nernst method to systems with small or intermediate Sc numbers raises additional doubts.

---

†Nevertheless, the method has proved to be relatively accurate. Hale [6] compared the results of his own calculations with those of Siver (for δ_d equal to the stationary values). The results differ little (about 8%) over the whole range of times considered.

Several methods have hitherto been developed for the calculation of nonstationary heat- and mass-transfer processes in the vicinity of a rotating disc. Here, however, the approximate method proposed by one of the present authors and Kiryanov [16] will be followed. The method is described in detail in § 2.8. In spite of some errors connected with this calculation, the approximate method suffices to describe the nonstationary convective diffusion at the rotating disc electrode in a consistent and relatively simple way.

§ 5.3. Establishment of a Stationary Regime for the Given Surface Concentration of the Transferred Species

The process of establishment of a stationary state in the case of a controlled change of surface concentration will now be considered. The value of surface concentration is set by a suitable choice of the electrode potential. A similar problem has been formulated in § 5.1.

It is assumed, for simplicity, that before the process started, the concentration of the diffusing species was uniform and equal to c^*. At the time $t = 0$, the surface concentration dropped to the value $c_S = 0$ and remained constant at $t > 0$. According to Eq. (5.12),

$$C(0, \eta) = F(\eta) = \begin{cases} 0 & \text{for} \quad \xi = 0, \quad \eta < 0, \\ 1 & \text{for} \quad \xi = 0, \quad \eta > 0. \end{cases} \tag{5.13}$$

The problem thus reduces to the solution of Eq. (5.6) with the boundary condition (5.13). Using the Laplace transform

$$L\left[C(\xi, \eta)\right] = \bar{C}(\xi, p) = p \int_0^\infty \exp\left(-p\eta\right) C(\xi, \eta)\, d\eta \tag{5.14}$$

we convert Eq. (5.6) into an ordinary differential equation:

$$p\bar{C} = \frac{d^2\bar{C}}{d\xi^2} + a\xi^2 \frac{d\bar{C}}{d\xi}. \tag{5.15}$$

Suitable transforms of the boundary conditions (5.8a) and (5.13) give

$$\text{for} \quad \xi \to \infty \qquad \bar{C} \to 0, \tag{5.16a}$$

$$\text{for} \quad \xi = 0 \qquad \bar{C}(0, p) = \bar{F}(p), \tag{5.16b}$$

where $\bar{F}(p)$ is the Laplace transform of the function $F(\eta)$.

The solution of the differential equation (5.15) is sought using the method described in § 2.8. The resulting approximate solution of Eq. (5.15) is as follows:

$$\bar{C}(\xi, p) \approx A Ai\left(\frac{p + a\xi}{a^{2/3}}\right) \exp\left(-a\xi^3/6\right). \tag{5.17}$$

The constant A is found from the condition (5.16b):

$$\bar{C}(\xi, p) = \bar{F}(p) \frac{Ai\,[(p + a\xi)/a^{2/3}]}{Ai\,(p/a^{2/3})} \exp\left(-a\xi^3/6\right). \tag{5.18}$$

The concentration distribution of the diffusing species at various times can be obtained by the inverse Laplace transform of function (5.18).

The following discussion is concerned only with the changes of the diffusional flux at the surface of the rotating disc. According to Eqs. (5.4) and (5.5), the diffusional flux equals

$$j = D\left(\frac{\partial c}{\partial z}\right)_S = -j_d\left(\frac{\partial C}{\partial \xi}\right)_S, \tag{5.19}$$

where $j_d = DC*/\delta_d$ is the limiting diffusion flux to the disc surface.

The Laplace transform of Eq. (5.19) is as follows:

$$L\,[j] = -j_d\left(\frac{\partial \bar{C}}{\partial \xi}\right)_S.$$

Using Eq. (5.18), we obtain

$$L\,[j] = -j_d\bar{F}(p) \frac{Ai'\,(p/a^{2/3})}{Ai\,(p/a^{2/3})} a^{1/3}. \tag{5.20}$$

The recovery of the original function from expression (5.20) is a complex mathematical operation. The latter will be simplified by utilizing the following interpolation formula for the logarithmic derivative of the Airy function:

$$\frac{Ai'(x)}{Ai(x)} \approx - \frac{1+x}{\sqrt{1.877+x}} \,. \tag{5.21}$$

For low and high values of the argument x, the above interpolation formula becomes identical with the logarithmic derivative of the Airy function. Furthermore, it was found by direct comparison that the values of the logarithmic derivative of the Airy function calculated from Eq. (5.21) coincide within 0.1% with the values calculated from tabulated data [17] over the whole range of values of the variable x.

Substitution of Eq. (5.21) into (5.20) gives

$$L[j] = j_d \bar{F}(p) \frac{p + a^{2/3}}{\sqrt{p + 1.877 a^{2/3}}} \,. \tag{5.22}$$

It can easily be shown [18] that the flux of the diffusing species to the rotating disc surface is given by

$$j = j_d \frac{d}{d\eta} \int_0^\eta F(\eta - \lambda) \left[\frac{\exp(-3.10\lambda)}{\sqrt{\pi\lambda}} + 0.94 \mathrm{erf} \sqrt{3.10\lambda} \right] d\lambda, \tag{5.23}$$

where the function $F(\eta)$ is determined by the boundary condition (5.13). Finally, the following expression describing the diffusional flux to a disc† is obtained:

$$j(t) = j_d \left[\frac{\exp(-3.10 Dt/\delta_d^2)}{\sqrt{\pi Dt/\delta_d^2}} + 0.94 \mathrm{erf}\left(\sqrt{3.10 Dt/\delta_d^2}\right) \right]. \tag{5.24}$$

Equation (5.24) makes possible the analysis of the kinetics of establishment of a stationary flux.

Initially (for $t \ll \delta_d^2/3.10D$) the flux is purely diffusional, convective contributions being insignificant. In fact, at low values of t, Eq. (5.24) can be transformed into

$$j(t) = j_d \frac{1}{\sqrt{\pi Dt/\delta_d^2}} = \frac{Dc^*}{\sqrt{\pi Dt}} \,. \tag{5.25}$$

---

†The character of changes of the diffusional flux to a disc obtained by Krylov [8] by analytical solution of Eq. (5.6) does not differ considerably from that described by Eq. (5.24).

Thus, in the initial stages of the process, a high concentration gradient arises at the interface, and the flux of species transported by diffusion is much larger than that by convection.

For long times $(t > \delta_d^2/3.10D)$, when the change of the original distribution occurs at distances commensurate with the thickness of the diffusion layer δ_d, diffusional and convective contributions become comparable, and a stationary limiting flux

$$j = 0.94 j_d \tag{5.26}$$

is established.

It can be seen from Eq. (5.26) that the approximate method of solving Eq. (5.6) gives flux values accurate to 6%.

The quantity

$$\tau_d \sim \delta_d^2/3.10D \tag{5.27}$$

is the characteristic time of establishment of a stationary state.

A result close to Eq. (5.27) was obtained by Bruckenstein and Prager [19], who calculated the time τ_d using a modified Nernst approach. Their calculations were verified experimentally by studies of Cu^{2+} ion reduction from KCl solution at a rotating

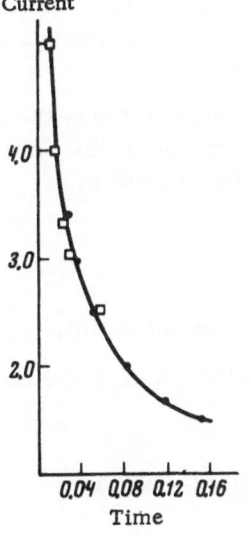

Fig. 5.1. Current transient (in dimensionless units) for a potential step from stationary value to a value in the limiting current region [19]. Reduction of Cu^{2+} ions ($\bullet$, $1.96 \cdot 10^{-3}$ M; $\square$, $3.91 \cdot 10^{-3}$ M) from 0.5 M KCl solution at a rotating gold electrode.

gold electrode. Applying a potential step from the stationary value to a constant value in the limiting current region, the authors obtained the transient shown in Fig. 5.1. It can be seen that the theory is in good agreement with experiment.

Transient processes were also investigated using reduction of $K_3Fe(CN)_6$ at a platinum disc electrode [20]. The transition time observed upon applying a potential step was close to that calculated according to Eq. (5.27).

§ 5.4. Establishment of a Stationary

Regime at a Given Density

of Diffusional Flux (Relaxation

of Potential upon Switching on

Constant Current)

It will be assumed that the flux of the species transferred to the disc surface varies according to a certain law. In this case, the boundary conditions are described by (5.8b). Let

$$\left(\frac{\partial C}{\partial \xi}\right)_s = \Phi(\eta) = \begin{cases} 0 & \text{for } \xi = 0, \quad \eta \leqslant 0, \\ -\Phi_0 & \text{for } \xi = 0, \quad \eta > 0. \end{cases} \qquad (5.28)$$

If $\Phi(\eta)$ is presented in the form of Eq. (5.28), it pertains to the case where electric current is absent up to the time t = 0. Beginning with t = 0, a constant current density $i = \Phi_0 i_d$ passes through the cell. The magnitude of Φ_0 is arbitrary.

The solution of the mathematical problem follows the method described in § 5.2. However, the determination of the constant A in expression (5.17) requires the use of the condition

$$\left(\frac{\partial C}{\partial \xi}\right)_s = \overline{\Phi}(p), \qquad (5.29)$$

where $\overline{\Phi}(p)$ is the Laplace transform of function (5.28).

Condition (5.29) results in the following form of the function $\overline{C}(\xi, p)$:

$$\overline{C}(\xi, p) = \overline{\Phi}(p) \frac{Ai\,[(p + a\xi)/a^{2/3}]}{a^{1/3} Ai'\,(p/a^{2/3})} \exp(-a\xi^3/6). \qquad (5.30)$$

In this case, the experimentally measurable quantity is the surface concentration of the diffusing species [or, more strictly, it is the electrode potential which is measured, which in turn depends on $c(0, t)$].

It follows from Eq. (5.30) that the Laplace transform of the surface concentration is of the form

$$L\,[\,c\,(0,\,t)] = c^{\bullet}\left[1 - \overline{\Phi}\,(p)\,\frac{Ai\,(p/a^{2/3})}{a^{1/3}Ai'\,(p/a^{2/3})}\right]. \tag{5.31}$$

Therefore, the surface concentration can be described by

$$c\,(0,\,\eta) = c^{\bullet}\left[1 - \frac{d}{d\eta}\int_{0}^{\eta}\Phi\,(\eta - \lambda)\,K\,(\lambda)\,d\lambda\right], \tag{5.32}$$

where $K(\lambda)$ is a function whose Laplace transform is given by

$$L\,[K\,(\lambda)] = \frac{Ai\,(p/a^{2/3})}{a^{1/3}Ai'\,(p/a^{2/3})}.$$

Presenting the logarithmic derivative of the Airy function in the form of the interpolation formula (5.21) previously described, we can express (5.32) in the form

$$c\,(0,\,\eta) = c^{\bullet}\left\{1 + \frac{d}{d\eta}\int_{0}^{\eta}\Phi\,(\eta - \lambda)\,[1.07\mathrm{erf}\,\sqrt{3.1\lambda} - \right.$$

$$\left. - 0.73\exp\,(-1.65\lambda)\,\mathrm{erf}\,\sqrt{1.45\lambda}\,]\right\}\,d\lambda. \tag{5.33}$$

The integral in Eq. (5.33) disappears when $\Phi\,(\eta)$ is substituted from Eq. (5.28):

$$c\,(0,\,t) = c^{\bullet}\left\{1 - \frac{i}{i_d}\,[1.07\,\mathrm{erf}\,\sqrt{3.1Dt/\delta_d^2} - \right.$$

$$\left. - 0.73\exp\,(-1.65Dt/\delta_d^2)\,\mathrm{erf}\,\sqrt{1.45Dt/\delta_d^2}\,]\right\}. \tag{5.34}$$

In the initial stages of the process (i.e., for $t \ll \delta_d^2/D$), effects of convection are negligibly small. In fact, in this limiting case, Eq. (5.34) becomes

$$c\,(0,\,t) = c^{\bullet} - \frac{2i\,\sqrt{t}}{nF\,\sqrt{\pi D}}; \tag{5.35}$$

i.e., the concentration changes are brought about by diffusion.

After a long time, a stationary regime is established and the surface concentration becomes constant and time-dependent; i.e.,

$$c\,(0,\ t) = c^*\,(1 - 1.07i/i_d).\qquad (5.36)$$

The above expression agrees within 7% with that derived for the surface concentration under stationary conditions.

The characteristic quantity for nonstationary convective diffusion during a galvanostatic transient is the transition time τ. In particular, chronopotentiometric studies of electrochemical systems utilize measurements of transition times to determine electrode-kinetic parameters [21].

The transition time is defined as that elapsed till the surface concentration becomes zero, i.e., $c(0,\ \tau) = 0$.

From Eq. (5.34) a transcendental equation describing τ at a given i can be derived:

$$i_d/i = 1.07\ \text{erf}\ \sqrt{3.1D\tau/\delta_d^2} - 0.73\ \text{exp}\,(-1.65D\tau/\delta_d^2)\ \text{erf}\ \sqrt{1.45D\tau/\delta_d^2}.$$
$$(5.37)$$

Further considerations are restricted to a semiquantitative analysis of Eq. (5.37). In this way, the character of the dependence of τ on i can be established. The analysis is best carried out by calculating i at various τ values. From the relation $i = i(\tau)$ obtained, the inverse function $\tau = \tau\,(i)$ can be easily derived.

For small τ ($\tau \ll \delta_d^2/D$), the i vs τ relation is similar to that found in the case of diffusion near a planar electrode, when convection can be neglected:

$$i = nF\,\frac{c^*\,\sqrt{\pi D}}{2\,\sqrt{\tau}}.\qquad (5.38)$$

For larger τ values ($\tau \gg \delta_d^2/D$), the transition time becomes independent of current density, a relation characteristic of stationary convective diffusion at a rotating disc.

The relations described above are represented in Fig. 5.2 by the solid line. Points correspond to the experimental results of Buck and Keller [1]. In spite of the rather low accuracy of experimental data, the agreement between theory and experiment is satisfactory.

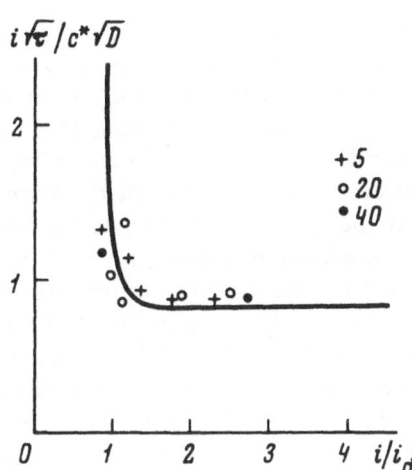

Fig. 5.2. Dependence of the transition time on current at a rotating disc electrode. The solid line is calculated; the points are experimental (constructed [31] from results of Buck and Keller [1]). Reduction of $K_3Fe(CN)_6$ (0.004 M) in KCl solution at a platinum electrode. Rotational speeds (rpm) are indicated on the plot.

The relation obtained between the transition time and diffusional flux (or current) has a simple physical meaning. If the current passing through the cell is higher than the limiting diffusion current at the rotating disc electrode, the surface concentration drops to zero in a time τ less than the time required for establishment of a stationary state (δ_d^2/D). Motions of the liquid have no significance during this period of time. However, a current $i > i_d$ cannot be passed for an indefinitely long time. Only at a current $i = i_d$ can a stationary zero surface concentration, $c(0, t) = 0$, be achieved.

The above dependence of current density on the transition time coincides with that derived by Levich [5] on the basis of qualitative considerations.

Transient processes under galvanostatic conditions were studied also by Kruglikov et al. [20, 22, 23]. They measured transition times of galvanostatic transients at a rotating platinum electrode in alkaline $K_3Fe(CN)_6$ solution (applying currents higher than the limiting current values). As predicted by Eq. (5.38), they found a linear relation between i and $1/\sqrt{\tau}$ at low τ values; for large τ values the transition time becomes dependent on the rotational speed, and its dependence on i disappears.

The characteristic properties of transients at the rotating disc can be observed in the processes of metal passivation.

Several passivation studies have been carried out using rotating discs made of metal to be passivated. Galvanostatic anodic pulses result in passivation (i.e., in an abrupt decrease of anodic dissolution). Just and Landsberg [24] investigated passivation of gold using this method, by measuring the time necessary to passivate the electrode (instead of the transition time proper). They found that the relation of the passivation current to passivation time and Cl^- ion concentration, at not too high rotational speeds, correspond to Eq. (5.38). From this, they concluded that the observed passivation time is in fact the transition time, and consequently that the passivation of gold is a diffusional phenomenon, not connected with the formation of a surface oxide layer. The passivation is caused by a decrease to zero of the surface concentration of Cl^- ions as a result of the reaction

$$Au + 4Cl^- \rightarrow AuCl_4^- + 3e^-.$$

Popova et al. [25] imposed polarization pulses on a rotating electrode to study diffusional processes accompanying passivation of zinc during its anodic dissolution in alkaline solutions. They found that at high currents the passivation time is independent of stirring. At low currents the passivation time increased with increasing rate of stirring. At rotation speeds of the order of 3000-5000 rpm in concentrated KOH solutions (0.5-2.5 N) no passivation occurred. This result agrees qualitatively with the prediction of Eq. (5.38). It shows that passivation of zinc can only occur after a certain necessary change of the base and zincate ion concentrations takes place in the vicinity of the electrode. Under conditions of intense stirring and low currents, the required changes do not occur even after a stationary state has become established, so that the electrode does not passivate. Similar results were obtained in a study of passivation of indium in citric and perchloric acids.

Transient processes connected with imposition of a constant current on a disc electrode were used by Doronin and Kabanova [26] to develop a chronopotentiometric method for analytical determination of univalent thallium ions. Hg^{2+} ions are added to the Tl^+ solution, and a preliminary cathodic co-deposition of mercury and thallium is carried out at a platinum electrode. Thereafter, the current is switched off, and thallium amalgam spontaneously decomposes according to

$$Tl\,(Hg) \rightarrow Tl^+ + e^-,$$
$$Hg^{2+} + 2\,e^- \rightarrow Hg.$$

The reaction involves dissolved Hg^{2+} ions. Their surface concentration drops to zero, and the rate of the overall process is controlled by diffusion from the bulk to the electrode. The measured transition time is close to that calculated from Eq. (5.38) for the given Tl^+ ion concentration. The method has a $\pm 10\%$ accuracy for solutions $5 \cdot 10^{-8} - 10^{-6}$ g-eq/liter in univalent thallium. A similar method was developed for determination of dissolved oxygen at concentrations of the order of 10^{-7} M. An important condition necessary for obtaining high accuracy of measurements is uniform accessibility of the disc surface. Nonuniformity results in variations of the thickness of the amalgam layer and of the transition times at various points on the surface and consequently results in lowered accuracy and sensitivity of the method.

§ 5.5. Establishment of a Stationary

Current for an Irreversible

Electrode Reaction

A theoretical analysis of the establishment of a stationary regime in the case of the mixed kinetics of a heterogeneous reaction proceeding at a rotating disc can be carried out using the methods described in §§ 5.3 and 5.4.

As an example, the current for an irreversible electrode reaction

$$Ox + ne^- \xrightarrow{k_f} Red$$

will be considered. Let the species Ox undergo no reduction up to the time t = 0 so that its concentration is thus uniform over the whole cell:

$$c_O = \overset{*}{c_O} \qquad \text{for } t = 0, \quad z > 0. \tag{5.39}$$

After a suitable change of the electrode potential, Ox starts to be reduced at the electrode. According to § 3.1, at the disc surface,

$$nFD\left(\frac{\partial c_0}{\partial z}\right)_s = k_f c_{0s} nF \qquad \text{for } t \geqslant 0. \qquad (5.40)$$

Using the dimensionless variables introduced in § 5.2, we can easily transform (5.39) and (5.40) into

$$C_0 = 0 \qquad \text{for } \eta = 0, \quad \xi \geqslant 0, \qquad (5.41a)$$

$$\left(\frac{\partial C_0}{\partial \xi}\right)_s = \frac{k_f \delta_d}{D}(C_{0s} - 1) \qquad \text{for } \eta > 0, \quad \xi = 0. \qquad (5.41b)$$

The latter conditions serve to determine the constant A [Eq. (5.17)] necessary for solving the equation of convective diffusion.

The complete solution of the problem has a very cumbersome form. A detailed analysis of the expression describing the current at a rotating disc electrode has, however, been carried out by one of the authors with Podgaetskii [27]. Their basic results are as follows.

After a long time ($t \to \infty$), a stationary state of convective diffusion is established. The stationary current was shown to be given by

$$i = i_{d0} \frac{0.94 k_f \delta_d/D}{0.94 + k_f \delta_d/D}. \qquad (5.42)$$

[The above expression agrees (with a 6% accuracy) with the usual expression for current under mixed kinetic and diffusion control (3.29).]

Theoretical analysis leads to the conclusion that in the initial stages the transient current varies with the ratio of the rate constants of diffusion and of the electrode reaction.

1. If the specific rate of the electrode reaction is lower than the rate of mass transport (i.e., $\delta_d \ll D/k_f$), the initial current is equal to the limiting kinetic current (3.19):

$$i \approx i_k = nF k_f c^*. \qquad (5.43)$$

Deviations from the value of kinetic current in time are small since diffusion compensates for the loss of the reactant due to the electrode reaction.

2. If the electrode reaction is fast (i.e., $\delta_d \gg D/k_f$), a stationary state becomes established in the following way. Initially, when $t < D/k_f^2$, the relaxation process is determined by the electrode kinetics; later, when $t > D/k_f^2$, the rate of establishment of a stationary state depends only on the process of convective diffusion.

By analyzing the time dependence of current, the rate constant of the electrode reaction, k_f, and the number of electrons, n, transferred in the electrode reaction can be simultaneously determined. In fact, the transient current (as $t \to 0$) depends in a relatively simple manner on k_f. The number n can be obtained from the expression for the limiting current (5.42) established at the end of the process. These advantages of transient measurements in stirred media are discussed again in § 5.7.

§ 5.6. Polarization of the Electrode

with Potentials Varying in Time:

General Equations

The approximate method of solving the equation of convective mass transport to the surface of a rotating disc presented in § 5.3 and the quantitative relations so obtained allow a theoretical analysis to be made of other, more complex cases of nonstationary diffusion [28].

A redox reaction

$$Ox + ne^- \underset{k_b}{\overset{k_f}{\rightleftarrows}} Red$$

proceeding at a disc electrode is now considered. Both Ox and Red species are regarded as being present in the solution.

The convective transport of the reactant to the electrode surface is described by equations of the form of Eq. (5.3):

$$\frac{\partial c_O}{\partial t} = D \frac{\partial^2 c_O}{\partial z^2} - w \frac{\partial c_O}{\partial z},$$

$$\frac{\partial c_R}{\partial t} = D \frac{\partial^2 c_R}{\partial z^2} - w \frac{\partial c_R}{\partial z},$$

where c_O and c_R are concentrations of the oxidized and reduced forms, respectively. The initial concentrations of both species are c_O^* and c_R^*, respectively. The same concentrations are maintained in the bulk solution. For simplicity, it is assumed that $D_O = D_R = D$.

The conditions at the electrode surface (i.e., for $z = 0$) are obtained from the condition of mass balance for the reactant and product

$$D\left(\frac{\partial c_O}{\partial z}\right)_S = -D\left(\frac{\partial c_R}{\partial z}\right)_S$$

and from Eq. (3.11) describing the rate of electrode reaction,

$$i = i_0\left\{\frac{c_{OS}}{c_O^*}\exp\left[-\alpha nF(\varphi - \varphi_e)/RT\right] - \frac{c_{RS}}{c_R^*}\exp\left[(1-\alpha)nF(\varphi - \varphi_e)/RT\right]\right\},$$

where i_0 is the exchange current density, φ_e is the equilibrium potential, and $\varphi = \varphi(t)$ is the electrode potential.

Using the dimensionless variables introduced in § 5.2, the equations of convective diffusion can be presented in the form of Eq. (5.6), and the boundary conditions can be rewritten as follows:

$$\left(\frac{\partial C_O}{\partial \xi}\right)_S = -\frac{c_R^*}{c_O^*}\left(\frac{\partial C_R}{\partial \xi}\right)_S \qquad (5.44a)$$

and

$$i = i_0\left\{(1 - C_{SO})\exp\left[-\alpha nF(\varphi - \varphi_e)/RT\right] - (1 - C_{SR})\exp\left[(1-\alpha)nF(\varphi - \varphi_e)/RT\right\}. \qquad (5.44b)$$

The yet unknown flux of the Ox species at the electrode surface is given by

$$\left(\frac{\partial C_O}{\partial \xi}\right)_S = Q(\eta). \qquad (5.45a)$$

It follows from condition (5.44a) that

$$\left(\frac{\partial C_R}{\partial \xi}\right)_S = -\frac{c_O^*}{c_R^*}Q(\eta). \qquad (5.45b)$$

It has been shown in § 5.4 that the present method applied to a potential transient occurring upon the application of a constant current allows the surface concentration of the oxidized species to be expressed in the following form [cf. Eq. (5.32)]:

$$\frac{c_{SO}}{\overset{\bullet}{c}_O} = 1 - C_{SO} = 1 - \frac{d}{d\eta}\int_0^{\eta} Q\,(\eta - \lambda)\,K\,(\lambda)\,d\lambda. \qquad (5.46a)$$

The form of the $K(\lambda)$ function was discussed in §5.4.

Similarly, the surface concentration of the reduced species can be described by

$$\frac{c_{SR}}{\overset{\bullet}{c}_R} = 1 - C_{SR} = 1 + \frac{\overset{\bullet}{c}_O}{\overset{\bullet}{c}_R}\frac{d}{d\eta}\int_0^{\eta} Q\,(\eta - \lambda)\,K\,(\lambda)\,d\lambda. \qquad (5.46b)$$

Substituting Eqs. (5.46a) and (5.46b) into Eq. (5.44b), we obtain an integral equation which describes the flux Q to the rotating disc surface for an arbitrary potential sweep:

$$-Q = \frac{i_0}{i_{dO}}\Big\{(\exp\,[-\alpha nF\,(\varphi - \varphi_e)/RT] - \exp\,[(1 - \alpha)\,nF\,(\varphi - \varphi_e)/RT]) -$$
$$- (\exp\,[\alpha nF\,(\varphi - \varphi_e)/RT] +$$
$$+ \frac{\overset{\bullet}{c}_O}{\overset{\bullet}{c}_R}\exp\,[(1 - \alpha)\,nF\,(\varphi - \varphi_e)/RT])\,\frac{d}{d\eta}\int_0^{\eta} Q\,(\eta - \lambda)\,K\,(\lambda)\,d\lambda\Big\}.$$
$$(5.47)$$

Equation (5.47) was derived using the following definition of the current density i at the electrode:

$$i = nFD\left(\frac{\partial c_O}{\partial z}\right)_s = -i_{dO}\left(\frac{\partial C_O}{\partial \xi}\right)_s = -i_{dO}Q\,(\eta).$$

The general equation (5.47) allows special cases of polarization of the rotating disc with varying potential to be analyzed.

§ 5.7. Polarization of the Electrode with a Linear Potential Sweep. The Case of a Reversible Reaction

In the case of a fast electrode reaction controlled by mass transport of the reactants to the electrode surface,† Eq. (5.47)

---

†The exchange current of the electrode reaction must satisfy the condition $i_0 \gg i_{dO}$.

simplifies to

$$\frac{d}{d\eta} \int_0^\eta Q\,(\eta - \lambda)\,K\,(\lambda)\,d\lambda = \frac{1 - \exp\,[nF\,(\varphi - \varphi_e)/RT]}{1 + \dfrac{c_O^\bullet}{c_R^\bullet}\exp\,[nF\,(\varphi - \varphi_e)/RT]}. \tag{5.48}$$

The electrode is polarized by potentials varying according to $\varphi = \varphi_i - \gamma t$, where φ_i is the initial potential and γ is the rate of potential change

The following designations will be used:

$$\Theta = \exp\,[nF\,(\varphi_i - \varphi_0)/RT] \quad \text{and} \quad \sigma = nF\gamma\delta_d^2/RTD, \tag{5.49}$$

where φ_0 is the standard reversible potential of the redox system.

Equation (5.48) can be conveniently rewritten in the form

$$\frac{d}{d\eta} \int_0^\eta Q\,(\eta - \lambda)\,K\,(\lambda)\,d\lambda = \frac{1 - \Theta\exp\,(-\sigma\eta)\,c_R^\bullet/c_O^\bullet}{1 + \Theta\exp\,(-\sigma\eta)}. \tag{5.50}$$

Solution of the above equation (by means of a Laplace transform) gives

$$Q\,(\eta) = -\frac{d}{d\eta} \int_0^\eta \frac{1 - \Theta\exp\,[-\sigma\,(\eta - \lambda)]\,c_R^\bullet/c_O^\bullet}{1 + \Theta\exp\,[-\sigma\,(\eta - \lambda)]} \times$$
$$\times \left\{ \frac{\exp\,(-3{,}10\lambda)}{\sqrt{\pi\lambda}} + 0.94\,\mathrm{erf}\,\sqrt{3.10\lambda} \right\} d\lambda. \tag{5.51}$$

It is convenient to present the expression describing the electrode current in a form similar to the corresponding expression for a planar electrode [29] (Sevcik–Randles formula). In the following discussion, $\eta_{1/2} = (\ln \Theta)/\sigma$.

Equation (5.51) can be rewritten as

$$i = i_{d0}\frac{\sigma}{4}\left(1 + \frac{1}{\Theta}\right) \int_0^\eta \left\{ \frac{\exp\,[-3{,}10\,(\eta - \lambda)]}{\sqrt{\pi\,(\eta - \lambda)}} + \right.$$
$$\left. + 0.94\,\mathrm{erf}\,\sqrt{3.10\,(\eta - \lambda)} \right\} \frac{d\lambda}{\cosh^2 \sigma\,\dfrac{(\lambda - \eta_{1/2})}{2}}. \tag{5.52}$$

A detailed analysis of Eq. (5.52) was carried out by Filinovskii with Girina and Feoktistov [28]. In particular, it was shown that after a sufficiently long time

$$i \rightarrow i_{d0}. \qquad (5.53)$$

The character of the time dependence of the current, $i(\eta)$, depends considerably on the magnitude of the parameter σ.

For fast sweeps ($\sigma \gg 1$), current transients exhibit a maximum located at $\eta \sim 1/\sigma$ (similarly to the location of maxima on Sevcik—Randles curves [29]) and independent of the angular velocity of the electrode. This is due to the fact that for $\sigma \gg 1$, the current varies with potential so rapidly that the effects of convection are insignificant.

Conversely, for $\sigma \ll 1$, the potential varies at a sufficiently slow rate so that diffusion processes are able to follow the changes of concentration at the electrode. The diffusion process can be considered to be in a quasi-stationary state, and the current transient can then be described by

$$i(\eta) = i_{d0} \frac{1 - \dfrac{\overset{\bullet}{c}_R}{\overset{\bullet}{c}_O} \exp[\sigma(\eta_{1/2} - \eta)]}{1 + \exp[\sigma(\eta_{1/2} - \eta)]}. \qquad (5.54)$$

The problem arising for linear potential sweep conditions at a disc electrode was solved also by Fried [15] and Vashchenko [14]

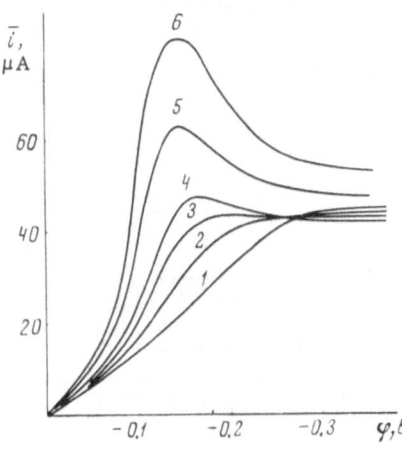

Fig. 5.3. Polarogram for the reduction of Tl^+ ions from $5 \cdot 10^{-2}$ N $LiClO_4$ solution at a rotating copper amalgam disc electrode [28]. Rotation speed 2700 rpm. Sweep rate (V/sec): 1) 0.0125; 2) 0.1; 3) 0.4; 4) 1.6; 5) 4.0; 6) 8.0. Potential referred to the mercury pool at the bottom of the cell.

using the Nernst model of the diffusion layer. Their expressions are very complex and hence difficult to analyze.

The experimental verification of the above calculations is described in [14, 15, 28, 30]. Polarograms obtained using the rotating disc electrode (Fig. 5.3) clearly show the effect of decreasing sweep rate; the characteristic maximum for nonstationary diffusion at a stationary electrode ($\sigma \gg 1$) progressively disappears, and the curve is eventually transformed into an ordinary polarographic wave observed under conditions of stationary diffusion at a rotating electrode ($\sigma \ll 1$). The critical value of the sweep rate γ corresponding to this change of behavior increases with increasing rotation speed m (rpm) (Fig. 5.4). Plotting the relative change of current i/i_{dO} against $(\gamma/m)^{1/2}$ (Fig. 5.5), we can clearly see that, in accordance with Eq.(5.49), the characteristic parameter which determines the transition from nonstationary to stationary diffusion is neither γ nor m seperately, but their ratio.

§ 5.8. Polarization of an Electrode

with a Periodically Oscillating

Potential

Here a rotating disc electrode polarized at a constant potential $\bar{\varphi}$, with a superimposed oscillating potential $\tilde{\varphi}$ of small ampli-

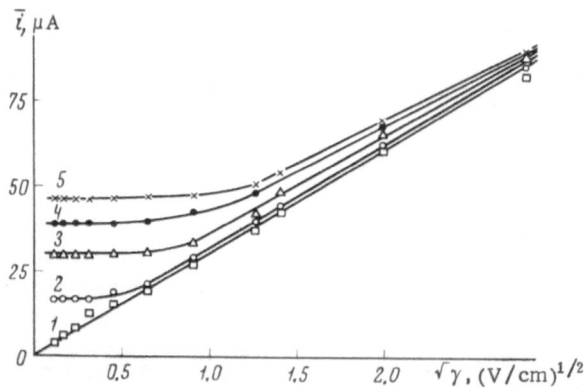

Fig. 5.4. Current (maximum or limiting) as a function of the square root of the potential sweep rate (from Fig. 5.3). Rotational speed (rpm): 1) 0; 2) 560; 3) 1400; 4) 2700; 5) 3700.

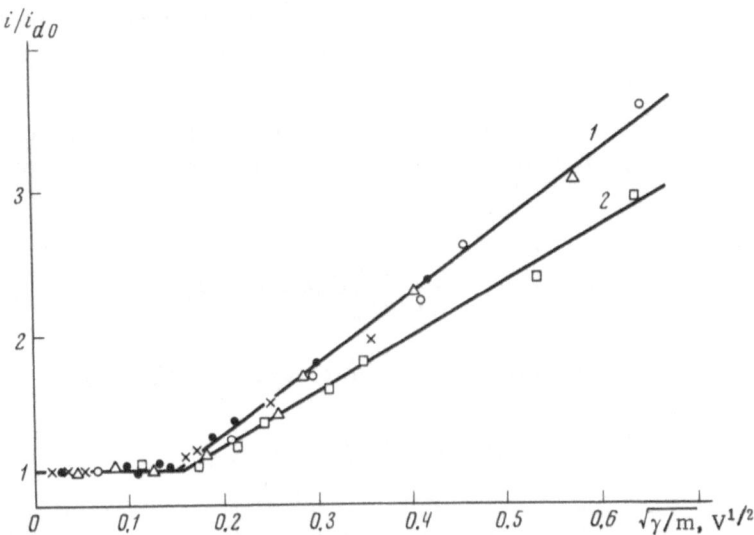

Fig. 5.5. Ratio of current (maximum or limiting) to its limiting values as a function of the square root of the sweep rate/rotation speed ratio. 1) Tl^+ reduction at an amalgamated electrode [28] (for symbols, see Fig. 5.4); 2) $K_3Fe(CN)_6$ reduction at a graphite electrode [15].

tude, will be considered, i.e.,

$$\varphi = \overline{\varphi} + \varepsilon_f \widetilde{\varphi}, \tag{5.55}$$

where ε_f is a small-valued parameter.

Periodic components appear in the Faradaic current passing through the cell. The Faradaic current can be described by

$$i = i_{d0} (\overline{Q} + \varepsilon_f Q_1 + \varepsilon_f^2 Q_2 + \ldots). \tag{5.56}$$

Determination of the relation between $\widetilde{\varphi}$ and Q_1, i.e., the determination of the Faradaic impedance of the cell, Z, allows the kinetic parameters of the electrode reaction to be established [31].

The impedance Z can be found using a method similar to that employed by Matsuda [32]. Expressions (5.55) and (5.56) for the polarizing voltage and current, respectively, are first substituted into Eq. (5.47). Then the factors $\exp[-\alpha n F \varepsilon_f \widetilde{\varphi}/RT]$ and $\exp[(1 - \alpha) n F \varepsilon_f \widetilde{\varphi}/RT]$ are expanded in a Taylor series. Gathering together terms of the same power of the parameter $\overline{\varepsilon}_f$, we obtain a system of equations which allows the quantities $\overline{Q}$, Q_1, Q_2, ... to be determined.

For terms in zero powers of ε_f (corresponding to the constant component of polarization), we obtain the ordinary integral equation

$$-\overline{Q} = \frac{i_0}{i_{d0}} \left\{ A - B \frac{d}{d\eta} \int_0^\eta \overline{Q}(\lambda) K(\eta - \lambda) d\lambda \right\}. \tag{5.57}$$

For terms in the 1st power of ε_f, we obtain the expression

$$Q_1 = \frac{i_0}{i_{d0}} \left\{ B \frac{d}{d\eta} \int_0^\eta Q_1(\lambda) K(\eta - \lambda) d\lambda + \right.$$

$$\left. + \frac{nF\widetilde{\varphi}}{RT} \left[A_1 + B_1 \frac{d}{d\eta} \int_0^\eta \overline{Q}(\lambda) K(\eta - \lambda) d\lambda \right] \right\}. \tag{5.58}$$

The following symbols are introduced to simplify the notation:

$$A = \exp[-\alpha nF(\overline{\varphi} - \varphi_e)/RT] - \exp[(1 - \alpha) nF(\overline{\varphi} - \varphi_e)/RT],$$

$$B = \exp[-\alpha nF(\overline{\varphi} - \varphi_e)/RT] + \frac{\overset{\bullet}{c}_O}{\overset{\bullet}{c}_R} \exp[(1 - \alpha) nF(\overline{\varphi} - \varphi_e)/RT],$$

$$A_1 = \alpha \exp[-\alpha nF(\overline{\varphi} - \varphi_e)/RT] +$$

$$+ (1 - \alpha) \exp[(1 - \alpha) nF(\overline{\varphi} - \varphi_e)/RT], \tag{5.59}$$

$$B_1 = \alpha \exp[-\alpha nF(\overline{\varphi} - \varphi_e)/RT] -$$

$$- (1 - \alpha) \frac{\overset{\bullet}{c}_O}{\overset{\bullet}{c}_R} \exp[(1 - \alpha) nF(\overline{\varphi} - \varphi_e)/RT].$$

In a similar way, integral equations can be found with terms in higher powers of ε_f in the expression for the Faradaic current.

Since the main interest is in the passage of the oscillating signal through the cell, it is convenient to choose the constant polarizing potential $\overline{\varphi}$ as the reversible value φ_e, i.e., $\overline{\varphi} = \varphi_e$. In this case,

$$\overline{Q} \equiv 0, \quad B = (1 + \overset{\bullet}{c}_O/\overset{\bullet}{c}_R), \quad A_1 = 1. \tag{5.60}$$

Equation (5.58) is then transformed into

$$Q_1 = \frac{i_0}{i_{dO}} \left\{ (1 + \overset{*}{c_O}/\overset{*}{c_R}) \frac{d}{d\eta} \int_0^\eta Q_1(\lambda) K(\eta - \lambda) \, d\lambda + \frac{nF\tilde{\varphi}}{RT} \right\}. \qquad (5.61)$$

The above integral equation can be solved using Laplace transforms. In fact, after Laplace transform of Eq. (5.61) according to
standard rules [18], a simple algebraic equation is obtained the
solution of which is

$$\bar{Q}_1 = \frac{nF\overline{\tilde{\varphi}}/RT}{i_{dO}/i_0 - (1 + \overset{*}{c_O}/\overset{*}{c_R}) \, \bar{K}}, \qquad (5.62)$$

where Q_1 and $\overline{\tilde{\varphi}}$ are the Laplace transforms of the functions Q_1
and $\tilde{\varphi}$, respectively; $\bar{K}$ is the Laplace transform of the kernel
of the integral equation (5.61). The transform $\bar{K}$ can be found
using the approximation described in §§ 2.8 and 5.3:

$$\bar{K} = \frac{Ai\,(p/a^{2/3})}{a^{1/3}Ai'\,(p/a^{2/3})}.$$

According to the general theory of electric circuits [33], the
transform impedance is chosen as the quantity equal to the ratio
of the transform voltage (in our case, $\tilde{\varphi}$) to the transform current
(in our case, $i_{dO}\bar{Q}_1$), i.e.,

$$Z = \left[\frac{i_{dO}}{i_0} - (1 + \overset{*}{c_O}/\overset{*}{c_R}) \, \bar{K} \right] RT/nFi_{dO}. \qquad (5.63)$$

If the variable potential component is a harmonic function of
time, i.e.,

$$\tilde{\varphi} = \tilde{\varphi}_0 \exp{(i\,\Omega t)} = \tilde{\varphi}_0 \exp{(i\,\Omega\delta_d^2/D)}, \qquad (5.64)$$

where Ω is the frequency of the harmonic signal, $\tilde{\varphi}_0$ is its amplitude, and $i = (-1)^{1/2}$, the magnitude of the impedance Z can be
easily obtained from Eq. (5.63) by a simple substitution of the
Laplace parameter p by the quantity $i\Omega\delta_d^2/D$ (cf. [33]).

Thus, polarization of the rotating disc electrode with a harmonic potential of small amplitude results in a Faradaic imped-

ance given by

$$Z\left(\frac{i\Omega\delta_d^2}{D}\right) = \frac{RT}{nFi_{dO}}\left[\frac{i_{dO}}{i_0} - \left(1 + \frac{c_O^*}{c_R^*}\right)Ai\left(\frac{i\Omega\delta_d^2}{Da^{2/3}}\right)\middle/ a^{1/3}Ai'\left(\frac{i\Omega\delta_d^2}{Da^{2/3}}\right)\right]. \quad (5.65)$$

It must be mentioned that application of the above method to calculations of the Faradaic impedance generally requires consideration of the problem of establishment of stationary-state conditions. In particular, the substitution of p by $i\Omega\delta_d^2/D$ should be accompanied by an additional analysis of the process of switching on the oscillating potential. For the case described, calculation of transient processes is very difficult. Similar processes were discussed in detail in §§ 5.3-5.5 where the establishment of a stationary regime of convective diffusion was analyzed.

In order to interpret the results obtained, the electrochemical cell will be represented in the usual way by a resistance R_s in series with a capacity C_s. The impedance of such a circuit can be written as

$$Z\left(i\,\Omega\delta_d^2/D\right) = R_s(\Omega) - i/\Omega C_s(\Omega). \quad (5.66)$$

It follows from Eq. (5.66) that

$$R_s(\Omega) = Re(Z) \quad \text{and} \quad \frac{1}{\Omega C_s} = -Im(Z), \quad (5.67)$$

where $Re(Z)$ and $Im(Z)$ are the real and imaginary parts in Eq. (5.65), respectively.

Separation of the real and imaginary parts in expression (5.65) is impossible owing to the complex dependence of the right side on Ω. However, this can be done using the approximate formula (5.21) for the logarithmic derivative of the Airy function. Since even this approximation [Eq. (5.21)] results in very cumbersome final expressions for $R_s(\Omega)$ and $C_s(\Omega)$, further considerations are restricted to two limiting cases. The latter allow the effects of convective diffusion on the Faradaic impedance to be evaluated.

As a preliminary remark, we note that the term $\Omega\delta_d^2/D$ in Eq. (5.65) can be interpreted as follows. The quantity $(D/\Omega)^{1/2}$, which appears in diffusion problems with harmonic boundary conditions, is usually called the effective thickness of the diffusion layer δ_f. Therefore, $\Omega \delta_d^2/D = \delta_d^2/\delta_f^2$ reflects the ratio of the thickness of the diffusion layer to the thickness of the layer over which the effects of the periodically oscillating boundary conditions extend.

1. Let $\delta_d \gg \delta_f$. In this case stirring affects little the passage of the oscillating signal. When $\Omega \delta_d^2/D \gg 1$, the following asymptotic expression is valid†:

$$\frac{Ai\,(i\,x/a^{2/3})}{a^{1/3}Ai'\,(i\,x/a^{2/3})} \approx -\frac{1}{\sqrt{i\,x}} + \frac{a}{4\,(i\,x)^2} + \cdots \tag{5.68}$$

Substituting Eq. (5.68) into Eq. (5.65) for the Faradaic impedance, and using expression (3.10) for the exchange current, we obtain the following expression:

$$R_s = \frac{RT}{(nF)^2}\left[\frac{1}{c_O^* k_f^0 \exp\,(-\alpha nF\varphi_e/RT)} + \right.$$
$$\left. + \left(\frac{1}{c_O^*} + \frac{1}{c_R^*}\right)\left(\frac{1}{(2\Omega D)^{1/2}} + \frac{\alpha D}{4\Omega^2\delta_d^3}\right)\right],$$
$$\frac{1}{\Omega C_s} = \frac{RT}{(nF)^2}\left(\frac{1}{c_O^*} + \frac{1}{c_R^*}\right)\frac{1}{\sqrt{2\Omega D}}. \tag{5.69}$$

In the case of a fast electrode reaction proceeding under diffusion control, expressions for the Faradaic impedance and pseudo-capacity are close to similar expressions obtained for semiinfinite linear diffusion [21, 29, 34, 35]. Stirring increases the cell resistance somewhat, but the pseudo-capacity remains practically unchanged.

The phase shift between current and potential, which in the case of semiinfinite linear diffusion is $\pi/4$, decreases upon stirring.

2. Let $\delta_d \ll \delta_f$. This inequality can be realized in practice for low frequencies of the harmonic signal and with intensive stirring. The condition $\delta_d \ll \delta_f$ can be interpreted physically as corresponding to the establishment of a quasi-stationary diffusion regime, when the convective transport can adjust itself to the slowly

---

†Equation (5.68) can be obtained by asymptotic expansion of $Ai(ix/a^{2/3})$ and $Ai'(ix/a^{2/3})$.

changing conditions at the electrode. When $\Omega \delta_d^2/D \ll 1$,

$$\frac{Ai\,(\mathrm{i}\,\Omega\delta_d^2/Da^{2/3})}{a^{1/3}Ai'\,(\mathrm{i}\,\Omega\delta_d^2/Da^{2/3})} \approx -1.07.$$

Then

$$R_s = \frac{RT}{(nF)^2}\left[\frac{1}{c_O^{\bullet} k_f^0 \exp\,(-\alpha nF\varphi_e/RT)} + 1.07\,\frac{\delta_d}{D}\left(\frac{1}{c_O^{\bullet}} + \frac{1}{c_R^{\bullet}}\right)\right],$$

$$\frac{1}{\Omega C_s} = 0.\tag{5.70}$$

In this case, the Faradaic impedance consists of the active resistance only, the phase displacement between the current and potential being zero.

The Faradaic impedance of a cell with a rotating disc electrode was discussed by Epelboin and Schuhmann et al. [36, 37]. Their calculations were based on the Nernst model of the diffusion layer. The frequency dependence in their expressions for the Faradaic impedance was described by the function $\tanh\,(\mathrm{i}\Omega\delta_d^2/D)^{1/2}/(\mathrm{i}\Omega\delta_d^2/D)^{1/2}$. This functional dependence coincides with Eq. (5.69) in the high-frequency region. However, the possibility of using Schuhmann's and Epelboin's expressions for the low-frequency region (for $\Omega < D/\delta_d^2$) is not immediately obvious. The authors proved the applicability of their equations in the low-frequency region by a special, very tedious calculation.

Experimental measurements of the impedance of a rotating platinum disc electrode were carried out [36] for the KI_3-KI system at frequencies up to 30 kHz. From the magnitude of the Faradaic impedance, the thickness of the diffusion layer was calculated and compared with the diffusion layer thickness δ_d calculated according to Eq. (2.20).

Both quantities were found to be quite close in value. Consequently, the theory quantitatively describes the behavior of the rotating disc, at least at sufficiently high frequencies.

Qualitative agreement with the theory was found in a study of the corrosion of rotating disc electrodes of lead, zinc, and iron subjected to alternating current [38]. At high ac frequencies, when $\delta_f \ll \delta_d$, the corrosion rate is low and independent of the rotation speed. At low frequencies (~ 1 Hz), corrosion proceeds at measurable rates, particularly at high rotation speeds. The effective thickness of the diffusion layer δ_f exceeds in this case the magnitude of δ_d. Therefore, metal ions which are formed during the anodic part of the cycle reach the boundary of the "stationary" diffusion layer and are transferred irreversibly into the bulk of the solution.

§ 5.9. Autooscillating Processes

Nonstationary phenomena which can appear in mass-transport studies at a rotating disc electrode include autooscillations. Spontaneous periodic oscillations of current at constant cell voltage may result from both surface (e.g., electrode passivation) and bulk (e.g., in presence of diffusion in the surface layer) processes.

Although autooscillations, for example, during anion reduction were first observed at a dropping mercury electrode, their properties are not connected with the actual properties of any given type of electrodes; but they have a general character. Autooscillations can also be observed at solid electrodes. It is necessary

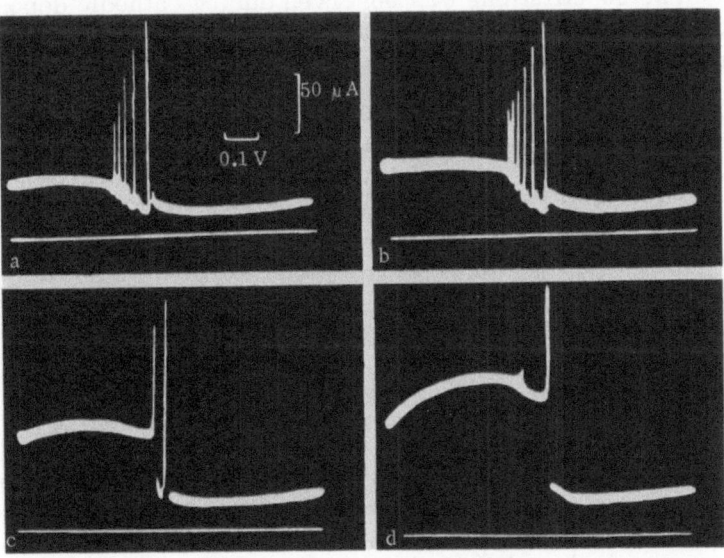

Fig. 5.6. Effect of rotation speed on autooscillations [39]. Solution $3.7 \cdot 10^{-3}$ M $K_2S_2O_8$; copper disc electrode ($r_0 = 0.95$ mm). Rotation speed (rev/sec): a) 5; b) 8; c) 15; d) 29. Rate of potential sweep 0.25 V/sec.

that the solution in the vicinity of the electrode be continuously
renewed, as is the case at a mercury dropping electrode owing to
the tangential movement of the surface. The surface of nonamalga-
mated solid electrodes is motionless, but the solution can be stirred
by electrode rotation. The rotating disc electrode was used
by Hochstein [39] to study autooscillating processes in passivation-
free electrochemical systems with a negative resistance. Hochstein showed that the existence of a region with negative resistance
together with concentration polarization due to the discharging ion
is a sufficient condition for the appearance of autooscillations.

Calculations [39] show that the potential range over which
autooscillations are observed should decrease with decreasing
thickness of the diffusion layer. The latter is determined at a
rotating disc by the rotation speed. In fact, it was found, in a
study of persulfate reduction at a rotating copper disc, that the
potential range over which autooscillations are observed becomes
narrowed with increasing angular velocity of the electrode, and
autooscillations degenerate eventually into a single peak, as shown
in the polarograms in Fig. 5.6. A similar effect of the rotation
speed on autooscillations was observed during cathodic deposition
of cadmium from a cyanide solution [40], as well as in anodic oxidation of ethylene at a platinum electrode [41].

It should be mentioned that the rotating disc electrode is particularly useful for such studies owing to the uniform accessibility
of its surface.

References

1.   R. P. Buck and H. E. Keller, Anal. Chem., 35:400 (1968).
2.   I. Shain and A. Z. Crittenden, Anal. Chem., 26:281 (1956).
3.   A. N. Frumkin et al., Élektrokhimiya, 1:17 (1965); N. A. Shumilova and
      V. S. Bogotzky, Electrochim. Acta, 13:285 (1968).
4.   L. A. Dorfman, Hydrodynamic Drag and Heat Transfer of Rotating Bodies,
      Fizmatgiz, Moscow (1960); L. G. Loitsyanskii, The Laminar Boundary Layer,
      GIFML, Moscow (1962); K. Stewartson, in: Problems in Mechanics, Vol. 4,
      H. Dryden and T. von Kármán, eds., IL, Moscow (1963), p. 9.
5.   V. G. Levich, Physical Thermodynamics, Prentice Hall, Inc. (1962).
6.   J. M. Hale, J. Electroanal. Chem., 6:187 (1963); 8:332 (1964).
7.   L. Nanis and I. Klein, Extended Abstracts, 20th CITCE Meeting, Strasbourg,
      1969, p. 52.
8.   V. S. Krylov, ibid., p. 135.

9. R. D. Cess and E. M. Sparrow, in: International Heat Transfer Conference, Boulder, 1961, p. 468.

10. R. D. Cess, Appl. Sci. Res., A13:233 (1964).

11. D. R. Olander, Intern. J. Heat Mass Transfer, 5:825 (1962).

12. R. D. Andrews and N. Riley, Quart. J. Mech. Appl. Math., 22:19 (1969).

13. Yu. G. Siver, Zh. Fiz. Khim., 33:2586 (1959); 34:577 (1960).

14. V. V. Vashchenko, N. G. Chovnyk, and L. L. Savvin, Ukr. Khim. Zh., 32:676 (1966).

15. I. Fried and P. J. Elving, Anal. Chem., 37:464, 803 (1965); I. Fried and E. Shamir, Electrochim. Acta, 14:941 (1969).

16. V. Yu. Filinovskii and V. A. Kir'yanov, Dokl. Akad. Nauk SSSR, 156:1412 (1964).

17. G. D. Yakovleva, Tables of Airy Functions and Their Derivatives, Nauka, Moscow (1969).

18. V. A. Dumkin and A. P. Prudnikov, Textbook of Operational Computations, Vysshaya Shkola, Moscow (1965).

19. S. Bruckenstein and S. Prager, Anal. Chem., 39:1161 (1967).

20. S. S. Kruglikov, N. T. Kudryavtsev, and R. P. Sobolev, Studies in the Field of Electroplating, Novocherkassk (1965), p. 66.

21. B. B. Damaskin, Principles of Modern Methods of Investigation of Electrochemical Reactions, Izd. MGU (1965).

22. S. S. Kruglikov, R. P. Sobolev, and N. T. Kudryavtsev, Tr. MKhTI im. D. I. Mendeleeva, 49:150 (1965).

23. S. S. Kruglikov et al., Dokl. Akad. Nauk SSSR, 149:911 (1963).

24. G. Just and R. Landsberg, Z. Phys. Chem., 226:183 (1964); Electrochim. Acta, 9:817 (1964).

25. T. I. Popova et al., Dokl. Akad. Nauk SSSR, 132:639 (1960); Studies in the Field of Electrical Power Sources, Novocherkassk (1966), p. 144; Izv. Akad. Nauk SSSR, Ser. Khim., 1963:1187; Élektrokhimiya, 6:104 (1970).

26. A. N. Doronin and O. L. Kabanova, Zh. Anal. Khim., 20:1321 (1965); 24:791 (1969).

27. V. Yu. Filinovskii and E. M. Podgaetskii, Élektrokhimiya, 4:671 (1968).

28. G. P. Girina, V. Yu. Filinovskii, and L. G. Feokitstov, Élektrokhimiya, 3:941 (1967); Rev. Polarogr. (Japan), 14:N3-6 (1967).

29. F. Delahay, New Instrumental Methods in Electrochemistry, Interscience, New York (1954).

30. Yu. S. Gorodetskii, Élektrokhimiya, 2:122 (1966).

31. V. Yu. Filinovskii, Dissertation, Moscow (1965).

32. H. Matsuda, Z. Elektrochem., 62:977 (1958).

33. A. Ango, Mathematics for Electrical Engineers and Radio Engineers, GIFML, Moscow (1967).

34. E. Warburg, Ann. Phys., 67:493 (1899); 6:125 (1901).

35. V. A. Kir'yanov, and V. Yu. Filinovskii, Zh. Fiz. Khim., 37:2122 (1963).

36. I. Epelboin, M. Keddam, and J. C. Lestrade, Compt. Rend., C263:1110 (1966); Rev. Gen. Electr., 76:777, 823 (1967).

37. J. M. Coueignoux and D. Schuhmann, J. Electroanal. Chem., 17:245 (1968).

38. Yu. N. Mikhailovskii, Zh. Fiz. Khim., 38:995 (1964).

39.   A. Ya. Gokhshtein et al., Dokl. Akad. Nauk SSSR, 140:1114 (1961); 144:821
       (1962).
40.   R. M. Vishomirskis and Yu. P. Shivitskis, Élektrokhimiya, 1:864 (1965).
41.   J. Wojtowicz et al., J. Chem. Phys., 48:4333 (1968); Electrochim. Acta,
       14:1119 (1969).

Studies of Chemical Reactions in a Bulk Phase (Kinetic and Catalytic Currents)

The methodological advantages of the rotating disc electrode, as well as the relative simplicity of the theory of mass transfer to the surface of a disc electrode, make this device a suitable tool for studies of transport processes accompanied by chemical reactions of the transferred species.

The present chapter is concerned primarily with chemical reactions of electrochemically active species, i.e., species also taking part in electrode reactions at the disc surface. In this case, the electric current measured allows the kinetics of chemical reactions to be studied with relatively high accuracy.

Electrode processes involving chemical steps differ considerably from each other [1-3]. Depending on the sequence of chemical and electrochemical steps, they can be divided into three groups, according to whether: 1) the chemical reaction precedes the electrochemical step, 2) the chemical and electrochemical steps are in a parallel relationship, and 3) the electrochemical step precedes the chemical reaction.

The magnitude of electric current is affected considerably in the case of "fast" chemical reactions. "Slow" chemical processes usually affect little the rate of diffusion to the disc surface and consequently the diffusional currents. A criterion for evaluating the rates of chemical reactions which affect the transport processes will be derived in the following sections.

§6.1. General Method of Calculation of Kinetic Currents at the Rotating Disc Electrode

Several methods of calculation, both approximate and more exact, have been developed for the theoretical description of polarographic kinetic currents [2-4]. These procedures can be applied without serious modifications to the case of the rotating disc electrode [5].

A single chemical reaction is assumed to occur in the solution involving N species A_1, ..., A_N. The reaction can be represented by

$$\sum_{\varkappa=1}^{N} \mu_{\varkappa} A_{\varkappa} = 0, \tag{6.1}$$

where μ_k are the respective stoichiometric coefficients. It is obvious that in the above equation some of the μ_k coefficients are positive, and the others negative.

One of the components of the chemical reaction (species A_1) is assumed to be electrochemically active and undergoes an electrode reaction:

$$A_1 + ne^- \rightarrow A_R. \tag{6.2}$$

The product of the electrode reaction, A_R, can be among the species A_2, ..., A_N.

If the substance A_1 were electrochemically inactive, all the A_1, ..., A_N species in solution would be in equilibrium, obeying the relation

$$\prod_{\varkappa} c_{\varkappa}^{*\mu_{\varkappa}} = \sigma, \tag{6.3}$$

where c_k^* is the bulk concentration of substance k and σ is the equilibrium constant.

The electrode reaction disturbs the equilibrium in the vicinity of the electrode, and the deviation from equilibrium thus caused can be characterized by the following function:

$$\rho_c F(c_1, \ldots, c_N) = \rho_c \prod_{\mu_{\varkappa} > 0} c_{\varkappa}^{\mu_{\varkappa}} - \rho_c \sigma \prod_{\mu_{\varkappa} < 0} c_{\varkappa}^{-\mu_{\varkappa}}, \tag{6.4}$$

where ρ_c is the rate constant of the bulk reaction (the polarographic designations are followed), and c_k is the concentration of the k-th substance in the vicinity of the electrode. The first product on the right side of Eq. (6.4) involves species with positive stoichiometric coefficients, and the second, those with negative values.

The chemical reactions which occur in the system result in the appearance (or disappearance) of some species in the bulk near the electrode. Therefore, the convective transport equations (2.6) must include additional terms characterizing these chemical processes. In the case discussed, the mass transport in the stationary state is described by a system of N ordinary differential equations

$$D_\varkappa \frac{d^2c_\varkappa}{dz^2} = w \frac{dc_\varkappa}{dz} + \mu_\varkappa \rho_c F(c_1, \ldots, c_N), \qquad (6.5)$$

where D_k is the diffusion coefficient of substance k.

The boundary conditions for the above system (6.5) will now be formulated. Since equilibrium is disturbed only in the vicinity of the electrode, the bulk concentrations retain their equilibrium values, i.e.,

$$c_\varkappa \to c_\varkappa^* \qquad \text{as } z \to \infty. \qquad (6.6)$$

For a fast electrode reaction (6.2), the surface concentration of the substance A_1 can be assumed to be zero, so that

$$c_1 = 0 \quad \text{for } z = 0. \qquad (6.7)$$

Substances A_2, ..., A_N do not participate in the electrode reaction and therefore†

$$\left(\frac{dc_2}{dz}\right)_S = \cdots = \left(\frac{dc_N}{dz}\right)_S = 0 \qquad \text{for } z = 0. \qquad (6.8)$$

For simplicity, the diffusion coefficients of the various species are considered equal. Consequently, the thickness of the diffusion layer δ_d at the disc is the same for all species [2.20].

---

†The boundary condition (6.8) may change in the case of electrochemical reactions.

All distances will be measured in units of δ_d,

$$z = \delta_d \xi. \tag{6.9}$$

and dimensionless concentrations

$$C_\varkappa = c_\varkappa / c_\varkappa^* \tag{6.10}$$

will be introduced, characterizing deviations of concentration from the equilibrium values.

Functions $F(c_1, \ldots, c_N)$ describing the chemical reaction can now be rewritten as follows:

$$\rho_c F(c_1, \ldots, c_N) = \rho_c \prod_{\mu_\varkappa > 0} (c_\varkappa^*)^{\mu_\varkappa} \left\{ \prod_{\mu_\varkappa > 0} C_\varkappa^{\mu_\varkappa} - \prod_{\mu_\varkappa < 0} C_\varkappa^{-\mu_\varkappa} \right\}$$

$$= \rho_c \prod_{\mu_\varkappa > 0} (c_\varkappa^*)^{\mu_\varkappa} \, \Phi(C_1, \ldots, C_N), \tag{6.11}$$

and the set of equations (6.5) becomes

$$\frac{d^2 C_\varkappa}{d\xi^2} + a\xi^2 \frac{dC_\varkappa}{d\xi} - \frac{\mu_\varkappa c_1^*}{|\mu_1| c_\varkappa^*} \frac{1}{\varepsilon_c^2} \Phi(C_1, \ldots, C_N) = 0, \tag{6.12}$$

where $a = 2.13$ and

$$\varepsilon_c^2 = c_1^* D \left[\delta_d^2 |\mu_1| \rho_c \prod_{\mu_\varkappa > 0} (c_\varkappa^*)^{\mu_\varkappa} \right]^{-1}. \tag{6.13}$$

The boundary conditions (6.6)-(6.8) retain their previous form, viz:

$$C_\varkappa \to 1 \quad \text{as} \quad \xi \to \infty, \tag{6.14}$$

$$C_1 = 0 \quad \text{for } \xi = 0, \tag{6.15}$$

$$\left(\frac{dC_2}{d\xi} \right)_S = \cdots = \left(\frac{dC_N}{d\xi} \right)_S = 0 \quad \text{for } \xi = 0. \tag{6.16}$$

First the parameter ε_c [Eq. (6.13)] will be considered. Its magnitude depends on the chemical reaction rate; ρ_c will be small in the case of fast chemical reactions ($\varepsilon_c \ll 1$) and large when ρ_c is low ($\varepsilon_c \gg 1$). In the latter case, the last term on the left side of Eq. (6.12) is small and can be calculated by sucessive approx-imations. This indicates also that slow chemical reactions affect little the convective transport to the rotating disc surface.

Thus, the magnitude of the parameter ε_c can serve as a criterion for determining the relative effect of chemical bulk reactions on the processes occurring at the rotating disc electrode.

It should be mentioned that the parameter ε_c can be presented in another way which reveals better its physical significance. The quantity μ_c defined by

$$\mu_c^2 = \frac{c_1^* D}{|\mu_1| P_c \prod_{\mu_\kappa > 0} c_\kappa^{* \mu_\kappa}} \tag{6.17}$$

has the dimensions of length and can be called the thickness of the kinetic reaction layer. From (6.13) and (6.17), we obtain

$$\varepsilon_c^2 = \mu_c^2 / \delta_d^2 . \tag{6.18}$$

It will be further shown, on the basis of the simplest cases, that the thickness of the kinetic reaction layer μ_c is a measure of the dimensions of the region at the disc electrode where chemical equilibrium is appreciably disturbed.

In the case of fast chemical reactions $(\mu_c \ll \delta_d)$, the thickness of the region is much less than that of the diffusion boundary layer. Electrode processes resulting in disturbance of chemical equilib-rium strongly depend on the rate constant ρ_c of the chemical reac-tion. When $\mu_c \gg \delta_d$, participation of chemical reactions has usually little effect.

In order to present the methods of calculation, the system (6.12) will first be considered. New functions

$$C_1 \equiv C_1,$$

$$\Psi_1 = C_1 - \frac{\mu_1 c_2^*}{\mu_2 c_1^*} C_2,$$

$$\Psi_2 = C_2 - \frac{\mu_2 c_3^*}{\mu_3 c_2^*} C_3,$$

$$\cdots \cdots \cdots \cdots \cdots \cdots$$

$$\Psi_{N-1} = C_{N-1} - \frac{\mu_{N-1} c_N^*}{\mu_N c_{N-1}^*} C_N \tag{6.19}$$

are introduced for this purpose.

The total number of unknown functions remains the same; however, the form of Eqs. (6.12) is changed to

$$\frac{d^2C_1}{d\xi^2} + a\xi^2 \frac{dC_1}{d\xi} - \frac{\mu_1}{|\mu_1| e_c^2} \Phi_1(C_1, \Psi_1, \ldots, \Psi_{N-1}) = 0,$$

$$\frac{d^2\Psi_1}{d\xi^2} + a\xi^2 \frac{d\Psi_1}{d\xi} = 0,$$

$$\frac{d^2\Psi_2}{d\xi^2} + a\xi^2 \frac{d\Psi_2}{d\xi} = 0,$$

$$\cdots \cdots \cdots \cdots \cdots \cdots \cdots \cdots \tag{6.20}$$

$$\frac{d^2\Psi_{N-1}}{d\xi^2} + a\xi^2 \frac{d\Psi_{N-1}}{d\xi} = 0.$$

The boundary conditions which should be satisfied by solutions of Eqs. (6.20) are as follows:

for $\xi \to \infty$

$$C_1 \to 1,$$

$$\Psi_1 \to 1 - \frac{\mu_1}{\mu_2} \frac{c_2^*}{c_1^*},$$

$$\Psi_2 \to 1 - \frac{\mu_2}{\mu_3} \frac{c_3^*}{c_2^*}, \tag{6.21}$$

$$\cdots \cdots \cdots \cdots \cdots \cdots \cdots$$

$$\Psi_{N-1} \to 1 - \frac{\mu_{N-1}}{\mu_N} \frac{c_N^*}{c_{N-1}^*};$$

for $\xi = 0$

$$C_1 = 0, \tag{6.22a}$$

$$\left(\frac{d\Psi_1}{d\xi}\right)_S = \left(\frac{dC_1}{d\xi}\right)_S, \tag{6.22b}$$

$$\left(\frac{d\Psi_2}{d\xi}\right)_S = \cdots = \left(\frac{d\Psi_{N-1}}{d\xi}\right)_S = 0. \tag{6.22c}$$

Solution of Eqs. (6.20) with boundary conditions (6.21) and (6.22a)–(6.22c) leads to the determination of the current density as

$$i = nFD\left(\frac{dc_1}{dz}\right)_S = \frac{nFDc_1^*}{\delta_d}\left(\frac{dC_1}{d\xi}\right)_S = i_{d1}X, \tag{6.23}$$

where i_{d1} is the limiting diffusion current of A_1 at the disc and

$$X = \left(\frac{dC_1}{d\xi}\right)_S \tag{6.24}$$

is the unknown factor describing the effect of the chemical reaction on the diffusion current.

It can be easily demonstrated from Eqs. (6.20), (6.21), and (6.22) that the functions $\Psi_1, \ldots, \Psi_{N-1}$ should be equal to

$$\Psi_1(\xi) = \left(1 - \frac{\mu_1 c_2^*}{\mu_2 c_1^*}\right) - X \int_{\xi}^{\infty} \exp\left(-\frac{ay^3}{3}\right) dy,$$

$$\Psi_2 = \Psi_2^* = 1 - \frac{\mu_2}{\mu_3} \frac{c_3^*}{c_2^*} ,$$

(6.25)

$$\cdot \quad \cdot \quad \cdot \quad \cdot \quad \cdot \quad \cdot \quad \cdot \quad \cdot \quad \cdot \quad \cdot \quad \cdot \quad \cdot \quad \cdot \quad \cdot$$

$$\Psi_{N-1} = \Psi_{N-1}^* = 1 - \frac{\mu_{N-1}}{\mu_N} \frac{c_N^*}{c_{N-1}^*} .$$

(6.26)

The functions $\Psi_2^*, \ldots, \Psi_{N-1}^*$, in fact, turn out to be constant and independent of the distance from the electrode.

The problem of solution of Eqs. (6.20) is now reduced to the solution of the following ordinary differential equation:

$$\frac{d^2 C_1}{d\xi^2} + a\xi^2 \frac{dC_1}{d\xi} - \frac{1}{\varepsilon_c^2} \frac{\mu_1}{|\mu_1|} \Phi_1 (C_1, \Psi_1, \Psi_2^*, \ldots, \Psi_{N-1}^*) = 0, \quad (6.27)$$

where the functions $\Psi_1, \Psi_2^*, \ldots, \Psi_{N-1}^*$ are defined by relations (6.25) and (6.26). The boundary conditions for Eq. (6.27) are as follows:

for $\xi \to \infty$

$$C_1 \to 1,$$

(6.28a)

for $\xi = 0$

$$C_1 = 0.$$

(6.28b)

No assumptions conerning the magnitude of the parameter ε_c have, as yet, been made. The choice of the method of solving Eq. (6.27) and the resulting concentration distribution C_1 depend, however, on the magnitude of ε_c.

When $\varepsilon_c \gg 1$, the function Φ_1 in Eq. (6.27) is multiplied by a small-valued parameter. The equation can then be solved by successive approximations. In the zeroth approximation, the concentration distribution is not distorted by the chemical reaction. It has already been mentioned that, in the case $(\mu_c \gg \delta_d)$, the chemical reaction little affects the limiting diffusion current at the disc.

When $\varepsilon_c \ll 1$, i.e., in the case of a fast chemical reaction, the small parameter is a factor multiplying the derivatives. Equations of this type were already discussed in Chap. 4. It was shown

CHAPTER 6

in § 4.3 that such equations can be solved using the "boundary layer" method. It will be shown later that this procedure is approximately equivalent to methods used by the Czechoslovakian authors [2, 4]. By analogy with § 4.3, the region in which diffusion occurs is divided into two parts: "the external region" and a narrow "boundary layer" region at the electrode.

Function $\bar{C}_1$ in the external region can be simply obtained by equating $\varepsilon_c = 0$ in Eq. (6.27). Then

$$\Phi_1(\bar{C}_1, \Psi_1, \Psi_2^\bullet, \ldots, \Psi_{N-1}^\bullet) = 0. \tag{6.29}$$

The above equation can be used together with Eqs. (6.25) and (6.26) to determine the concentration distribution of A_1 far from the electrode. Obviously, $\bar{C}_1 = \bar{C}_1(\xi)$ satisfies the boundary condition as $\xi \to \infty$, i.e., $\bar{C}_1 \to 1$ as $\xi \to \infty$. Thus, far from the electrode, concentrations of all species are in chemical equilibrium (6.3). However, the distribution $\bar{C}_1$ thus found does not satisfy the boundary condition (6.28b) at the electrode ($\bar{C}_1 \neq 0$ for $\xi = 0$).

Distribution $\bar{C}_1$ must be corrected in order to eliminate the above contradiction. By analogy with § 4.4, the concentration distribution of A_1 in the vicinity of the electrode is sought in the form

$$C_1 = \bar{C}_1 - \tilde{C}_1, \tag{6.30}$$

where the function $\tilde{C}_1$ is the correction to the concentration distribution $\bar{C}_1$ close to the disc.

Introducing Eq. (6.30) into Eq. (6.27), we obtain an equation which should be satisfied by the function $\tilde{C}_1$:

$$\frac{d^2\tilde{C}_1}{d\xi^2} + a\xi^2 \frac{d\tilde{C}_1}{d\xi} - \left(\frac{d^2\bar{C}_1}{d\xi^2} + a\xi^2 \frac{d\bar{C}_1}{d\xi}\right) +$$
$$+ \frac{\mu_1}{|\mu_1| \varepsilon_c^2} \Phi_1 [(\bar{C}_1 - \tilde{C}_1), \Psi_1, \Psi_2^\bullet, \ldots, \Psi_{N-1}^\bullet] = 0. \tag{6.31}$$

Since the detailed behavior of the function $\tilde{C}_1$ is of interest, the scale of distance measurements should be extended, e.g., by defining

$$\xi = \varepsilon_c t. \tag{6.32}$$

Introducing the new variable in Eq. (6.31), we obtain the following expression describing the $\widetilde{C}_1$ function in the boundary layer as $\varepsilon_c \to 0$:

$$\frac{d^2\widetilde{C}_1}{dt^2} + \frac{\mu_1}{|\mu_1|}\, \Phi\,[(\bar{C}_{1S} - C_1),\ \Psi_{1S},\ \Psi_2^{\bullet},\, \ldots,\ \Psi_{N-1}^{\bullet}] = 0. \qquad (6.33)$$

It is necessary to note a peculiarity in the derivation of Eq. (6.33). The new variable t is also introduced in functions $\bar{C}_1$ and Ψ_1 which previously depended on the variable ξ. Thus, by fixing $\varepsilon_c \to 0$, we obtain in place of functions $\bar{C}_1(\xi)$ and $\Psi_1(\xi)$ the values of these functions at the electrode, i.e., as $\varepsilon_c \to 0$, $\bar{C}_1 \to \bar{C}_1(0) = \bar{C}_{1S}$, and $\Psi_1 \to \Psi_1(0) = \Psi_{1S}$. Thus, in the limits, Eq. (6.33) contains new parameters, $\bar{C}_{1S}$ and Ψ_{1S}, and the dependence on the variable t is considerably simplified.

The physical meaning of the above transformations is quite clear. In fact, functions $\bar{C}_1(\xi)$ and $\Psi_1(\xi)$ describe the concentration changes of the electrochemically active species within the diffusion boundary layer of thickness δ_d. The equilibrium is disturbed at distances of the order of $\mu_c \ll \delta_d$, i.e., when $\xi \ll 1$. Therefore, it can be assumed that in the very narrow region close to the electrode, functions C_1 and Ψ_1 change little.

Equation (6.33) can be solved with the boundary conditions

$$\bar{C}_1 \to 0 \qquad \text{as } t \to \infty, \qquad (6.34a)$$

$$\bar{C}_1 = \bar{C}_{1S} \qquad \text{for } t = 0. \qquad (6.34b)$$

Integration of Eq. (6.33) is easy and leads to the following relation between the function $\widetilde{C}_1$ and its derivative $d\widetilde{C}_1/dt$†:

$$\frac{1}{2}\left(\frac{d\widetilde{C}_1}{dt}\right)^2 = -\frac{\mu_1}{|\mu_1|}\int_0^{\widetilde{C}_1} \Phi_1\,[(\bar{C}_{1S} - y),\ \Psi_{1S},\ \Psi_2^{\bullet},\, \ldots,\ \Psi_{N-1}^{\bullet}]\, dy. \qquad (6.35)$$

In order to obtain the concentration distribution within the kinetic layer, another integration must be carried out. The latter operation usually results in complex expressions from which

---

†The obvious conditions $\widetilde{C}_1 \to 0$ and $(d\widetilde{C}_1/dt) \to 0$ in the region far from the electrode have been used.

$\tilde{C}_1(t)$ cannot be recovered in an explicit form. However, in the case discussed, this integration is not necessary. In fact, it can be seen from Eqs. (6.35) and (6.34b) that, at t = 0,

$$\left(\frac{d\tilde{C}_1}{dt}\right)_S = \pm\left\{-\frac{2\mu_1}{|\mu_1|}\int_0^{\overline{C}_{1S}} \Phi_1\left[(\overline{C}_{1S}-y),\ \Psi_{1S},\ \Psi_2^*,\ \ldots,\ \Psi_{N-1}^*\right]dy\right\}^{1/2};\quad (6.36)$$

Eq. (6.36) suffices to complete the solution of the problem.

It can be easily shown that all functions depend on the unknown quantity X as a parameter. The quantity X, which describes the flux of the electroactive reactant A_1, enters Eq. (6.25) for the function Ψ_1 and, consequently, in the subsequent expressions for $\overline{C}_1$ and $\tilde{C}_1$.

The flux X can be obtained using Eq. (6.22b), from which

$$X = \left(\frac{d\Psi_1}{d\xi}\right)_S = \left(\frac{dC_1}{d\xi}\right)_S.$$

Since $(dC_1/d\xi)_S = (d\overline{C}_1/d\xi)_S - (d\tilde{C}_1/d\xi)_S$, and the derivatives $(d\overline{C}_1/d\xi)_S$ and $(d\tilde{C}_1/d\xi)_S$ were already calculated, the following algebraic expression can be easily obtained:

$$X = \left(\frac{d\overline{C}_1}{d\xi}\right)_S - \frac{1}{\varepsilon_c}\left(\frac{d\tilde{C}_1}{dt}\right)_S.\quad (6.37)$$

The right side of the above equation depends only on X, ε_c, c_1^*, c_2^*, ..., c_N^*, and μ_1, μ_2, ..., μ_N. Solution of Eq. (6.37) allows the magnitude of the kinetic current to be found. It should be stressed, however, that the simple solution of Eq. (6.37) is only possible in the case of a first-order chemical reaction.

The general method of calculation of kinetic currents at the rotating disc electrode will be further clarified in the discussion of particular examples, to be treated below.

§ 6.2. Preceding Chemical Reaction

A. An electrochemical system in which a first-order chemical reaction precedes the charge transfer† will be considered:

$$A_2 \underset{\rho_c\sigma}{\overset{\rho_c}{\rightleftharpoons}} A_1 \quad (\text{or}\quad A_2 - A_1 = 0,\quad \mu_1 = -1,\quad \mu_2 = 1).\quad (6.38)$$

---

†As opposed to [2, 4] and following [1, 3, 6], the rate constant of the chemical reaction resulting in the formation of the electroactive product is designated by ρ_c.

The electrode reaction (6.2)

$$A_1 + ne^- \rightarrow \text{product}$$

is assumed to be fast.

Obviously, in this case, when the reactant in the electrode reaction is formed in a preceding chemical step, the kinetic current is observed if A_1 is practically absent in the solution, i.e., if the chemical equilibrium is strongly shifted towards the electrochemically inactive substance A_2. Under such conditions, A_1 is supplied basically at a rate equal to that of the chemical reaction (6.38), and the effect of the latter is then most strongly pronounced.

The following considerations are also valid for pseudo-monomolecular reactions. The system of differential equations (6.5) for the present case is as follows:

$$D \frac{d^2 c_1}{dz^2} = w \frac{dc_1}{dz} - \rho_c (c_2 - \sigma c_1),$$

$$D \frac{d^2 c_2}{dz^2} = w \frac{dc_2}{dz} + \rho_c (c_2 - \sigma c_1), \qquad (6.39)$$

with the boundary conditions

$$c_1 \rightarrow \overset{*}{c_1}, \quad c_2 \rightarrow \overset{*}{c_2}, \quad \sigma = \overset{*}{c_2}/\overset{*}{c_1} \qquad \text{as } z \rightarrow \infty, \qquad (6.40a)$$

$$c_1 = 0, \quad \left(\frac{dc_2}{dz}\right)_s = 0 \qquad \text{for } z = 0. \qquad (6.40b)$$

Eq. (6.39) can be rewritten in terms of dimensionless variables [(6.9) and (6.10)] as follows:

$$\frac{d^2 C_1}{d\xi^2} + a\xi^2 \frac{dC_1}{d\xi} - \frac{1}{e_c^2}(C_1 - C_2) = 0,$$

$$\frac{d^2 C_2}{d\xi^2} + a\xi^2 \frac{dC_2}{d\xi} + \frac{1}{\sigma e_c^2}(C_1 - C_2) = 0. \qquad (6.41)$$

The boundary conditions (6.40a)-(6.40b) become

$$C_1 = C_2 \rightarrow 1 \qquad \text{as } \xi \rightarrow \infty, \qquad (6.42a)$$

$$C_1 = 0; \quad \left(\frac{dC_2}{d\xi}\right)_s = 0 \qquad \text{for } \xi = 0, \qquad (6.42b)$$

and the parameter ε_c^2 in Eq. (6.41) is given by

$$\varepsilon_c^2 = \frac{Dc_1^*}{\delta_d^2 \rho_c c_2^*} = \frac{D}{\delta_d^2 \rho_c \sigma} \,, \tag{6.43}$$

while the thickness of the reaction layer is

$$\mu_c = \sqrt{D/\rho_c\sigma}. \tag{6.44}$$

The physical meaning of the quantity μ_c is often interpreted as the distance covered by the diffusing particle of the electroactive substance in the mean lifetime $1/\rho_c\sigma$.

Defining

$$\Psi_1 = C_1 + \sigma C_2, \tag{6.45}$$

we easily find [cf. Eq. (6.25)] that

$$\Psi_1 = (1 + \sigma) - X \int_{\xi}^{\infty} \exp\left(- ay^3/3\right) dy, \tag{6.46}$$

where $X = (d\Psi_1/d\xi)_S$ is an unknown constant.

In order to find the distribution of the electroactive species A_1, it is necessary to solve the equation

$$\frac{d^2 C_1}{d\xi^2} + a\xi^2 \frac{dC_1}{d\xi} + \frac{1}{\varepsilon_c^2}\left[\frac{\Psi_1}{\sigma} - C_1 \frac{(1 + \sigma)}{\sigma}\right] = 0. \tag{6.47}$$

Since the chemical reaction is assumed to be fast $(\varepsilon_c \ll 1)$, the method described in § 6.1 can be used.

According to Eq. (6.29), the distribution of A_1 beyond the limits of the kinetic reaction layer is given by

$$\bar{C}_1 = \frac{1}{1 + \sigma}\Psi_1 = 1 - \frac{X}{1 + \sigma}\int_{\xi}^{\infty} \exp\left(- ay^3/3\right) dy, \tag{6.48}$$

which corresponds to the chemical equilibrium between A_1 and A_2.

The deviation from equilibrium within the kinetic layer is described [cf. Eq. (6.30)] by the function

$$\tilde{C}_1 = \bar{C}_1 - C_1, \tag{6.49}$$

which can be transformed, as previously, into

$$\frac{d^2\tilde{C}_1}{dt^2} - \frac{1+\sigma}{\sigma} C_1 = 0 \qquad (6.50)$$

using the boundary conditions

$$\tilde{C}_1 \to 0 \qquad \text{as } t \to \infty, \qquad (6.51a)$$

$$\tilde{C}_1 = \tilde{C}_{1S} \qquad \text{for } t = 0. \qquad (6.51b)$$

The solution of Eq. (6.50) satisfying the boundary conditions (6.51a) and (6.51b) is as follows:

$$\tilde{C}_1 = \tilde{C}_{1S} \exp\left(-t \sqrt{(1+\sigma)/\sigma}\right). \qquad (6.52)$$

From Eq. (6.48) †, we obtain

$$\tilde{C}_{1S} = 1 - X/(1+\sigma). \qquad (6.53)$$

Substitution of the quantities obtained in Eq. (6.37) gives the following expression for X:

$$X = \frac{X}{1+\sigma} + \frac{1}{\varepsilon_c} \sqrt{\frac{1+\sigma}{\sigma}} \left(1 - \frac{X}{1+\sigma}\right), \qquad (6.54)$$

from which

$$X = \frac{(1+\sigma) \sqrt{(1+\sigma)/\sigma}}{\varepsilon_c \sigma + \sqrt{(1+\sigma)/\sigma}}. \qquad (6.55)$$

It can be seen from Eq. (6.55) that for $\sigma = c_2^*/c_1^* \gg 1$, i.e., in the case where the chemical equilibrium is shifted in favor of the electrochemically inactive species A_2,

$$X \approx \frac{\sigma}{1 + \varepsilon_c \sigma} \qquad (6.56)$$

and the kinetic current at the rotating disc electrode is, according

---

†By definition [cf. Eq. (2.47)], $\int\limits_0^\infty \exp\left(-ay^3/3\right) dy = 1$.

to Eq. (6.23),

$$i = i_{d1} \frac{\sigma}{1 + \varepsilon_c \sigma}. \tag{6.57}$$

In the opposite limiting case, when the solution contains an excess of A_1, $\sigma \ll 1$ and $X \approx 1$. Kinetic effects are then not observed in this case, and the current is equal to the limiting diffusion current of A_1, i.e., $i \approx i_{d1}$, as has already been mentioned at the begining of this section.

Equation (6.57) for the kinetic current at the disc electrode will again be considered. Replacing the limiting diffusion current of A_1 (i_{d1}) by the limiting diffusion current of A_2(i_{d2}), we can transform Eq. (6.57) into

$$i = \frac{i_{d2}}{1 + \sigma \mu_c / \delta_d}, \tag{6.58}$$

since $i_{d2} = \sigma i_{d1}$. Since the solution contains an excess of the electrochemically inactive substance A_2 ($c_2^* \gg c_1^*$), the kinetic current is considerably larger than the limiting diffusion current which would be observed if the discharge of A_1 were not preceded by the chemical reaction (6.38).

Expression (6.58) corresponds to the equations derived for the polarographic kinetic currents resulting from a preceding monomolecular chemical reaction [2-4]. The changes which must be introduced in the polarographic expressions consist in the replacement of the parameters characteristic of the convective transport to the rotating disc electrode.

For the analysis of experimental results, Eq. (6.58) is used in the form

$$\frac{(1 - i/i_{d2})}{(i/i_{d2})} = \sigma \mu_c / \delta_d. \tag{6.59}$$

The right side of Eq. (6.59) is a linear function of $\sqrt{\omega}$ [cf. Eq. (2.20)]. The plot of $(1 - i/i_{d2})/(i/i_{d2})$ vs $\sqrt{\omega}$ is hence a straight line with a slope which is a measure of the rate constant ρ_c of the chemical reaction (6.38). In the absence of the latter ($\rho_c = 0$), the quantity $(1 - i/i_{d2})/(i/i_{d2})$ is independent of the rotation speed ω.

Equation (6.59) can be rewritten in another form which is often used for graphical processing of experimental data, viz.,

$$1 - A \frac{i}{\sqrt{\omega}} = B \frac{i}{\sqrt{P_c}} , \qquad (6.59a)$$

where $A = 1.61 \nu^{1/6}/nF \sigma c_1^* D^{2/3}$ and $B = 1/nFc_1^* \sqrt{\sigma}$. A plot of $i/\sqrt{\omega}$ vs i is shown in Fig. 6.1. The slope of the line depends on the rate of the preceding chemical reaction. When $\mu_c = 0$ (i.e., for a very fast reaction), the line is parallel to the axis of abscissas.

For simplicity, it has been assumed hitherto that the diffusion coefficients of A_1 and A_2 are equal ($D_1 = D_2 = D$). Dogonadze [7] discussed the general case of arbitrary diffusion coefficients ($D_1 \neq D_2$). He obtained the following expression for the disc current in the case of a preceding monomolecular reaction:

$$i = \frac{i_d}{1 + \sigma D_2 \mu_c / D_1 \delta_d} , \qquad (6.60)$$

where

$$i_d = \frac{nFD (c_1^* + c_2^*)}{\delta_d} , \qquad \mu_c = \frac{1}{\sqrt{P_c \sigma / D_1 + P_c / D_2}} ,$$

$$\delta_d = 1.6 \left(\frac{D}{\nu}\right)^{1/3} \left(\frac{\nu}{\omega}\right)^{1/2} , \qquad D = \frac{\sigma D_2 + D_1}{1 + \sigma} .$$

Both the present calculation and that of Dogonadze [7] were based on the assumption of a fast chemical reaction (6.38), i.e., for $\mu_c \ll \delta_d$. The general case of a preceding chemical reaction of arbitrary rate was analyzed by one of the present authors and

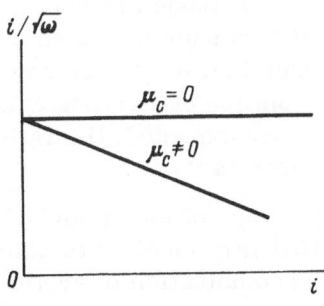

Fig. 6.1. Schematic representation of the dependence of kinetic current on rotation velocity [Eq. (6.59a)].

Kiryanov [8]. Calculations were based on the approximate method of evaluating the mass transport to the disc surface described in § 2.8. The following expression was obtained:

$$i = \frac{i_{d_1}(1+\sigma)}{1.06 + \dfrac{\sigma\sqrt{3.10 + \delta_d^2 \rho_c (1+\sigma)/D}}{1.65 + \delta_d^2 \rho_c (1+\sigma)/D}}. \tag{6.61}$$

When $\delta_d^2 \gg D/\rho_c (1+\sigma)$ (fast chemical reactions), expression (6.61) reduces to Eq. (6.57) with an accuracy of 6% (determined by the accuracy of the approximation).

Hale [9] calculated numerically the effect of a preceding monomolecular chemical reaction on the galvanostatic current at the disc electrode. His calculations of the establishment of the stationary state under conditions of kinetic limitation are valid for a wide range of bulk reaction rate constants.

One of the examples of an electrode reaction, complicated by a preceding monomolecular bulk process, is the reduction of H^+ ions from a weak acid solution (e.g., acetic acid). In aqueous solutions, the acid is in equilibrium with its ions:

$$H^+ + CH_3COO^- \underset{\rho_c}{\overset{\rho_c\sigma}{\rightleftharpoons}} CH_3COOH,$$

where ρ_c and $\rho_c\sigma$ are the rate constants for ionic dissociation and recombination, respectively. Hydrogen ions are consumed in the electrode reaction so that, in the vicinity of the electrode, the above equilibrium becomes shifted to the left side. The magnitude of the limiting current is determined by the rate of dissociation. Polarization curves of hydrogen evolution from a buffered acetic acid solution obtained at a platinum disc electrode are shown in Fig. 6.2b. The dependence of $\bar{i}/(\omega/2\pi)^{1/2}$ on $\bar{i}$ for the same system is shown in Fig. 6.3b [cf. Eq. (6.59a) and Fig. 6.1]. For comparison, analogous curves obtained with a solution of a strong acid (HCl) are shown in Figs. 6.2a and 6.3a. As expected, the ratio $\bar{i}/(\omega/2\pi)^{1/2}$ is independent of $\bar{i}$ in the latter case.

Vielstich and Jahn [10–13] used the slope of the straight line in Fig. 6.3b and the value of $\sigma = 3.1 \cdot 10^{-5}$ liter $\cdot$ mole^{-1} to calculate the rate constants of dissociation and recombination of acetic acid

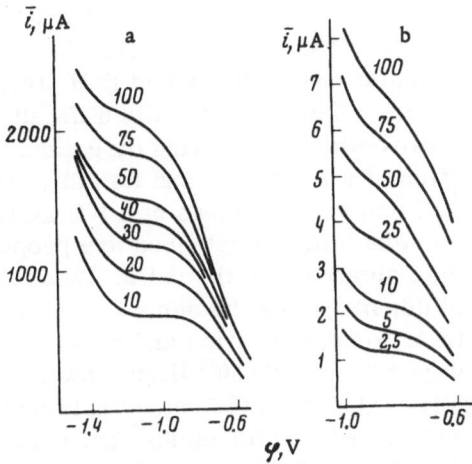

Fig. 6.2. Polarization curves for hydrogen evolution at a platinum electrode from $5 \cdot 10^{-2}$ N HCl + 1 N KCl (a) and from an acetate buffer solution ($2 \cdot 10^{-2}$ N, pH = 6.2) in 1 N KCl (b) [10, 11]. The rotation velocity of the electrode is indicated on the curves.

in 1 N KCl: $\rho_c = 3 \cdot 10^5$ sec^{-1}, $\rho_c \sigma = 1 \cdot 10^{10}$ liters $\cdot$ mole$^{-1} \cdot$ sec^{-1}. The values agree with those obtained using a dropping mercury electrode. The rate constants of dissociation and recombination of citric acid were calculated from the magnitude of the limiting current of hydrogen reduction from a 0.09 M solution of sodium citrate in 1 N NaClO$_4$: $\rho_c = 10^5$ sec^{-1}, $\rho_c \sigma = 0.7 \cdot 10^{10}$ liters $\cdot$ mole$^{-1} \cdot$ sec^{-1} (25°C).

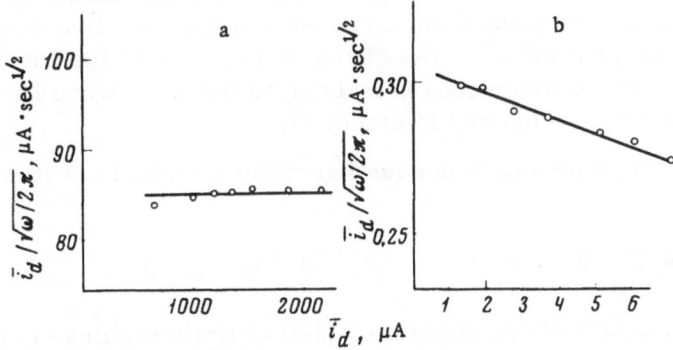

Fig. 6.3. $i / (\omega / 2\pi)^{1/2}$ as a function of $\bar{i}$ (from data in Fig. 6.2).

It can be seen from Fig. (6.2a and b) that the limiting current is poorly defined in the case of a weak acid, thus decreasing the accuracy of the calculations. Albery and Bell [14], however, proposed a method for circumventing this difficulty by simultaneously operating two cells with rotating disc electrodes. The cells, containing a solution of a weak acid and its salt, and a strong acid, respectively form two resistors of a Wheatstone bridge network. The two remaining resistances are proportional to the surfaces of the disc electrode in the cells. When the bridge is balanced, current densities are the same in both cells. The measured rate constants for dissociation and recombination of acetic acid were $9.1 \cdot 10^5$ sec^{-1} and $5.2 \cdot 10^{10}$ liters $\cdot$ mole$^{-1} \cdot$ sec^{-1}, respectively (i.e., slightly higher than the values obtained by Vielstich and Jahn). The rate constants for dissociation and recombination of trimethylacetic acid are $1.42 \cdot 10^5$ sec^{-1} and $1.53 \cdot 10^{10}$ liters $\cdot$ mole$^{-1} \cdot$ sec^{-1}, respectively. The accuracy was estimated by the authors to be within $\pm 10\%$ (the accuracy of limiting current measurements by the above differential method is $\pm 2\%$).

Albery [15] pointed out that the experimental reduction of H$^+$ ions from weak acids cannot, strictly speaking, be described by the Levich and Koutecký theory [5, 6]. In fact, in the range of relatively low concentrations investigated, the effect of the electric field in the diffuse part of the double layer on the dissociation constant, as well as on migration of protons, must be taken into account.

B. As a second example, the case of an electrode reaction preceded by decomposition of a dimer in the solution bulk may be considered. It is assumed as previously, that the chemical equilibrium is shifted toward the electrochemically inactive dimer (in this case, the effect of the chemical reaction on the current is a strong one). Calculations pertaining to this case were first carried out by Levich and Koutecký [5].

Let the stoichiometric equation of the chemical reaction be

$$A_2 \underset{\rho_c{}^\sigma}{\overset{\rho_c}{\rightleftharpoons}} 2A_1 \quad (\text{or} \quad A_2 - 2A_1 = 0, \quad \mu_2 = 1, \quad \mu_1 = -2), \quad (6.62)$$

where the electrode reaction step (6.2) is again assumed to be fast.

Equations (6.5) describing the behavior of A_1 and A_2 are in the form

$$D \frac{d^2 c_1}{dz^2} = w \frac{dc_1}{dz} - 2\rho_c (c_2 - \sigma c_1^2),$$

$$D \frac{d^2 c_2}{dz^2} = w \frac{dc_2}{dz} + \rho_c (c_2 - \sigma c_1^2). \tag{6.63}$$

As opposed to the case of monomolecular reaction, $\sigma = c_2^* / c_1^{*2}$. The boundary conditions (6.40a) and (6.40b) retain their form.

In dimensionless form, Eqs. (6.63) are in the forms

$$\frac{d^2 C_1}{d\xi^2} + a\xi^2 \frac{dC_1}{d\xi} + \frac{1}{\varepsilon_c^2} (C_2 - C_1^2) = 0,$$

$$\frac{d^2 C_2}{d\xi^2} + a\xi^2 \frac{dC_2}{d\xi} - \frac{\nu_c}{\varepsilon_c^2} (C_2 - C_1^2) = 0, \tag{6.64}$$

where

$$\varepsilon_c^2 = D / 2\rho_c \sigma c_1^* \delta_d^2, \tag{6.65}$$

$$\nu_c = c_1^* / 2c_2^*. \tag{6.66}$$

Introducing the function

$$\Psi_1 = C_1 + C_2 / \nu_c, \tag{6.67}$$

we obtain a system of two equations (6.20) for C_1 and Ψ_1.

According to Eq. (6.25),

$$\Psi_1 = \frac{1 + \nu_c}{\nu_c} - X \int_\xi^\infty \exp(- ay^3/3) \, dy. \tag{6.68}$$

Then from Eqs. (6.67) and (6.64), we obtain

$$\frac{d^2 C_1}{d\xi^2} + a\xi^2 \frac{dC_1}{d\xi} + \frac{1}{\varepsilon_c^2} (\nu_c \Psi_1 - \nu_c C_1 - C_1^2) = 0. \tag{6.69}$$

In the case considered for a fast electrode reaction $(\varepsilon_c^2 \ll 1)$, the solution of Eq. (6.69) is sought using the "boundary layer" method described in § 6.1.

In the "external region," A_1 is in equilibrium and its distribution can be found from the condition

$$\bar{C}_1^2 + v_c \bar{C}_1 - v_c \Psi_1 = 0 \qquad (6.70)$$

as

$$\bar{C}_1 = \sqrt{\frac{v_c^2}{4} + v_c \Psi_1} - \frac{v_c}{2}, \qquad (6.71)$$

where $\Psi_1(\xi)$ is defined by Eq. (6.68).

Deviations from the equilibrium distribution $\tilde{C}_1 = \bar{C}_1 - C_1$ which take place inside the kinetic reaction layer are described [cf. Eq. (6.33)] by

$$\frac{d^2\tilde{C}_1}{dt^2} - [(2\bar{C}_{1S} + v_r)\,\tilde{C}_1 - \tilde{C}_1^2] = 0, \qquad (6.72)$$

where $\bar{C}_{1S} = \bar{C}_1(0)$ is the surface concentration of A_1 obtained from the equilibrium distribution (6.71).

The boundary conditions for Eq. (6.72) are as follows:

$$\tilde{C}_1 \to 0 \qquad \text{as } t \to \infty, \qquad (6.73a)$$
$$\tilde{C}_1 = \tilde{C}_{1S} \qquad \text{for } t = 0. \qquad (6.73b)$$

Integration of Eq. (6.72) results in

$$\frac{d\tilde{C}_1}{dt} = \pm\left[(2\bar{C}_{1S} + v_c)\tilde{C}_1^2 - \frac{2}{3}\tilde{C}_1^3\right]^{1/2}. \qquad (6.74)$$

The magnitude of the additional flux to the electrode, which results from the chemical reaction, can be easily obtained by substituting the boundary condition (6.73b) into Eq. (6.74):

$$\left(\frac{d\tilde{C}_1}{dt}\right)_S = \pm\left(\frac{4}{3}\tilde{C}_{1S}^3 + v_c\tilde{C}_{1S}^2\right)^{1/2}. \qquad (6.75)$$

The total flux, described by Eq. (6.37), requires calculation of $(d\bar{C}_1/d\xi)_S$. The latter quantity can be obtained by differentiation of Eq. (6.71):

$$\left(\frac{d\bar{C}_1}{d\xi}\right)_S = \frac{v_c X}{2\bar{C}_{1S} + v_c}. \qquad (6.76)$$

Substituting Eqs. (6.75) and (6.76) into Eq. (6.37) enables

$$X = \frac{v_c X}{2\bar{C}_{1S} + v_c} \mp \frac{1}{\varepsilon_c} \left(\frac{4}{3} \bar{C}_{1S}^3 + v_c \bar{C}_{1S}^2 \right)^{1/2} \qquad (6.77)$$

to be obtained.

The magnitude of $\bar{C}_{1S}$ depends only on X and v_c and can be found from Eq. (6.71), viz.,

$$\bar{C}_{1S} = \left[\left(1 + \frac{v_c}{2} \right)^2 - v_c X \right]^{1/2} - v_c/2. \qquad (6.78)$$

From Eqs. (6.77) and (6.78), an expression for the unknown flux of the electroactive monomer can be obtained.

Instead of the general analysis of Eq. (6.77), only the appearance of kinetic currents will be considered; i.e., the chemical equilibrium in the bulk is assumed to be shifted in favor of the electrochemically inactive dimer A_2 ($v_c = c_1^*/2c_2^* \ll 1$). In this case, the equations obtained are considerably simplified.

First of all, it is to be noted that the kinetic current at the disc electrode, $i = i_{d1} X$, should now be proportional to the bulk concentration of A_2, i.e.,

$$X \sim c_2^*/c_1^* \sim 1/v_c. \qquad (6.79)$$

The magnitude of C_{1S} depends only on X and v_c and can be found from Eq. (6.71), viz.

$$\bar{C}_{1S} \approx \sqrt{1 - v_c X}. \qquad (6.80)$$

Equation (6.77) reduces to†

$$X = \frac{1}{\varepsilon_c} \left[\frac{4}{3} (1 - v_c X)^{3/2} \right]^{1/2}. \qquad (6.81)$$

Equation (6.81) is an algebraic expression with respect to X. It can be rewritten in the form

$$\frac{(1 - i/i_{d2})^{3/4}}{i/i_{d2}} = \frac{2}{\sqrt{3}} v_c \mu_c/\delta_d, \qquad (6.82)$$

---

†Since X is a positive quantity, Eq. (6.77) must be taken with the plus sign.

where

$$\mu_c = \sqrt{D/2\rho_c \sigma c_1^*}, \qquad i_{d_2} = 2nFDc_2^*/\delta_d.$$

Equation (6.82) demonstrates that the dependence of current on the rotational speed ($\delta_d \sim 1/\sqrt{\omega}$) differs from that describing a monomolecular reaction [cf. Eq. (6.59)]. This difference can be used for determining the order of a preceding chemical reaction.

Equation (6.81) will be transformed into a form obtained first by Levich and Koutĕcky [5, 6]. Taking into account that $(d\Psi_1/d\xi)_S = X$ and $\Psi_{1S} = (1 - \nu_C X)/\nu_c$ when $\xi = 0$ and $\nu_c \ll 1$, we can rewrite Eq. (6.81) in the form

$$\left(\frac{d\Psi_1}{d\xi}\right)_S = K\Psi_{1S}^{3/4} \qquad (6.83)$$

given by Levich and Koutĕcky, where

$$K = \frac{\nu_c^{3/4}}{\varepsilon_c}\left(\frac{4}{3}\right)^{1/4}. \qquad (6.84)$$

Expressing $(d\Psi_1/d\xi)_S$ in terms of Ψ_{1S}, we can again rewrite Eq. (6.83) in the form

$$\Psi_{1S}^{3/4} + \frac{1}{K}\Psi_{1S} - \frac{1}{\nu_c K} = 0, \qquad (6.85)$$

and it is seen that Eq. (6.85) is equivalent to Eq. (6.81). However, it has now the form of an equation already discussed in § 3.2. The latter section contains also a nomogram (cf. Fig. 3.1) which allows solutions of Eq. (6.85) to be found for any given values of parameters.

The kinetic disc current can be represented by

$$i = i_{d1}(K\Psi_{1S}^{3/4}) = i_{d1}\left(\frac{1}{\nu_c} - \Psi_{1S}\right). \qquad (6.86)$$

Dogonadze [7] discussed the case of arbitrary diffusion coefficients ($D_1 \neq D_2$). His results are as follows: if the kinetic current is sought in the form (6.86), determination of Ψ_{1S} requires solution of the equation

$$\Psi_{1S}^{\prime 3/4} + \frac{1}{K}\, \Psi_{1S}^{\prime} - \frac{D_2/D_1}{\nu_c K} = 0, \tag{6.87}$$

where

$$K = \left[\frac{2\,(\rho_c \sigma)^3\,(1 - D_1/3D_2)^2}{\rho_c D_1 D_2} \right]^{1/4} \frac{\delta_d}{c_1^{*1/4}}. \tag{6.88}$$

Hale [9] numerically integrated equations for convective diffusion to the rotating disc electrode under conditions of a preceding chemical reaction (6.62). He calculated the time for establishing the stationary state under galvanostatic conditions for various values of the rate constant ρ_c of the chemical reaction.

§ 6.3. Catalytic Processes

The catalytic electrode processes to be considered below can be schematically presented as follows: a product of the electrode reaction undergoes a chemical transformation which regenerates the initial electroactive reactant. As a result, the measured limiting current can considerably exceed the current which would be observed in the absence of a chemical reaction.

A. As the first example, a catalytic process of the first order will be considered.

Let a fast electrode reaction

$$A_1 + ne^- \to A_2 \tag{6.89}$$

be followed by a pseudo-monomolecular chemical reaction

$$A_2 \underset{\rho_c \sigma}{\overset{\rho_c}{\rightleftharpoons}} A_1 \quad (\text{or} \quad A_2 - A_1 = 0, \quad \mu_1 = -1, \quad \mu_2 = 1). \tag{6.90}$$

It can easily be established that the system of differential equations describing the behavior of A_1 and A_2 in the solution is identical with the system (6.89).

The boundary conditions for distances far removed from the rotating disc electrode also retain their previous form [cf. Eq.

(6.40a)]:

$$c_1 \to \overset{*}{c_1}, \quad c_2 \to \overset{*}{c_2}, \quad \sigma = \overset{*}{c_2}/\overset{*}{c_1} \qquad \text{as } z \to \infty. \tag{6.91a}$$

However, the boundary conditions at the electrode surface (6.40b) must be changed to

$$\left(\frac{dc_1}{dz}\right)_S = -\left(\frac{dc_2}{dz}\right)_S \quad \text{and } c_1 = 0 \qquad \text{for } z = 0. \tag{6.91b}$$

The first of conditions (6.91b) reflects the mass conservation in the electrode reaction; the second corresponds to fast discharge of A_1 at the electrode according to reaction (6.89).

The method of solving the problem formulated above is the same as that described in § 6.2. However, the boundary condition for the function Ψ_1 resulting from Eq. (6.91b) differs from Eq. (6.22b); it is

$$\left(\frac{d\Psi_1}{d\xi}\right)_S = 0 \qquad \text{for } \xi = 0. \tag{6.92}$$

This new boundary condition considerably simplifies the problem. In particular, the function Ψ_1 is everywhere constant, i.e.,

$$\Psi_1 = 1 + \sigma. \tag{6.93}$$

The concentration of the electroactive substance, $\bar{C}_1$, is also constant beyond the boundaries of the kinetic reaction layer:

$$\bar{C}_1 = 1. \tag{6.94}$$

Inside the kinetic reaction layer μ_c, the concentration of A_1 varies according to the following relation:

$$C_1 = \bar{C}_1 - \tilde{C}_1 = 1 - \exp\left(-\frac{\xi}{\varepsilon_c}\sqrt{\frac{1+\sigma}{\sigma}}\right), \tag{6.95}$$

The disc current density is given by

$$i = i_{d1}\frac{1}{\varepsilon_c}\sqrt{\frac{1+\sigma}{\sigma}}, \tag{6.96}$$

or

$$i = i_{d1} \frac{\delta_d}{\mu_c} \sqrt{\frac{1+\sigma}{\sigma}}. \tag{6.96a}$$

It follows from Eq. (6.96a) that $i/i_{d1} \sim \omega^{-1/2}$. The latter dependence can be used to study the kinetic process.

Substituting Eq. (6.43) in Eq. (6.96), we find that

$$i = nF(c_1^* + c_2^*) \sqrt{\rho_c D/(1+\sigma)}, \tag{6.97}$$

a relation first derived by Levich and Koutecky [5, 6].

It can be seen from Eq. (6.97) that the catalytic current is independent of the rotation speed, being determined by the rate of the chemical transformation, in the bulk, of the electrode reaction product A_2 into the initial reactant A_1.

The fact that the catalytic current is independent of the rotational velocity is a consequence of the assumption of a high rate for the chemical reaction. The condition $(\varepsilon_c \ll 1)$ characterizing a fast chemical reaction was used in the solution of the problem; catalytic processes were assumed to occur close to the electrode surface, within the diffusion layer, unaffected by convection.

One of the authors, in collaboration with Kiryanov, solved the problem of catalytic currents for first-order chemical reactions without assuming a high rate for the chemical reaction. Approximate calculations (cf. § 2.8) showed that the catalytic current at a rotating disc electrode is given by

$$i = i_{d1} \frac{1.65 + \rho_c(1+\sigma)\delta_d^2/D}{\sqrt{3.10 + \rho_c(1+\sigma)\delta_d^2/D}}. \tag{6.98}$$

For fast chemical reactions $[\rho_c(1+\sigma)\delta_d^2/D \gg 1]$ Eq. (6.98) reduces to Eq. (6.97). For slow chemical reactions $[\rho_c(1+\sigma) \times \delta_d^2/D \ll 1]$ the currents calculated from Eq. (6.98) agree within 6% with the magnitude of the diffusion–limiting current of A_1.

The above example of a catalytic reaction (6.89)–(6.90) corresponds, for example, to the $Fe^{3+}-H_2O_2$ system. Fe^{3+} ions are

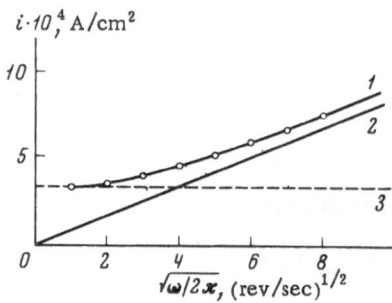

$i \cdot 10^4,\ \mathrm{A/cm^2}$

Fig. 6.4. The limiting current density as a function of the square root of rotation speed. Reduction of 10^{-3} M Fe^{3+} in 1 M $KCl + 2.5 \cdot 10^{-2}$ M H_2O_2 [16]. 1) Experimental; 2) diffusion-limited current; 3) behavior calculated according to Eq. (6.97).

$\sqrt{\omega/2\pi}$, $(\mathrm{rev/sec})^{1/2}$

reduced at the electrode

$$Fe^{3+} + e^- \rightarrow Fe^{2+},$$

and the product, Fe^{2+}, is reoxidized in the solution by hydrogen peroxide:

$$Fe^{2+} + {}^1/_2 H_2O_2 \rightleftarrows Fe^{3+} + OH^-.$$

At high H_2O_2 concentrations, the reaction behaves as a pseudo-monomolecular one (hydrogen peroxide, it is to be noted, does not enter directly into any electrode reactions at the potential for Fe^{3+} reduction).

It was experimentally shown [16] [in agreement with Eq. (6.98)] that the current is proportional to the square root of the rotation velocity under conditions of intensive stirring (low δ_d values); for high δ_d values, the rotation speed has practically no effect on the observed current which tends to its limiting value (Fig. 6.4).

An autocatalytic process of a similar type was found to take place during reduction of a chloride complex of divalent copper at a graphite electrode in the presence of dissolved oxygen. The electrode reaction product (a complex ion of univalent copper) is reoxidized in the solution by oxygen to the divalent copper complex. However, this reaction is so fast that it was not possible to measure its rate by the method described [17].

B. As a second example, a catalytic reaction with partial regeneration of the initial reactant will be considered†:

$$A_1 + n\,e^- \rightarrow A_2 \ (\text{at the electrode}), \tag{6.99a}$$

---

†Calculations pertaining the disproportionation reactions at the rotating disc electrode were carried out by one of the authors in collaboration with B. C. Potapov.

$$2A_2 \underset{\rho_c^{\sigma}}{\overset{\rho_c}{\rightleftharpoons}} A_1 \quad (\text{or} \quad 2A_2 - A_1 = 0, \quad \mu_1 = -1, \quad \mu_2 = 2) \quad \text{(in the solution).}$$

$$(6.99b)$$

An example of this type of reaction sequence is the dispropor- tionation of pentavalent uranium ions. The experimental results obtained in a polarographic study of this reaction were reported by Heyrovski and Kuta [3].

The concentrations of A_1 and A_2 are described by

$$D \frac{d^2c_1}{dz^2} = w \frac{dc_1}{dz} - \rho_c (c_2^2 - \sigma c_1),$$

$$D \frac{d^2c_2}{dz^2} = w \frac{dc_2}{dz} + 2\rho_c (c_2^2 - \sigma c_1),$$

$$(6.100)$$

where $\sigma = c_2^{*2}/c_1^*$ is the equilibrium constant, and c_1^* and c_2^* are the bulk concentrations of A_1 and A_2, respectively.

Equations (6.100) must be supplemented by the following bound- ary conditions:

$$c_1 \to c_1^*, \quad c_2 \to c_2^* \qquad \text{as } z \to \infty, \qquad (6.101a)$$

$$c_{1S} = 0, \quad \left(\frac{dc_1}{dz}\right)_S = -\left(\frac{dc_2}{dz}\right)_S \qquad \text{for } z = 0. \qquad (6.101b)$$

[The electrochemical reaction (6.99a) is, as previously, considered to be fast.] Using the dimensionless variables (6.9) and (6.10), we can transform Eqs. (6.100) into

$$\frac{d^2C_1}{d\xi^2} + a\xi^2 \frac{dC_1}{d\xi} + \frac{1}{\varepsilon_c^2} (C_2^2 - C_1) = 0,$$

$$\frac{d^2C_2}{d\xi^2} + a\xi^2 \frac{dC_2}{d\xi} - \frac{\nu_c}{\varepsilon_c^2} (C_2^2 - C_1) = 0,$$

$$(6.102)$$

where

$$\nu_c = 2c_1^*/c_2^* \qquad \text{and} \qquad \varepsilon_c^2 = D/\rho_c \sigma \delta_d^{\circ}. \qquad (6.103)$$

The boundary conditions (6.101a) and (6.101b) can be rewritten as

$$C_1 = C_2 \to 1 \qquad \text{as } \xi \to \infty, \qquad (6.104a)$$

$$C_{1S} = 0, \quad \frac{\nu_c}{2} \left(\frac{dC_1}{d\xi}\right)_S = -\left(\frac{dC_2}{d\xi}\right)_S \qquad \text{for } \xi = 0. \qquad (6.104b)$$

Following the general methods previously described, a function

$$\Psi_1 = v_c C_1 + C_2 \qquad (6.105)$$

is now introduced, which satisfies Eq. (6.20).

Designating again by X the factor characterizing the changes of the diffusion current effected by the chemical reaction, i.e., $(dC_1/d\xi)_S = X$, we can easily show that

$$\Psi_1 = (1 + v_c) - \frac{v_c X}{2} \int_\xi^\infty \exp\left(- a y^3/3\right) dy. \qquad (6.106)$$

The concentration distribution of the substance A_1 is described by

$$\frac{d^2 C_1}{d\xi^2} + a\xi^2 \frac{dC_1}{d\xi} + \frac{1}{\varepsilon_c^2} [(\Psi_1 - v_c C_1)^2 - C_1] = 0. \qquad (6.107)$$

Following the general methods of solving equations of this type, described in § 6.1, we can find the distribution of A_1 outside the kinetic reaction layer from the solution of the algebraic equation

$$\bar{C}_1 - (\Psi_1 - v_c \bar{C}_1)^2 = 0 \qquad (6.108)$$

in the form

$$\bar{C}_1 = \frac{(1 + 2v_c \Psi_1) \pm \sqrt{1 + 4v_c \Psi_1}}{2 v_c^2}. \qquad (6.109)$$

In particular, at the electrode surface ($\xi = 0$) the reduced concentration is given by

$$\bar{C}_{1S} = \frac{(1 + v_c)^2 + v_c^2 (1 - X) \pm \sqrt{(1 + 2v_c)^2 - 2v_c^2 X}}{2 v_c^2}, \qquad (6.110)$$

since, according to Eqs. (6.106) and (2.47), it follows that

$$\Psi_{1S} = (1 + v_c) - \frac{v_c X}{2}. \qquad (6.111)$$

The distribution of the substance A_1 within the kinetic reaction layer μ_c can be found by solution of the differential equation (6.33) for the auxiliary function $\widetilde{C}_1$ which describes the deviations from the equilibrium distribution $\widetilde{C}_1 = \overline{C}_1 - C_1$. In the case discussed, Eq. (6.33) has the following form:

$$\frac{d^2\widetilde{C}_1}{dt^2} - [(1 + 2v_c\Psi_{1S} - 2v_c^2\overline{C}_{1S})\,\widetilde{C}_1 + v_c^2\widetilde{C}_1^2] = 0 \qquad (6.112)$$

with the boundary conditions

$$\begin{aligned}\widetilde{C}_1 &\to 0 & \text{as } t \to \infty, \\ \widetilde{C}_1 &= \widetilde{C}_{1S} & \text{for } t = 0.\end{aligned} \qquad (6.113)$$

The first integral of Eq. (6.112) is in the form

$$\left(\frac{d\widetilde{C}_1}{dt}\right) = \pm\left[(1 + 2v_c\Psi_{1S} - 2v_c^2\overline{C}_{1S})\,\widetilde{C}_1^2 + \frac{2v_c^2}{3}\,\widetilde{C}_1^3\right]^{1/2} \qquad (6.114)$$

The value of $(d\widetilde{C}_1/dt)_S$ at the electrode surface (t = 0) is obtained after substitution in Eq. (6.114) of the corresponding boundary condition (6.113):

$$\left(\frac{d\widetilde{C}_1}{dt}\right)_S = \pm\left[(1 + 2v_c\Psi_{1S})\,\overline{C}_{1S}^2 - \frac{4}{3}\,v_c^2\overline{C}_{1S}^3\right]^{1/2}. \qquad (6.115)$$

All the equations derived above depend parametrically on X. The latter quantity can be found using Eq. (6.37). Substitution into Eq. (6.37) of the values of $(d\overline{C}_1/d\xi)_S$ [obtained by differentiation of Eq. (6.109)] and $(d\widetilde{C}_1/d\xi)_S$ gives

$$X = \frac{X}{2}\left[1 \pm \frac{1}{(1 + 4v_c\Psi_{1S})^{1/2}}\right] - \\ - \frac{1}{\varepsilon_c}\left\{\pm\left[(1 + 2v_c\Psi_{1S})\,\overline{C}_{1S}^2 - \frac{4}{3}\,v_c^2\overline{C}_{1S}^3\right]^{1/2}\right\}. \qquad (6.116)$$

The solution of Eq. (6.116) together with Eqs. (6.110) and (6.111) is a complex problem. Further discussion will therefore be restricted to the practically important case where the catalyst in the electrode process (substance A_2) is virtually absent in the bulk solution phase (i.e., $v_c = 2c_1^*/c_2^* \gg 1$). The relationship derived above obtains, of course, for this case. However, a number of expressions reduce to much simpler forms when $v_c \gg 1$. In particular, Eqs. (6.110) and (6.111) can be replaced by

$$\bar{C}_{1S} \approx 1 - X/2 \qquad (6.110a)$$

and

$$\Psi_{1S} \approx v_c(1 - X/2). \qquad (6.111a)$$

Equation (6.116), from which X can be obtained, simplifies now to

$$\frac{X}{2}\left(1 - \frac{X}{2}\right)^{-3/2} = \frac{v_c}{\varepsilon_c}\left(\frac{2}{3}\right)^{1/2}. \qquad (6.117)$$

It is convenient to present Eq. (6.117) in a form suitable for analysis of experimental data:

$$\frac{(i/2i_{d_1})}{(1 - i/2i_{d_1})^{3/2}} = \left(\frac{2}{3}\right)^{1/2} \frac{v_c \delta_d}{\mu_c}, \qquad (6.117a)$$

where

$$\mu_c = \sqrt{D/\rho_c \sigma}.$$

A numerical solution of Eq. (6.117) can be conveniently carried out using the nomogram shown in § 3.2. In fact, using the function Ψ_{1S} [cf. Eq. (6.111)], Eq. (6.117) can be rewritten in the form

$$\Psi_{1S}^{3/2} + \frac{\Psi_{1S}}{K} - \frac{v_c}{K} = 0, \qquad (6.118)$$

where

$$K = (2v_c/3)^{1/2}/\varepsilon_c.$$

When the rate constant ρ_c of the chemical reaction (6.99a) is very high and the recombination of the intermediates (A_2) occurs at the electrode surface, the maximum current density $i = 2i_{d_1}$ is observed.

According to Ulstrup [18], the rotating disc electrode can be successfully used in studies of disproportionation processes whose rate constants ρ_c are less than 10^8 liters $\cdot$ mole^{-1} $\cdot$ sec^{-1}.

Equation (6.117) [or (6.117a)] is similar to the equation derived by Brdicka, Hanus, and Koutečky [2] for the analogous electrochemical processes occurring at a dropping mercury electrode. An equation of the same type was also obtained by Ulstrup [18] for the case of a disproportionation reaction at a rotating disc electrode. His derivation was based on the Nernst model of convective diffusion.

The partial bulk regeneration of the reactant described by Eqs. (6.99a)-(6.99b) corresponds to the gaseous reaction of the etching of a germanium disc in iodine vapor investigated by Olander [19]:

$$Ge_{(solid)} + 2I_{2(gas)} \rightarrow GeI_{4(gas)}.$$

The reaction product partially decomposes near the disc surface:

$$GeI_4 \rightleftarrows GeI_2 + I_2,$$

The latter reaction results in a somewhat increased rate of the overall etching process. However, at the temperature attained (460°C) the decomposition rate remains low and the effect could not be quantitatively evaluated.

Disproportionation is observed in electrooxidation of iodide ions:

$$2I^- \rightarrow I_2 + 2e^- \quad \text{(at the electrode)},$$
$$I_2 + H_2O \rightleftarrows I^- + IOH + H^+ \quad \text{(in the solution)}.$$

The latter reaction was studied at an azobenzene−graphite electrode by Landsberg et al. [20, 21]. A quantitative analysis, similar to that described above, was not carried out; however, the concentrations of I^-, I_2, and IOH within the boundary layer were calculated numerically with the assumption $\mu_c \ll \delta_d$.

C. A more complex case of a catalytic current is observed in iodine reduction at a platinum electrode from a solution containing iodate ions [21, 22]. The electroreduction $I_2 + 2e \rightarrow 2I^-$ is followed by regeneration of iodine in a homogeneous reaction involving iodate ions:

$$5I^- + IO_3^- + 6H^+ \underset{\rho_c^{\sigma}}{\overset{\rho_c}{\rightleftarrows}} 3I_2 + 3H_2O.$$

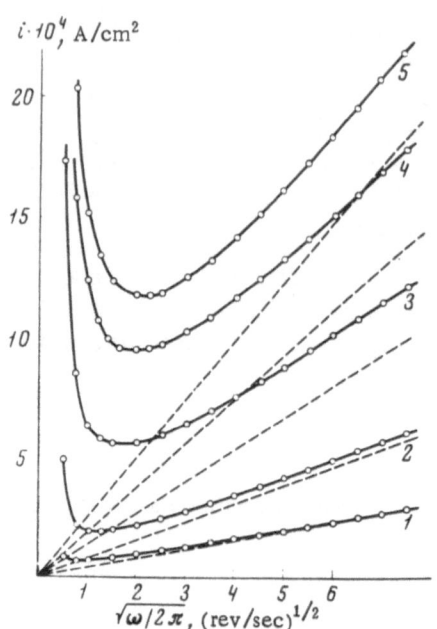

Fig. 6.5. The limiting current density
for iodine electroreduction in 0.016 M
KIO_3 (pH = 1.95) at platinum as a function of the square root of the rotation
speed of the electrode [21]. Iodine concentration (moles/liter): 1) $1.18 \cdot 10^{-4}$;
2) $2.45 \cdot 10^{-4}$; 3) $4.33 \cdot 10^{-4}$; 4) $6.0 \cdot$
10^{-4}; 5) $7.9 \cdot 10^{-4}$. The dashed lines
are calculated diffusion limiting currents for curves 1-5.

It can be seen from the above equations that iodine is formed in
a larger amount than that consumed in the electrode reaction.
This results in an unusual dependence of current on the rotation
speed of the electrode; the relevant plot has a minimum (Fig. 6.5).
As opposed to the previously considered case (A), weak stirring
is accompanied not by a constant, but by a sharply increasing,
current.

A relatively simple mathematical analysis of the set of equations describing the above process was given by Beran and
Bruckenstein [22]. The authors used an original method of solving
the convective mass-transport problem for the vicinity of the rotating disc. The method combines the Nernst model with the integral method described in § 2.8 (Beran and Bruckenstein call this
the "moments method"). Assuming a linear distribution of iodine
in the kinetic reaction layer μ_c, these authors obtained an equation
from which the thickness μ_c can be calculated. By analyzing the
experimental dependence of the catalytic reduction current on the
rotation speed, they determined the rate constants of the homo-
geneous chemical reaction and found $\rho_c = 4 \cdot 10^5$ liter$^4 \cdot$ mole$^{-4} \cdot$
sec^{-1} and $\rho_c \sigma = 5.1 \cdot 10^9$ liter$^5 \cdot$ mole$^{-5} \cdot$ sec^{-1}.

Complex cases of kinetic currents and their coupling with catalytic ones are observed during electroreduction of chromic acid at gold rotating electrodes in the presence of sulfates [23], and of vanadic acid at platinum electrodes in the presence of di- and trivalent iron salts [24].

Catalytic electrode reactions encountered in practice are very diversified, as can be seen even from this short review. A review of the wide range of such processes is beyond the scope of this book. Analysis of the experimental results can often be helped by the calculations of Brdicka, Hanus, and Koutecký [2] carried out for a number of catalytic processes at the dropping mercury electrode. As previously mentioned, their formulas can be used for the description of similar reactions proceeding at the rotating disc electrode.

§ 6.4. The Effect of Chemical Reactions
Following the Electrochemical Processes

A. The effect of consecutive chemical steps on the kinetics of an electrode process will be exemplified by considering monomolecular reactions.

Let the product A_1 of the electrode reaction

$$A_3 + n e^- \rightleftarrows A_1 \qquad (6.119a)$$

enter a chemical step in the vicinity of the electrode

$$A_1 \underset{\rho_c}{\overset{\rho_c \sigma}{\rightleftarrows}} A_2. \qquad (6.119b)$$

Obviously, the effect of this step on the kinetics of the electrode process is most pronounced when the latter process is fast [reaction (6.119a) is reversible] and the equilibrium in the kinetic reaction (6.119b) is shifted toward the substance A_2, which does not directly participate in the electrochemical step ($\sigma \gg 1$).

The set of equations describing the behavior of A_1 and A_2 near the rotating disc electrode takes the form of (6.39). However, the distribution of A_3 obeys the usual equation of convective diffusion to the rotating disc

$$D \frac{d^2 c_3}{dz^2} = w \frac{dc_3}{dz} . \qquad (6.120)$$

The boundary conditions corresponding to the situation considered are as follows:

$$c_1 \to c_1^*, \quad c_2 \to c_2^*, \quad (\sigma = c_2^*/c_1^*), \quad c_3 \to c_3^* \qquad \text{as } z \to \infty;$$

(6.121a)

$$\left(\frac{dc_3}{dz}\right)_S = -\left(\frac{dc_1}{dz}\right)_S, \quad \left(\frac{dc_2}{dz}\right)_S = 0, \quad \text{and} \quad c_{3S} = c_{1S} \exp\left[nF(\varphi-\varphi_0)/RT\right].$$

(6.121b)

The latter boundary condition (6.121b) is the Nernst equation for the reversible reaction (6.119a).

The calculations described in detail in § 6.2 for an electro-chemical process accompanied by a chemical reaction of the first order can be applied without significant changes to the analysis of the present case. It should only be remembered that the concentration of A_1 at the disc surface is, as opposed to § 6.2, not equal to zero but is given by the Nernst equation.

A strict mathematical discussion [8] of the above problem is replaced here by a qualitative analysis based on the concept of the reaction layer. A similar approach was first used by Adams and Galus [25].

In order to simplify calculations, the substances A_1 and A_2 will be considered to be virtually absent in the solution, i.e., $c_1^* = c_2^* = 0$. The substance A_3 is supplied to the disc by convective diffusion. According to the results previously obtained [cf. Eqs. (2.23) and (4.103)], the disc current is given by

$$i = i_{d3}(1 - c_{3S}/c_3^*),$$

(6.122)

where i_{d3} is the diffusion limiting current of A_3.

According to the first boundary condition in (6.121b), the diffusional flux of the substance A_3 results in the appearance of a similar flux of A_1. However, the distance which A_1 diffuses depends no longer on the stirring rate but on the lifetime of A_1 before it disappears in the chemical reaction (6.119b). The lifetime of A_1 is approximately given by $1/\rho_c \sigma$, and the distance which A_1 travels from the disc surface is of the order of magnitude of the thickness of the kinetic reaction layer, $\mu_c = (D/\rho_c \sigma)^{1/2}$. The surface concentration of A_1 is given by

$$c_{1S} = \frac{i\mu_c}{nFD}.$$

(6.123)

By combining Eqs. (6.122) and (6.123) with the last of the conditions (6.121b), the following relation is obtained:

$$\frac{1 - i/i_{d3}}{i/i_{d3}} = \varepsilon_c \exp\left[nF(\varphi - \varphi_0)/RT\right], \qquad (6.124)$$

where, as previously, $\varepsilon_c = \mu_c/\delta_d$.

Expression (6.124) can be represented in the form of a polarographic wave

$$\varphi = \varphi_{1/2} + \frac{RT}{nF} \ln \frac{i_{d3} - i}{i}, \qquad (6.125a)$$

where the half-wave potential is given by

$$\varphi_{1/2} = \varphi_0 + \frac{RT}{2nF} \ln \frac{p_c \sigma \delta_d^2}{D}, \qquad (6.125b)$$

and consequently depends considerably on the rate $\rho_c \sigma$ of the chemical reaction (6.119b). A fast following chemical step results in a shift of the half-wave potential $\varphi_{1/2}$ which in the absence of the latter reaction is equal to φ_0.

The value of $\varphi_{1/2}$ calculated by one of the authors together with Kir'yanov [8] for a following chemical reaction of the first order proceeding with an arbitrary rate constant $\rho_c \sigma$ is

$$\varphi_{1/2} = \varphi_0 - \frac{RT}{nF} \ln \frac{1.06 + \sigma \dfrac{\left[3.10 + p_c(1+\sigma)\delta_d^2/D\right]^{1/2}}{1.65 + p_c(1+\sigma)\delta_d^2/D}}{(1+\sigma)}. \qquad (6.126)$$

It is easy to verify that for fast chemical reactions expression (6.126) reduces to (6.125b). For slow reactions, the value of $\varphi_{1/2}$ given by (6.126) virtually coincides with φ_0.

Tong et al. [26] obtained a somewhat different expression for $\varphi_{1/2}$ using the Nernst model for convective diffusion to the rotating disc. Their expression reduces, however, to Eq. (6.125b) for chemical reactions.

The oxidation of N,N-dialkyl-p-phenylenediamines at a platinum electrode [26] belongs to reactions of the type represented

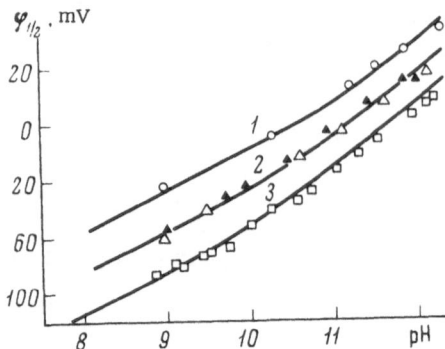

Fig. 6.6. The half-wave potential for oxidation of N,N-dialkyl-p-phenylene-diamines at platinum (against sat. cal. electrode) as a function of pH [26]. Substituents: $R_1 = C_2H_5$; $R_2 = C_2H_5$ (1); $R_1 = C_2H_5$; $R_2 = C_2H_4-NH-SO_2CH_3$ (2); $R_1 = CH_3$, $R_2 = CH_3$ (3). The solid lines are calculated.

by (6.119a)-(6.119b):

$$NH_2-\langle\rangle-N{\overset{R_1}{\underset{R_2}{}}} + OH^- \rightleftarrows NH=\langle\rangle=N^+{\overset{R_1}{\underset{R_2}{}}} + H_2O + 2e^-$$

The oxidation product participates in a homogeneous chemical reaction and becomes deaminated to quinonemonoimine $NH=\langle\rangle=0$. The rate constant of this reaction depends on pH, namely, $\rho_c\sigma = const \cdot c_{OH^-}$. The experimental dependence of the half-wave potential $\varphi_{1/2}$ on pH is shown in Fig. 6.6 for three derivatives of phenylenediamine.

Apart from the simple consecutive monomolecular reaction scheme described by (6.119a) and (6.119b), Tong et al. [26] discussed a more complex scheme of chemical reactions, including the case of partial regeneration of intermediates.

B. The above qualitative approach to the problem of the effect of consecutive reactions on the kinetics of the preceding electrochemical step can be applied to the analysis of irreversible chemical reactions of higher orders, e.g., when A_1 decomposes into two or more components.

Analysis of the effect of a dimerization reaction

$$2A_1 \xrightarrow{\rho_c\sigma} A_2 \qquad\qquad (6.127)$$

is only a little more complex than that for monomolecular reactions.† The surface concentration of A_1 (c_{1S}) is given by

---

†This case was also considered by Adams and Galus [25].

Eq. (6.123). For dimerization reactions, we have

$$c_{1S} = \left(\frac{i}{nFD}\right)^{2/3}\left(\frac{3D}{2\rho_c\sigma}\right)^{1/3}.$$

(6.128)

Substituting c_3 and the new value of c_{1S} into the Nernst equation, we obtain the expression

$$\frac{1 - i/i_{d3}}{(i/i_{d3})^{2/3}} = \left(\frac{3D}{2\rho_c\sigma c_3^* \delta_d^2}\right)^{1/3}\exp\left[\frac{nF}{RT}(\varphi - \varphi_0)\right],$$

(6.129)

which replaces Eq. (6.124). The equation of the polarographic wave for dimerization reactions is of the form

$$\varphi = \varphi_{1/2} + \frac{RT}{nF}\ln\frac{i_{d3} - i}{i_{d3}^{1/3} i^{2/3}},$$

(6.130a)

and the half-wave potential is given by

$$\varphi_{1/2} = \varphi_0 + \frac{RT}{3nF}\ln\frac{\rho_c\sigma c_3^* \delta_d^2}{3D}.$$

(6.130b)

Similar to the case of a monomolecular reaction, the half-wave potential depends on the rate constant $\rho_c\sigma$ of dimerization and on the intensity of convective stirring (δ_d). However, in the case of dimerization (following the electrode step), $\varphi_{1/2}$ depends on the concentration of the depolarizer, c_3^*. The shift of $\varphi_{1/2}$ accompanying chemical reactions of the products of electrode processes may be used for analysis of the kinetics of chemical reactions and for establishment of the rate constant $\rho_c\sigma$. The general dependence of $\varphi_{1/2}$ on the characteristics of various chemical processes is similar to that obtained for the dropping electrode [2, 3].

Gray and Harrison [27] calculated numerically steady-state polarization curves for polymerization of the product A_1 of the electrode reaction, as well as the ratio of the measured "polymerization current" to the diffusion limiting current for A_1 formation as a function of the rate constants of polymerization and of breaking of the chain.

C. The last part of this section involves chemical reactions whose products [in the reaction scheme (6.119a)-(6.119b), the substance A_2] can discharge at the disc electrode according to the scheme

$$A_3 + n_1 e^- \xrightarrow[\substack{\text{at the elec-}\\\text{trode}}]{} A_1 \xrightarrow[\substack{\text{in the}\\\text{bulk}}]{\rho_c \sigma} A_2 + n_2 e^- \xrightarrow[\substack{\text{at the elec-}\\\text{trode}}]{} A_4. \tag{6.131}$$

The above sequence, consisting of an electrochemical, chemical, and second electrochemical reaction is often called an ECE mechanism and represents an important class of electrochemical processes proceeding with unstable intermediates. The application of the rotating disc electrode to studies of such processes was discussed by Adams et al. [28-30], Karp [31], and Filinovskii [32].

The specific role of convective stirring in reactions of type (6.131) will now be considered. Up to now, the rate of the chemical reaction was assumed to be relatively high, and the effect of stirring was neglected in the region where chemical reactions take place. The convective mass transport caused by the rotation of the disc electrode played, in a certain sense, a secondary role, complicating the study of chemical reactions.

The application of the rotating disc electrode to ECE reactions has a different basis. If the chemical reaction is fast and its rate constant $\rho_c \sigma$ is much higher than that of the convective transport of A_1 into the bulk solution, the latter substance becomes converted into A_2 which in turn fully reduces at the electrode to A_4. The electrode current corresponds in this case to the discharge of $n_1 + n_2$ electrons in the electrochemical reaction. It is neither possible to establish whether the intermediates decompose at the surface or near the electrode, nor to determine the rate constant $\rho_c \sigma$ of this reaction.

Conversely, in the presence of sufficiently intensive convective stirring, when the substance A_1 is partly removed into the bulk solution, the amount of A_2 which reduces at the electrode is somewhat less than the amount of A_1 formed in the first electrochemical step. The current observed corresponds to the transfer of a somewhat smaller charge.

In other words, the apparent, or effective, number of electrons transferred is $n_{eff} < n_1 + n_2$. The deviation of n_{eff} from the sum $(n_1 + n_2)$ depends on the amount of A_2 carried by the convective diffusion into the bulk solution. At very high speeds of rotation of the disc electrode, $n_{eff} = n_1$ since the second electrochemical step is much slower than the diffusion of A_2 away from the electrode. Equations describing the concentration distributions of A_1, A_2, and A_3 retain their previous form† [cf. (6.39) and (6.120)].

The boundary conditions, however, at the disc surface are different; for simplicity, those given below refer to the case for which A_1 and A_2 are absent from the bulk,

$$c_1^* \to 0, \quad c_2^* \to 0, \quad c_3 \to c_3^* \qquad \text{as } z \to \infty, \qquad (6.132a)$$

and both substances undergo a fast reduction at the electrode surface (limiting current conditions):

$$c_{3S} = 0, \quad \left(\frac{dc_3}{dz}\right)_S = -\left(\frac{dc_1}{dz}\right)_S, \quad c_{2S} = 0 \qquad \text{for } z = 0. \qquad (6.132b)$$

Adams et al. [28] analyzed the behavior of system (6.131). They concluded that when $n_1 = n_2$ the disc current is given by

$$i = i_{d3}[2 - \exp(-\rho_c \sigma \delta_d^2 / D\pi)], \qquad (6.133)$$

where i_{d3} is the diffusion limiting current of A_3. It should be noted that Eq. (6.133), which describes the main characteristic effects of convection on the kinetics of the processes in question, somewhat overestimates convective effects. In the case of fast chemical reactions ($\rho_c \sigma > D/\delta_d^2$), Eq. (6.133) results in an incorrect dependence of i on the rate constant $\rho_c \sigma$. Karp [31], who apparently based his calculations on the Nernst model of convective diffusion, obtained in the analogous case a more correct expression

$$i = i_{d3}\left[2 - \frac{\tanh(\rho_c \sigma \delta_d^2 / D)^{1/2}}{(\rho_c \sigma \delta_d^2 / D)^{1/2}}\right]. \qquad (6.134)$$

The approximation described in § 2.8 was used by Filinovskii [32] to derive a somewhat simpler expression

$$i = 0.94 i_{d3}\left[2 - \frac{(1 + \delta_d^2 \rho_c \sigma / 1.9 D)^{1/2}}{1 + \delta_d^2 \rho_c \sigma / D}\right]. \qquad (6.135a)$$

---

†All diffusion coefficients are, as previously, considered to be equal.

When $n_1 \neq n_2$, Eq. (6.135a) must be replaced by

$$i = 0.94 i_{d3} \left\{ 1 + \frac{n_2}{n_1} \left[1 - \frac{(1 + \delta_d^2 \rho_c \sigma/1.9 D)^{1/2}}{1 + \delta_d^2 \rho_c \sigma/D} \right] \right\}. \qquad (6.135b)$$

Equations (6.135a) and (6.135b) can be used for arbitrary ratios of the chemical and diffusional rate constants. It is easy to reduce them to the limiting cases $n_{eff} = n_1$ and $n_{eff} = n_1 + n_2$ discussed at the beginning of this section. The errors involved are caused by the approximations used.

Applying Eq. (6.135b) to the case of fast chemical reactions ($\rho_c \sigma \gg D/\delta_d^2$), we easily obtain

$$i/i_{d3} \approx 0.94 \left[\left(1 + \frac{n_2}{n_1} \right) - \frac{n_2}{n_1} (D/1.9 \rho_c \sigma \delta_d^2)^{1/2} + \cdots \right]. \qquad (6.136)$$

Substitution into Eq. (6.136) of the expression (2.20) for the thickness of the diffusion boundary layer δ_d at the rotating disc gives

$$i/i_{d3} \approx 0.94 \left[\left(1 + \frac{n_2}{n_1} \right) - \frac{n_2}{n_1} \left(\frac{D^{1/3}}{4.9 \rho_c \sigma \nu^{1/3}} \right)^{1/2} \sqrt{\omega} + \cdots \right]. \qquad (6.137)$$

The plot of i/i_{d3} vs $\sqrt{\omega}$ is linear for low rotation speeds (or for fast chemical reactions of the intermediate). The intercept on the ordinate axis is a measure of the total number of electrons transferred in reaction (6.131), and the slope is proportional to $(\rho_c \sigma)^{-1/2}$.

The discussion of the utilization of the rotating disc electrode in studies of unstable intermediates in electrochemical reactions (considered at the beginning of this section) remains valid even if the chemical reactions of intermediates proceed according to a more complex mechanism than (6.131). However, the quantitative dependence of n_{eff} on the stirring intensity becomes more complex. Adams et al. [29, 30] calculated numerically the dependence of n_{eff} on the rotation speed ω for a series of cases corresponding to the ECE mechanism of electrochemical reactions. In particular, scheme (6.131) was complemented by the case where disappearance of intermediates occurs by dimerization:

$$2A \xrightarrow{\rho_c \sigma} A_2.$$

The computer-calculated results for the dependence of n_{eff} on ω

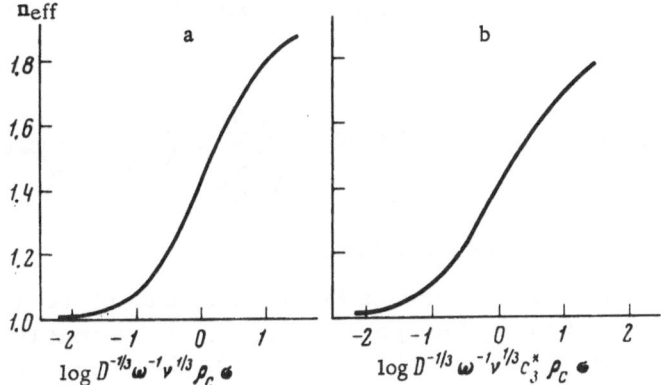

Fig. 6.7. Calculated dependence of n_{eff} on the rate constant of the chemical reaction $\rho_c \sigma$, rotation speed ω, and concentration of the electrochemically active species c_3^* [29]. a) Chemical reaction of the first order ($n_1 = n_2 = 1$); b) chemical reaction of the second order ($n_1 = 1$, $n_2 = 2$).

are shown in Fig. 6.7 for chemical reactions of the first and second order. The calculations are based on the general principles of programming diffusional problems in a rotating disc system formulated by Feldberg [33].

Adams et al. studied the oxidation of aromatic hydrocarbons in nitrobenzene solutions [28] and of diphenylamine derivatives in acetonitrile solutions [29] at platinum electrodes. Both reactions proceed according to the ECE mechanism. In the latter case, for example, triphenylamine oxidizes ($n_1 = 1$) at the electrode to a cation-radical

which dimerizes in the solution

The product, tetraphenylbenzidine, is oxidized at the electrode ($n_2 = 2$) to

Both electrode reactions are fast.

The experiments consisted in measurements of the limiting current. From these, the apparent number of electrons transferred, n_{eff}, was determined as a function of the triphenylamine concentration c_3 and of the angular velocity of the disc ω. The experimentally obtained n_{eff} vs log ω^{-1} curves were compared with theoretical n_{eff} vs log $D^{1/3}\omega^{-1}\nu^{1/3}c_3^* \rho_c\sigma$ curves calculated for a reaction of the second order (Fig. 6.8), to obtain the rate constant $\rho_c\sigma$. In this way, the rate constants for dimerization of triphenylamine ($3 \cdot 10^3$ liters $\cdot$ mole^{-1} $\cdot$ sec^{-1}) and its 4-acetyl ($2.6 \cdot 10^3$), 4-cyano ($1.1 \cdot 10^4$), 4-chloro ($8 \cdot 10^2$), and 4-nitro ($1 \cdot 10^4$) derivatives were measured. The errors were estimated [29] to be 10-30%.

§ 6.5. Investigation of Chemical Reactions of Electrochemically Inactive Species

The previous discussion concerned chemical reactions involving an electrochemically active species. The rotating disc electrode, however, can obviously be applied to the case where all reaction components are electrochemically inactive (in particular, to the studies of kinetics of homogeneous gaseous reactions).

Since the study of these reactions involves certain special methodological characteristics, and the kinetics of the processes are different from those considered in the previously discussed cases, the application of the rotating disc to the study of homogeneous reactions will now be discussed in detail. The calculations described below are based on the work of Litt and Serad [34].

Let it be assumed that the substance A diffuses away from the disc surface to which species B is supplied from the bulk. The two substances react at a certain distance from the disc surface

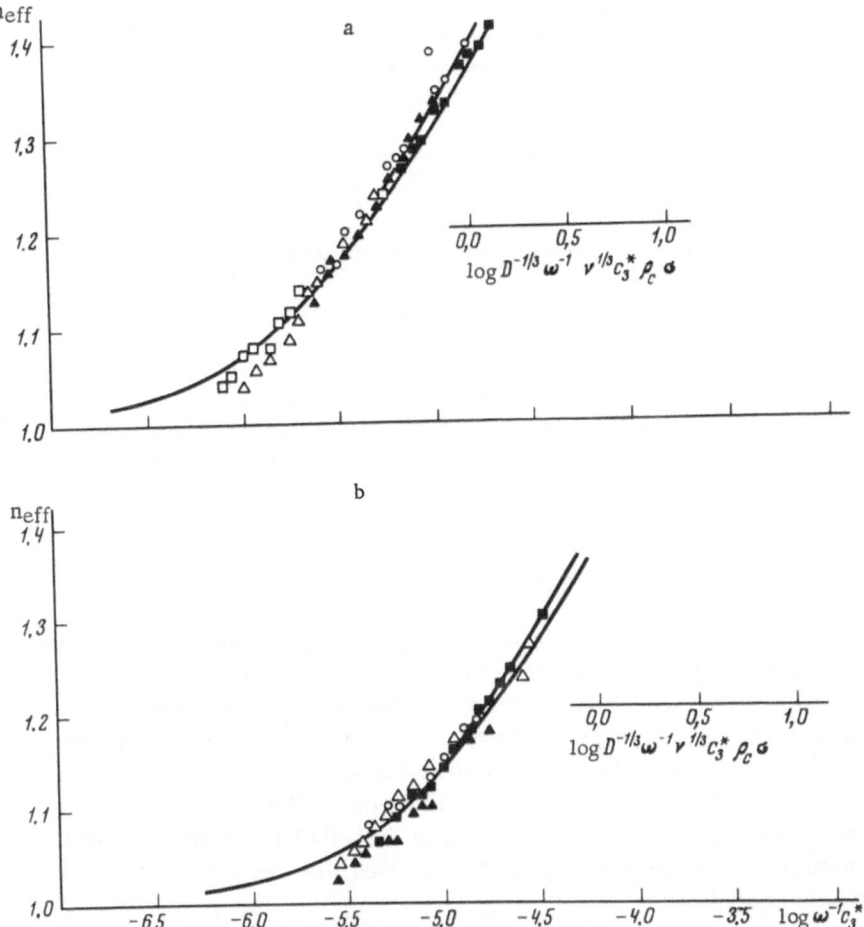

Fig. 6.8. Comparison of the experimental n_{eff} vs log $\omega^{-1} c_3^*$ curves with the calculated n_{eff} vs log $D^{-1/3} \omega^{-1} \gamma^{1/3} c_3^* \rho_c \sigma$ curves (upper scales) for an ECE reaction [29]. a) 4-Cyanotriphenylamine; b) 4-acetyltriphenylamine. Points correspond to various concentrations of the species investigated.

according to a fast irreversible reaction

$$\mu_A A + \mu_B B \to product. \tag{6.138}$$

The region where they react is called the reaction zone.

It follows from the previous considerations that the thickness of the reaction zone and its location relative to the disc surface

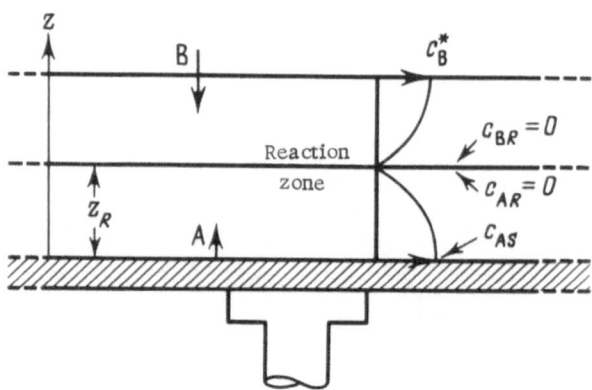

Fig. 6.9. Schematic diagram of the reaction zone and the concentration profile of A and B species at the disc surface. z) Distance from the rotating disc; c_{AS} and c_{AR}) concentrations of A at the disc surface and in the reaction zone, respectively; c_B^* and c_{BR}) concentrations of B in the bulk and in the reaction zone, respectively; z_R) distance of the reaction zone from the disc surface.

depend on the rate of reaction (6.138), the initial concentrations of the reacting species, the rate of the supply of the latter to the reaction zone, and other factors. If the rate of reaction (6.138) is sufficiently high, the reaction can be assumed to occur instantaneously. Then the concentrations of A and B in the reaction zone are virtually equal to zero, and the thickness of the reaction zone is negligibly small. The concentration distribution of the species transferred in the vicinity of the rotating disc is shown for this case in Fig. 6.9.

If the effects of the components of the chemical reaction (6.138) on the motion of the fluid at the rotating disc† are neglected, the concentration distribution of A and B at a uniform disc obeys (according to § 2.3) the following transport equations:

$$D_A \frac{d^2 c_A}{dz^2} - w(z)\frac{dc_A}{dz} = 0, \qquad (6.139a)$$

$$D_B \frac{d^2 c_B}{dz^2} - w(z)\frac{dc_B}{dz} = 0, \qquad (6.139b)$$

†The assumption that the fluid motion is independent of the composition of the fluid may be invalid in the gas phase. A strict solution of the problem then requires consideration of a set of gasdynamic equations [35].

where $w(z)$ is the normal component of the velocity of the fluid. The boundary conditions complementing Eqs. (6.139a) and (6.139b) are as follows:

$$c_A = c_{AS} \qquad \text{for } z = 0, \qquad (6.140a)$$

$$c_{AR} = c_{BR} = 0 \qquad \text{for } z = z_R, \qquad (6.140b)$$

$$c_B \to c_B^* \qquad \text{as } z \to \infty. \qquad (6.140c)$$

The diffusion coefficients D_A and D_B can differ; therefore, it is convenient to use the hydrodynamic variables

$$\zeta = z \sqrt{\omega/\nu}. \qquad (6.141)$$

Equations (6.141), (6.139a), and (6.139b) can be combined to yield

$$\frac{d^2 c_A}{d\zeta^2} - Sc_A H(\zeta) \frac{dc_A}{d\zeta} = 0, \qquad (6.142a)$$

$$\frac{dc^2_B}{d\zeta^2} - Sc_B H(\zeta) \frac{dc_B}{d\zeta} = 0, \qquad (6.142b)$$

where $Sc = \nu/D$ are the Schmidt numbers for species A and B. The function $H(\zeta)$ used in the above equations was introduced in §1.2, and defined by $w(z) = (\nu\omega)^{1/2} H(\zeta)$. The values of the function are given in Table 1.1.

Integration of Eqs. (6.142a) and (6.142b) with the boundary conditions (6.140a)-(6.140b) results in expressions for the concentration distribution of species A and B:

$$c_A = c_{AS} \left\{ 1 - \int_0^{\zeta} \exp\left[Sc_A \chi(y)\right] dy \Big/ \int_0^{\zeta_R} \exp\left[Sc_A \chi(y)\right] dy \right\} \quad 0 \leqslant \zeta \leqslant \zeta_R,$$

$$(6.143a)$$

$$c_B = c_B^* \left\{ 1 - \int_{\zeta}^{\infty} \exp\left]Sc_B \chi(y)\right] dy \Big/ \int_{\zeta_R}^{\infty} \exp\left[Sc_B \chi(y)\right] dy \right\} \quad \zeta_R \leqslant \zeta < \infty,$$

$$(6.143b)$$

where ζ_R is the distance of the reaction zone from the disc surface [in dimensionless units $\chi(y) = \int_0^y H(x)\,dx$].

Under steady-state conditions the rate of the species A supplied must be equal to the flux of B consumed in reaction (6.138).
This obvious condition can be used to determine the location of
the reaction zone. In fact, using the equality

$$- \mu_A D_A \left(\frac{dc_A}{dz}\right)_{z_R} = \mu_B D_B \left(\frac{dc_B}{dz}\right)_{z_R} \qquad \text{for } z = z_R,$$

together with Eqs. (6.143a) and (6.143b), we obtain the following
expression:

$$\frac{\mu_A D_A c_{AS}}{\mu_B D_B \dot{c}_B} = \exp\left[(Sc_B - Sc_A)\,\chi(\zeta_R)\right] \frac{\int_0^{\zeta_R} \exp\left[Sc_A \chi(y)\right]\,dy}{\int_{\zeta_R}^{\infty} \exp\left[Sc_B \chi(y)\right]\,dy}. \qquad (6.144)$$

Equation (6.144) is the required expression which allows the location of the reaction zone (i.e., ζ_R) to be determined.

The flux of the species A transported from the disc surface is
easily obtained from Eq. (6.143a):

$$j_A = - D_A \left(\frac{dc_A}{dz}\right)_S = D_A c_{AS}\,\sqrt{\omega/\nu}\left(\int_0^{\zeta_R} \exp\left[Sc_A \chi(y)\right]\,dy\right)^{-1}. \qquad (6.145)$$

Introduction of the dimensionless diffusion flux of A (cf. § 2.4)
results in the following expression for the Nusselt number for the
case considered:

$$Nu_d = j_A r_0 / D_A c_{AS} = r_0\,\sqrt{\omega/\nu}\left(\int_0^{\zeta_R} \exp\left[Sc_A \chi(y)\right]\,dy\right)^{-1}. \qquad (6.146)$$

In the absence of a chemical reaction, when $\zeta_R \to \infty$,

$$Nu_d^0 = r_0\,\sqrt{\omega/\nu}\left(\int_0^{\infty} \exp\left[Sc_A \chi(y)\right]\,dy\right)^{-1}. \qquad (6.147)$$

The effect of the chemical reaction is conveniently expressed by the so-called "reaction factor"

$$F_R = \frac{\mathrm{Nu}_d}{\mathrm{Nu}_d^0} = \int_0^\infty \exp\left[\mathrm{Sc}_A \chi\,(y)\right] dy \Big/ \int_0^{\zeta_R} \exp\left[\mathrm{Sc}_A \chi\,(y)\right] dy \,. \quad (6.148)$$

When the diffusion coefficients of both reacting species are equal $(D_A = D_B = D)$, Eq. (6.144) reduces to

$$\frac{\displaystyle\int_0^{\zeta_R} \exp\left[\mathrm{Sc}_A \chi\,(y)\right] dy}{\displaystyle\int_{\zeta_R}^\infty \exp\left[\mathrm{Sc}_B \chi\,(y)\right] dy} = \frac{\mu_A c_{AS}}{\mu_B c_B^*} \,. \quad (6.149)$$

The reaction factor is given in this case by

$$F_R = 1 + \mu_B c_B^* / \mu_A c_{A\,S} \,. \quad (6.150)$$

If the bulk concentration of B is much higher than the surface concentration of A $(c_B^* \gg c_{AS})$, the reaction zone approaches the disc surface $(\zeta_R \to 0)$. The reaction factor is expressed then by [34]:

$$F_R = \frac{\displaystyle\int_0^\infty \exp\left[\mathrm{Sc}_A \chi\,(y)\right] dy}{\displaystyle\int_0^\infty \exp\left[\mathrm{Sc}_B \chi\,(y)\right] dy}\left[1 + \left(\frac{\mathrm{Sc}_A}{\mathrm{Sc}_B}\right)\frac{\mu_B c_B^*}{\mu_A c_{A\,S}}\right]. \quad (6.151)$$

The theoretical dependence of F_R on the concentrations of reacting species is shown in Fig. 6.10, reprinted from [34]. Calculations were made for various values of the Schmidt numbers Sc_A and Sc_B.

It can be seen from the above formulas that the "reaction factor" is independent, in particular, of the rotation speed. This was verified experimentally by Litt and Serad [34] in a study of neutralization of moderately soluble acids in dilute alkaline solutions. An acid sample pressed in the form of a disc was dissolved upon rotation in water $(F_R = 0)$ or in alkaline solutions of various concentrations. Dissolution of the benzoic acid in water and in NaOH_{aq} was most extensively studied.

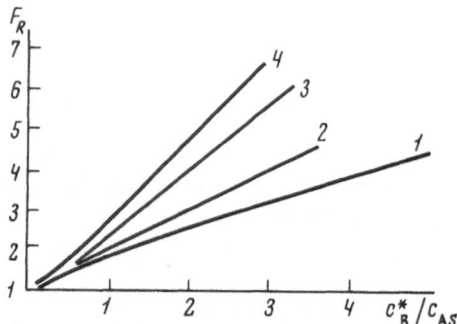

Fig. 6.10. The reaction factor F_R as a function of the concentration ratio of reactants A and B [34]. Values of the Schmidt numbers: 1) $Sc_A = 0.7$; $Sc_B = 2$; 2) $Sc_A = Sc_B$; 3) $Sc_A = 2$; $Sc_B = 0.7$; 4) $Sc_A = 900$; $Sc_B = 300$.

The experimental results [34], together with the theoretical calculations (solid lines), are shown in Fig. 6.11. It can be seen from the figure that within the limits of the experimental error the reaction factor is independent of the Reynolds number, i.e., of the rotation speed. The experimental and calculated values of F_R are in good agreement.

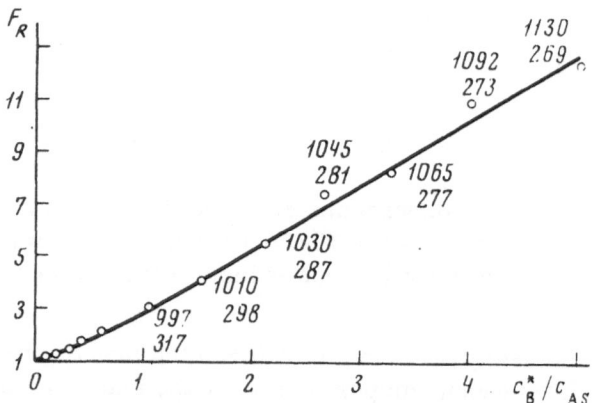

Fig. 6.11. The reaction factor as a function of the ratio c_B^* / c_{AS}. Dissolution of benzoic acid (A) in NaOH (B) solution [34]. The solid line is calculated according to Eq. (6.151) and corrected for the dependence of the Schmidt numbers on the ratio c_B^* / c_{AS}. The points are experimental. Reynolds numbers are shown on the diagram.

For further information, the reader is referred to [34]; one fact, however, which has been discussed in detail by Litt and Serad, should be considered here. It was assumed in the above calculations that the expressions for the concentration distribution at the rotating disc derived in § 1.2 are valid. In particular, the normal component of the fluid motion at the disc surface was assumed to be zero. This is valid only in the case where the flux of the species transferred from the disc surface $[j = -D(dc/dz)_s + w(0)c]$ does not appreciably affect the hydrodynamic mass-transport regime. At high fluxes (e.g., in the gas phase) the medium has a certain velocity at the disc surface [i.e., $w(0) \neq 0$]. The evaluation of this velocity requires simultaneous solution of the hydrodynamic and diffusional problems.

It is obvious that the scheme considered above can be applied also to electrochemical systems. For example, the substance A, a product of a fast electrochemical reaction at the disc, enters a homogeneous reaction with an electrochemically inactive species B. A similar combination of electrochemical and chemical reactions will be considered, for example, in § 8.7, where the application of the rotating disc electrode with a ring to titration with an electrochemically generated titrating agent is considered.

References

1. S. G. Mairanovskii, Catalytic and Kinetic Waves in Polarography, Nauka, Moscow (1965).
2. R. Brdicka, V. Hanus, and J. Koutecky, Progress in Polarography, Vol. 2, Interscience Publishers, New York (1962), p. 146.
3. J. Heyrovsky and J. Kuta, Principles of Polarography, Academic Press (1966).
4. J. Koutecky and J. Koryta, Electrochim. Acta, 3:318 (1961).
5. J. J. Koutecky and V. G. Levich, Dokl. Akad. Nauk SSSR, 117:441 (1957); Zh. Fiz. Khim., 32:1565 (1958).
6. V. G. Levich, Physicochemical Hydrodynamics, Prentice Hall, Inc. (1962).
7. T. P. Dogonadze, Zh. Fiz. Khim., 32:2437 (1958).
8. V. A. Kir'yanov and V. Yu. Filinovskii, 3rd Symposium on Polarography, Abstracts, Kiev (1965), p. 42; V. Yu. Filinovskii, Dissertation, Institute of Electrochemistry, Academy of Sciences of the USSR, Moscow (1965).
9. J. M. Hale, J. Electroanal. Chem., 8:332 (1964).
10. W. Vielstich and D. Jahn, Advances in Polarography, Vol. 1, I. Longmuir, ed., Pergamon Press, Oxford (1960), p. 281.
11. W. Vielstich and D. Jahn, Z. Elektrochem., 64:43, 129 (1960).
12. W. Vielstich, Z. Anal. Chem., 173:84 (1960).

13.   M. Stackelberg, W. Vielstich, and D. Jahn, Anal. Real. Soc. Espan. Fis. Quim., B56:475 (1960).

14.   W. J. Albery and R. P. Bell, Proc. Chem. Soc., 1963:169.

15.   W. J. Albery, Trans. Faraday Soc., 61:2063 (1965).

16.   D. Haberland and R. Landsberg, Ber. Bunsenges., 70:724 (1966).

17.   F. Hine and K. Yamakawa, Electrochim. Acta, 13:2119 (1968).

18.   J. Ulstrup, Electrochim. Acta, 13:1717 (1968).

19.   D. R. Olander, Ind. Eng. Chem. Fundament., 6(2):178 (1967).

20.   W. Geissler, R. Nitzsche, and R. Landsberg, Electrochim. Acta, 11:389 (1966).

21.   D. Haberland and R. Landsberg, Ber. Bunsenges., 71:219 (1967).

22.   P. Beran and S. Bruckenstein, J. Phys. Chem., 72:3630 (1968).

23.   E. Budevskii and S. Toshev, Dokl. Akad. Nauk SSSR, 130:1047 (1960); Izv. Inst. Fiz. Khim. Bolg. Akad. Nauk, 1:183 (1961).

24.   M. A. Loshkarev and I. P. Chernovaev, Tr. Dnepropetr. Khim.-Tekhnol. Inst., 12(1):73, 91 (1959).

25.   Z. Galus and R. N. Adams, J. Electroanal. Chem., 4:248 (1962); Z. Galus, Chem. Anal. (Polska), 10:803 (1965); Bull. Acad. Polon. Sci., 13:63 (1965).

26.   L. K. J. Tong, Kai Liang, and W. R. Ruby, J. Electroanal. Chem., 13:245 (1967).

27.   D. G. Gray and J. A. Harrison, J. Electroanalyt. Chem., 24:187 (1970).

28.   P. A. Malachesky, L. S. Marcoux, and R. N. Adams, J. Phys. Chem., 70:4068 (1966); L. S. Marcoux, I. M. Fritsch, and R. N. Adams, J. Amer. Chem. Soc., 89:5766 (1967).

29.   L. S. Marcoux, R. N. Adams, and S. W. Feldberg, J. Phys. Chem., 73:2611 (1969).

30.   G. Manning, V. D. Parker, and R. N. Adams, J. Amer. Chem. Soc., 91:4584 (1969).

31.   S. Karp, J. Phys. Chem., 72:1082 (1968).

32.   V. Yu. Filinovskii, Élektrokhimiya, 5:635 (1969).

33.   S. W. Feldberg, Electroanalytical Chemistry, Vol. 3, A. J. Bard, ed., Marcel Dekker, New York (1969), p. 199.

34.   M. Litt and G. Serad, Chem. Eng. Sci., 19:867 (1964).

35.   L. G. Loitsyanskii, Laminar Boundary Layer, Fizmatgiz, Moscow (1962), p. 390.

Application of the Rotating Disc Electrode to Various Electrochemical Problems

§ 7.1. Separation of Two Simultaneous Electrode Processes

Electrochemical studies often require the separation of two simultaneous processes occurring at the electrode over the same potential range. Thus, for example, the polarographic waves for reduction or oxidation of various substances are distorted at highly negative potentials by hydrogen evolution or by discharge of ions present in the supporting electrolyte, and at sufficiently positive potentials by dissolution of the electrode metal itself. Separation of the principal and side processes by electrochemical methods is rather difficult.

However, as shown by Pleskov [1], this separation can be effected using the rotating disc electrode in the case where the rate of one of the simultaneous processes is diffusion limited. In this case one of the processes depends on the rotational speed, while that of the second one is independent of stirring. Electrode processes of the latter type include anodic metal dissolution, evolution of hydrogen from acid solutions, and of oxygen from alkaline solutions at moderate current densities.

The disc current density under conditions of mixed kinetics is given by (cf. § 3.4)

$$i = \frac{i_d}{1 + D/k\delta_d} \, , \tag{7.1}$$

255

where i_d is the diffusion–limiting current density and k is the rate constant of the electrochemical reaction. When two independent reactions simultaneously occur at the electrode, the disc current consists of two parallel currents:

$$i = i_1 + i_2 = \frac{i_{d1}}{1 + D_1/k_1\delta_{d1}} + \frac{i_{d2}}{1 + D_2/k_2\delta_{d2}} . \qquad (7.2)$$

The separation of the total current i into two partial currents i_1 and i_2 is difficult in the general case. However, it is possible in the case where one of the reactions proceeds under diffusional control $(D_1/k_1\delta_{d1} \ll 1)$ and the other, under kinetic control $(D_2/k_2\delta_{d2} \gg 1)$. The general equation (7.2) reduces in this case to

$$i = i_{d1} + i_{k2}, \qquad (7.3)$$

where i_{k2} is the kinetic current for the second reaction.

Taking into account that, according to § 2.4, the diffusion limiting current $i_{d1} \sim \omega^{1/2}$ and the kinetic current i_{k2} is independent of stirring, a method can be developed which allows the partial currents to be separated. In fact, the extrapolation of the experimental current values i plotted at constant potential against $\omega^{1/2}$ into the region of low rotation speeds $(\omega \rightarrow 0)$ results in currents consisting almost exclusively of the partial current i_{k2}.

It must be stressed that this separation requires extrapolation of the experimental results into the region of low rotation speeds; it cannot be obtained by a direct measurement of current in absence of stirring, since Eq. (7.1) becomes invalid at low ω values owing to natural convection.

The conditions imposed when Eq. (7.2) is simplified to Eq. (7.3) have a simple physical meaning. Since the magnitude of D_1 and D_2 (and consequently of δ_{d1} and δ_{d2}) differ little in common electrolytes, the imposed conditions require that at the given potential $k_1 \gg k_2$, i.e., the rate of reaction 1 is much higher than that of reaction 2.

a. Ionization of Oxygen at a Gold Electrode. The overpotential for hydrogen evolution on gold is low and the experimental polarization curves for ionization of molecular oxygen are distorted[†] by the parallel process of hydrogen evolution (Fig. 7.1a,

---

[†]At high overpotentials. – Editor.

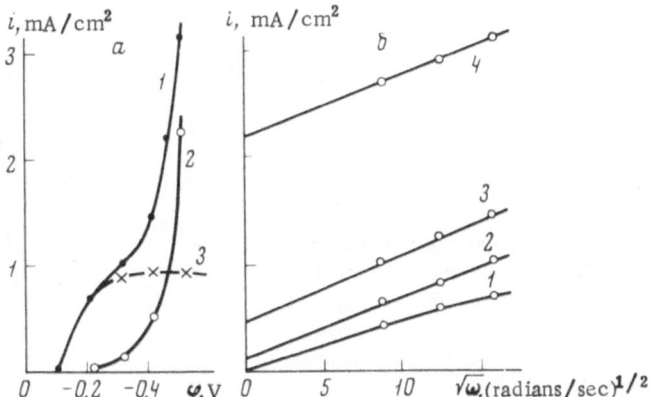

Fig. 7.1. Cathodic reduction of oxygen at a disc electrode. a) Polarization curves obtained at a gold electrode in air-saturated 1 N H_2SO_4 solution at 2550 rpm [1]: 1) experiment; 2) calculated curve for hydrogen solution; 3) calculated curve for oxygen ionization. b) Current density (Fig. 7.1a, curve 1) as a function of $\omega^{1/2}$ at potentials: 1) 0.22 V; 2) 0.32 V; 3) 0.42 V; 4) 0.52 V.

curve 1). Under the experimental conditions, the rate of O_2 ionization is diffusion controlled and proportional to $\omega^{1/2}$, whereas H_2 evolution (at moderate current densities) is independent of stirring.

In Fig. 7.1b the current i is plotted as a function of $\omega^{1/2}$. The stirring-independent part of the total current can be separated by extrapolation of the straight lines (Fig. 7.1b) to $\omega \to 0$. The values obtained are shown in Fig. 7.1a (curve 2). The calculated curve 2 corresponds to hydrogen evolution. Curve 3 (for oxygen ionization) was obtained by subtracting current densities corresponding to curve 2 from total current densities (curve 1).

b. Oxidation of Univalent Silver Ions at a Gold Electrode in KOH Solution. Oxidation of Ag^+ ions occurs simultaneously with oxygen evolution. Therefore, the Ag^+ oxidation wave is distorted (Fig. 7.2, curve 1). The total current density varies linearly with $\omega^{1/2}$. Using the method described above, we obtain the polarization curves for oxygen evolution (stirring-independent) and for Ag^+ oxidation (curves 2 and 3, respectively). Curve 2 has two distinct waves, whose height (limiting current) is proportional to $\omega^{1/2}$ and

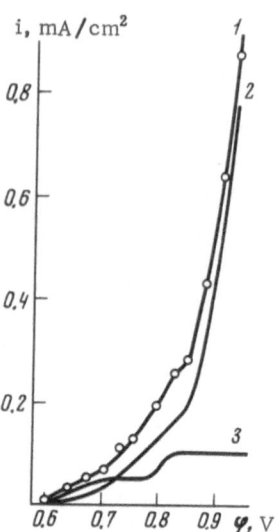

Fig. 7.2. Anodic polarization curves obtained in $2.3 \cdot 10^{-4}$ N Ag_2O + 13.4 N KOH solution at a gold electrode; 3100 rpm [1]. 1) Experimental; 2) calculated curve for oxygen evolution; 3) calculated curve for Ag^+ oxidation.

to the Ag_2O concentration in the solution. The height of the first wave is equal to the height of the cathodic Ag^+ reduction wave from the same solution. Consequently, the first wave coresponds to the one-electron process of oxidation of Ag^+ ions to AgO. The second wave is two times higher than the first one, corresponding to the oxidation of Ag^+ to trivalent silver.

The accuracy of calculations is about 5% for curve 3; the method can also be applied in cases where the rate of oxygen evolution exceeds that of silver ion oxidation by 10-15 times.

c. Oxidation of Divalent Vanadium Ions at a Germanium Electrode [2]. Oxidation of V^{2+} ions at a germanium anode is accompanied by dissolution of the latter. The method described above led to a separation of the stirring-independent polarization curve for germanium dissolution from that corresponding to the diffusion-limited oxidation of V^{2+} ions.

The above examples illustrate sufficiently the possibilities of the method. It was used recently in the determination of the background current during copper [3], iron [4], and thallium [5] deposition, as well as in microdeterminations of gold at rotating electrodes [6]. The method can be successfully applied in studies of iron and copper corrosion in acid solutions containing dissolved oxygen and in studies of the two parallel cathodic reactions −

reduction of molecular oxygen (diffusion-limited and stirring-dependent) and hydrogen evolution (stirring-independent in highly acidic solutions) [7]. Finally, the method can be used to separate currents for parallel reduction of the depolarizer from solution and from its adsorbed state [8].

§ 7.2. Study of the Type of Polarization

in Metal Deposition

The electrodeposition of metals at a rotating disc electrode serves two general purposes: 1) the electrodeposition process itself can be studied, and 2) the mechanism of the effects of admixtures of surface-active substances can be investigated.

A series of papers by Vishomirskis and Matulis et al. [9] was aimed at elucidating the type of polarization in metal deposition from complex salt solutions. Cathodic polarograms obtained in cyanide solutions of copper, cadmium, and zinc, and in phosphoric acid and alkaline solutions of palladium salts, exhibit a wave with a well-defined limiting current. Cathodic polarograms obtained in cyanide solutions of gold salts, in ammoniacal solutions of copper salts, and in borate solutions of nickel salts consist of two waves. In all these cases, the limiting currents are proportional to the concentration of the discharging ions and to $\omega^{1/2}$, and consequently they are diffusion controlled (cf. also [10]).

The reduction of bismuth from perchlorate solutions [11], of iron from alkaline [4] and sulfate [12] solutions, and of vanadium, manganese, chromium, and iron ions from molten salts [13] proceeds also under diffusion control. The same is true of electrodeposition of copper from Cu^{2+} solutions [14]; the latter process, however, is complicated by formation of intermediate Cu^+ ions [15, 16]. Kruglikov et al. [17] and others [15, 18-20] used the rotating disc electrode to explain the effect of leveling agents such as coumarin (in nickel deposition), thiourea (in copper deposition), and other systems.

The leveling agents increase the overpotential of cathodic metal deposition. The leveling effect is usually explained in terms of adsorption of the agent occurring to a larger extent at protrusions than at the hollows of the surface (owing to the higher adsorbability on protrusions, or to restriction of diffusion in the hollows); therefore, the growth of protrusions is slowed down, and

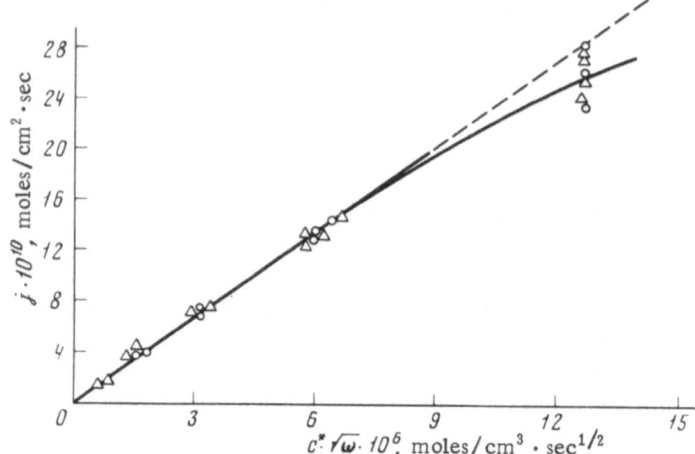

Fig. 7.3. The leveling effect of thiourea on copper deposition [17]. The dependence of the diffusional flux of thiourea on its concentration $c*$ and $\omega^{1/2}$. Points correspond to various experiments.

metal deposition proceeds with a higher rate on the hollows than on protrusions, resulting in a smoother surface.

The leveling agent passes to some extent into the deposit, and its concentration at the surface is lower than that in the bulk. In all the examples listed, diffusion was shown to play a considerable part in the transfer of the leveler to the metal surface. It was shown that the leveling capability of coumarin (in nickel deposition) is proportional to the product $c_{B}^{*}\omega^{1/2}$ (where c_{B}^{*} is the bulk concentration of coumarin and ω is the rotation speed of the electrode). Thus, the leveling effect is determined by the diffusional flux of coumarin to the electrode. When $\omega^{1/2}$ exceeds a certain critical value, the kinetics of the process passes from the diffusional into the mixed region. The coverage with the coumarin at protrusions then reaches a certain limiting value and remains constant with further increase of the diffusional flux. Since coumarin adsorption continues to increase at the hollows in the surface, the leveling effect (connected with the difference of adsorption at protrusions and hollows) decreases.

The consumption of thiourea during copper deposition is also diffusion controlled (Fig. 7.3).† The results obtained using the

---

†In some cases, the process is complicated by complex formation and slow adsorption and occurs in the mixed or kinetic region.

disc are in good agreement with direct measurements of the rate of thiourea codeposition using radioactively labeled atoms [22].

§ 7.3. Anodic Metal Dissolution

and Electropolishing

Studies of anodic dissolution are usually concerned with the nature of the rate-determining step and with the establishment of the type (diffusional or kinetic) of conditions under which the process occurs. The character of the polarization behavior can be established in a simple and quantitative way using a rotating disc electrode on the basis of the dependence of the rate of dissolution on the rotation speed. Thus, it has been shown [23] that the limiting current for copper dissolution in dilute phosphoric acid solution is proportional to $\omega^{1/2}$ and is determined by the rate of diffusion of nondissociated H_3PO_4 molecules to the electrode surface.

Electropolishing of aluminum in alcoholic solutions of magnesium perchlorate is diffusion controlled [24]; the same was found in the cases of anodic dissolution of cadmium in aqueous solutions [25], zinc in alkaline media [26], copper in H_3PO_4 solutions [27], and of iron and sulfur in molten salts [28]. Conversely, titanium anodes dissolve in fluoride baths under mixed kinetic control [29].

The anodic current of nickel dissolution in sulfuric acid is proportional to $\omega^{1/2}$ in the active and transpassive regions (at not too low concentrations of sulfuric acid) and is diffusion controlled. Conversely, in the passive region, current depends little on stirring, the reaction being kinetically controlled [31-34].

Electropolishing of metals is usually carried out at high current densities which results in high ohmic potential drops in the solution. Under these conditions the surface of the disc is not equipotential (cf. § 4.10). Therefore, polarization conditions differ in the center and at the edges of the disc, so that electropolishing may proceed in a different way at various regions of the disc surface.

Fedash [35] studied the electropolishing of copper in phosphoric acid and of steel in a sulfuric−phosphoric−chromic acid bath using a disc constructed of several concentric rings. Electropolishing starts after a certain critical current density is reached. This critical value is proportional to $\omega^{1/2}$ and is consequently of a diffusional nature. At lower current densities, the whole surface of the disc dissolves uniformly. With increasing current density, the critical value is reached at the edges of the disc where electropolishing starts. Further increase of the current density results in the critical value being also reached at the center of the disc. This effect is obviously connected with the fact that the disc does not present an equipotential surface at high current densities.

The measurements of the impedance of the rotating copper disc electrode electropolished in phosphoric acid led to the conclusion that the increase of the differential electrode resistance with increasing potential is connected not with the formation of thick insulating oxides but with decreasing conductivity of the electrolyte layer at the electrode which arises under diffusion-limiting current conditions. Rotation of the electrode decreases the thickness of the diffusion layer and, consequently, the resistance of the latter [36].

§ 7.4. Corrosion of Metals.

Dissolution of Nonmetals

Immersion of a metal in an electrolyte can, in principle, result in the passage† of the metal ions into the solution. The necessary quantity of electricity can be supplied externally (cf. § 7.3) or as a result of another (cathodic) reaction occurring in parallel and at the same rate as the anodic dissolution. In the latter case, the process corresponds, of course, to corrosion of the metal. The cathodic reaction usually consists in hydrogen evolution (dissolution of metals in acids) or in reduction of some other oxidizing agent (e.g., molecular oxygen).

The detailed nature of corrosion and problems connected with the latter are beyond the scope of this book. The following section is restricted to consideration of certain self-dissolution processes which are diffusion controlled. Thus, for example, the rate-determining step in the dissolution of metals in dilute acid solutions can be controlled by diffusion of hydrogen ions to the surface of the corroding sample. The first example concerns the case of a metal with a uniform surface.‡

---

†Continuous passage. — Editor.
‡In practice, metal surfaces are always rough and contain inclusions; however, as long as the dimensions of nonuniformities are small as compared to the thickness of the diffusion layer, the surface behaves as a uniform one with respect to diffusion processes.

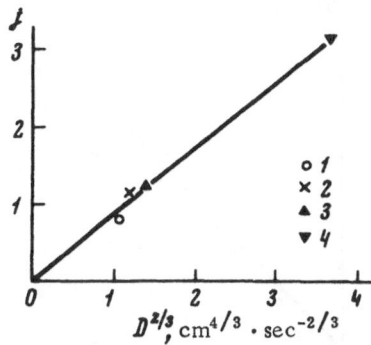

Fig. 7.4. Rate of dissolution of a magnesium disc in acids as a function of the
diffusion coefficient of H^+ ions [38]. 1)
CH_3COOH; 2) HCOOH; 3) $(COOH)_2$; 4)
HCl.

The rotating disc electrode is an effective tool in studies of
diffusion steps in corrosion. The rate of the overall process is
then usually determined by nonelectrochemical means (by the
decrease in weight of the electrode, or by changes in the electrolyte composition). The partial anodic and cathodic currents can
also be studied at the disc electrode under external polarization
conditions.

The rotating disc electrode was first applied to corrosion of
magnesium and manganese in hydrochloric, acetic, formic, and
oxalic acids [38]. The dissolution rate was found to be metal-
independent and proportional to the acid concentration and $\omega^{1/2}$,
clearly indicating diffusion-controlled kinetics. The corrosion
rate was also found to be proportional to $(D_{H^+})^{2/3}$ (where D_{H^+}
is the diffusion coefficient of hydrogen ions), as predicted by
Eq. (2.37). The dependence of the corrosion rate (expressed in
terms of the volume of gaseous hydrogen evolved by 1 cm^2 of
disc†) on $D_{H^+}^{2/3}$ is shown in Fig. 7.4.

Shortly after this first paper appeared, the rotating disc electrode became a common tool in corrosion studies. Zembura et al.
[7, 39-42] described a few characteristic cases of cathodic and
anodic control and their kinetics.

Cathodic Control; Diffusion-Controlled Kinetics. The rate of
zinc corrosion in 0.1 M Na_2SO_4 (pH = 4.5) is virtually equal to the
current density of the diffusion limiting current for hydrogen
evolution measured in the same solution at a rotating platinum

---

†Per unit time. — Editor.

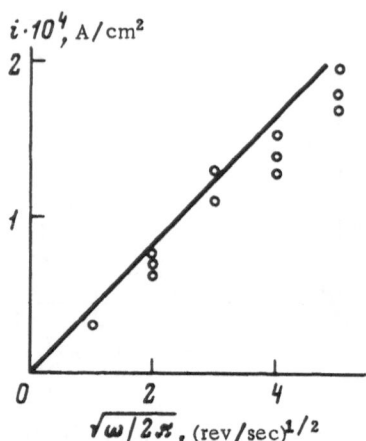

Fig. 7.5. Rate of dissolution of a rotating zinc disc in 0.1 M Na_2SO_4 at pH = 4.5 (circles) and the diffusion-limiting current density of H^+ ion reduction at platinum in the same solution (solid lines) as a function of $\omega^{1/2}$ [39].

cathode (cf. Fig. 7.5). Thus, the rate of the cathodic reaction and consequently that of corrosion, must be controlled by the convective diffusion of H^+ ions.

Cathodic Control; Two Depolarizers. This was observed in the case of iron and copper dissolution in air-saturated dilute acid solutions. The cathodic current consists of O_2 and H^+ reduction currents. At low pH values, hydrogen evolution is practically independent of stirring, as opposed to the oxygen reduction process. Using the method described in § 7.1, we can separate the two partial cathodic currents (at the given pH and potential).

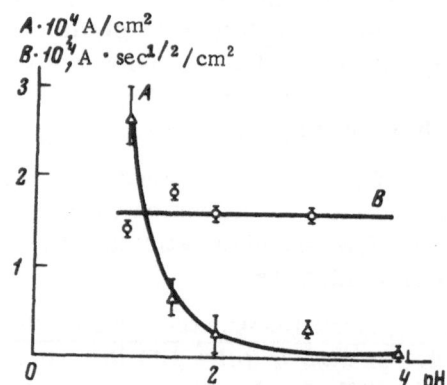

Fig. 7.6. Hydrogen ion (A) and oxygen (B) depolarization of an armco iron electrode corroding in air-saturated 0.1 M Na_2SO_4 solutions as a function of pH [42].

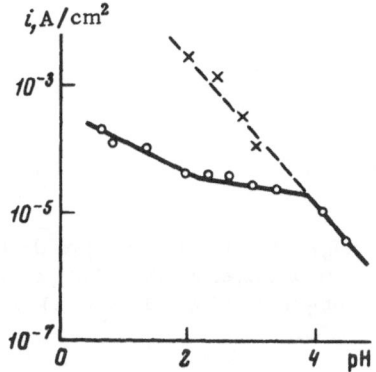

Fig. 7.7. Corrosion rate of a copper disc as a function of pH (circles) in acidified Na_2SO_4 solutions. ×) Diffusion-limiting current of H^+ ions [39].

When pH > 2, the contribution of H^+ ions to the cathodic current becomes insignificant in the case of corrosion of iron. The rate of the total process can be described by

$$i = A + B\sqrt{\omega/2\pi}, \tag{7.4}$$

where the first term pertains to hydrogen and the second to oxygen depolarization. The experimentally measured values of A and B are shown in Fig. 7.6.

Transition from Anodic Control with Chemical Kinetics to Cathodic Control with Diffusion-Controlled Kinetics. Copper corrosion in dilute H_2SO_4 solutions is controlled by diffusion of H^+ ions. With decreasing pH, the diffusion, and consequently the corrosion rate, increases until it becomes commensurate (at pH

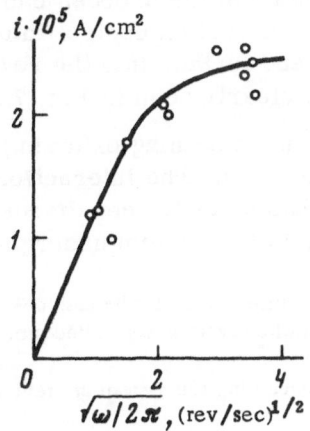

Fig. 7.8. Corrosion rate of a copper disc in 0.1 M Na_2SO_4 + 9.3 · 10^{-5} M H_2SO_4 as a function of $\omega^{1/2}$ [39].

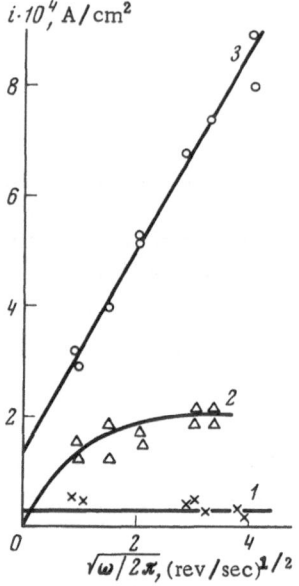

$i \cdot 10^4$, A/cm^2

Fig. 7.9. Effect of stirring on the corrosion rate of copper in 0.1 M H$_2$SO$_4$ at various temperatures [39]: 1) 25°C; 2) 50°C; 3) 75°C.

$\sqrt{\omega/2\pi}$, (rev/sec)$^{1/2}$

3.8, 25°C) with the rate of the anodic process (Fig. 7.7). The overall process occurs in the mixed kinetic control region, as indicated by the characteristic nonlinear dependence of the corrosion rate on $\omega^{1/2}$ (Fig. 7.8). Further decrease of pH affects little the rate of corrosion which then proceeds in the region of chemical kinetic control.

The rate of chemical processes is usually much more strongly affected by temperature changes than is the rate of diffusion. Therefore, at a given pH value lower than 3.8, the process can be shifted by increase of temperature from activation control into the mixed kinetic control region (50°C) and further into the region of diffusion control (75°C). This can be clearly seen in Fig. 7.9.†

The dissolution of metals in solutions containing oxidizing agents was studied quantitatively by Riddiford. The interaction of zinc with aqueous iodine solutions [43] has already been discussed in § 2.5.‡ Landsberg [45] found that the latter system undergoes

---

†When $\omega \to 0$, the corrosion rate tends to a certain finite value. In the case described, this is explained in terms of another parallel activation-controlled process, whose rate is stirring independent (cf. §7.1).

‡The earlier papers of Bircumshaw and Riddiford concerning this system are reviewed in Levich's monograph [44].

a transition from diffusion to mixed kinetic control in the presence of surface contaminants, which obviously inhibit the electrochemical reaction.

Gregory and Riddiford [46] measured the rate of dissolution of copper discs in sulfuric acid solutions of potassium dichromate. In the presence of a large excess of sulfuric acid, the rate of corrosion is controlled by diffusion of potassium dichromate. In this case, neither the chemical composition of the disc (copper, brass), nor the state of its surface affect the dissolution rate. The experimental results obtained for the system described are shown in Fig. 7.10 (circles). The solid line is calculated according to Eq. (2.38). The diffusion coefficients were independently measured (diffusion through a porous membrane). The experimental data agree with the theory to an accuracy of $\pm 2\%$, i.e., within the limits of the experimental error.

The rate of copper dissolution in a solution containing Fe^{3+} ions at low stirring rates is controlled by diffusion of Fe^{3+} ions to the copper surface. The cathodic reaction was separately studied using a platinum rotating disc electrode, while the overall corrosion process was examined at rotating copper discs [47]. At high rotation speeds ($Re > 2 \cdot 10^{-4}$), the process is activation controlled and stirring independent [48].

The corrosion of copper and nickel in acidified ethanol solutions was studied by Heitz [49]. The dependence of the corrosion rate on $\omega^{1/2}$ shows that the former is controlled by diffusion of H^+ ions. In the presence of a large excess of acid, oxygen adsorption

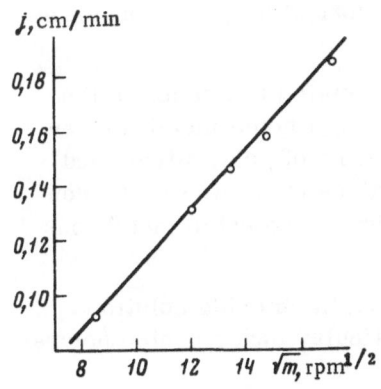

Fig. 7.10. Dissolution of a copper disc in $1\ M\ H_2SO_4 + 3 \cdot 10^{-2}\ M\ K_2Cr_2O_7$. Diffusion flux of $Cr_2O_7^{2-}$ ions as a function of $\omega^{1/2}$ [46].

at the metal surface, or the electrochemical reaction, seems to be the rate-determining step.

The dissolution of lead and nickel in nitric and trichloroacetic acids shows interesting features [50]. The kinetics is activation controlled in dilute and moderately concentrated HNO_3 solutions (when cathodic reduction of NO_3^- is the slow step), and diffusion controlled in concentrated ($\geq 40\%$) solutions (when the slow step is the diffusion of lead nitrate away from the disc surface). The corrosion rate is sensitive to admixtures in lead and in the solution bulk if they affect the rate of the heterogeneous reduction of HNO_3. For example, arsenic increases the reaction rate to the extent that in dilute HNO_3 solutions dissolution becomes diffusion controlled (by the rate of supply of HNO_3 to the electrode). Thus, for acid concentrations < 20% the reaction rate is controlled by diffusion of solution substrates, for 20-40% acid by activation control, and for 40-57% acid by the rate of diffusion of the products from the surface.

Kakovskii et al. investigated the corrosion of copper, zinc, and noble metals in alkaline cyanide solutions containing dissolved oxygen. The corrosion rate of silver and copper is diffusion limited: at low cyanide concentrations — by diffusion of CN^- ions; in concentrated KCN solutions — by diffusion of molecular oxygen [51].

Dissolution of gold occurs at higher rotation speeds in the chemical kinetic control region, possibly owing to passivation of gold with oxygen. Certain organic substances depassivate gold and increase the dissolution rate without, however, transferring the process into the diffusion regime [52]. A similar passivation of the surface of dissolving discs by adsorbed oxygen is observed for palladium [53] and zinc [54].

The rotating disc method was also applied to studies of the reverse process — electrodeless plating of a noble metal, driven by dissolution of the disc material. Plating of gold, silver, and copper by dissolution of iron and zinc discs is limited in the concentration range investigated by diffusion of the noble metal ions to the dissolving sample [55].

Corrosion of steel discs in acids and in chloride solutions [56-58] was extensively studied, in particular under nonisothermal conditions [58].

Novakovskii [59] suggested the use of a rotating disc electrode in studies of corrosion of pipelines. The experimental study of diffusion processes, and thus of diffusion, in a pipe is difficult, since convection in the liquid flowing through the pipe is relatively small, and a significant diffusion flux, easily reachable at a rotating disc under laminar conditions, can only be obtained under turbulent conditions. Novakovskii derived a relation between the flux in a pipe and the angular velocity of the disc, at which the density of the diffusion fluxes to the pipe walls and to the rotating disc is the same. A study of corroding discs then yields information concerning corrosion of pipelines.

According to Rozenfel'd [60], the corrosion of ship hulls moving in seawater can be modeled similarly.

The rotating disc can be used not only for studies of metals with uniform surfaces but also for investigating dissolution of inclusions and protection of metals. Dogonadze, Levich, and Chizmadzhev [61] calculated the potential distribution at a disc containing a protecting disc-shaped center (inclusion of a metal having a higher rate of the anodic process than that of the outer part of the disc). It was assumed that the rate of the cathodic processes (reduction of dissolved oxygen) at the part of the disc adjacent to the protector anode is diffusion limited and depends on the stirring rate, whereas at the much less polarized edge of the disc it is activation controlled. The results of calculations are in a good agreement with experiment [62].

The rotating disc method was also applied to studies of self-dissolution of nonmetallic substances resulting from redox processes at the solid—liquid interface. This concerns, for example, corrosion of anthracene in oxidizing solutions [63], of graphite in molten metal oxides containing FeO [64], and dissolution of sulfides in cyanide solutions containing molecular oxygen [65].

Finally, the dissolution of solids in liquids in the absence of chemical reactions can be very conveniently studied using the rotating disc. The nature of the rate-controlling step (diffusion or the heterogeneous process itself) was established by this method in the case of metal dissolution in melts [66], and of borates in water [67]. Some other processes are considered in the next section.

Thus, the rotating disc method can be used to establish the character of the rate-controlling step in a complex corrosion process and to determine the kinetic characteristics of the latter (the apparent activation energy and rate constant).

§ 7.5. The Rotating Disc in Systems with Variable Physical Properties

It was mentioned in § 2.1 that the diffusion coefficient depends on the solution composition, temperature, and other parameters. However, until now all relevant quantities (ρ, μ, D, etc.) characterizing the physical properties of a system were considered constant. This approximation, fully justified in the case of dilute solutions, is insufficient for highly concentrated solutions.

A number of practically important processes (e.g., electroplating, metal dissolution, etc.) are carried out in concentrated solutions whose physical properties depend considerably on composition. On the other hand, a detailed analysis of kinetic properties allows information on the medium structure and on the interactions between the particles to be obtained. Therefore, a study of the kinetic characteristics of the system as a function of composition, temperature, pressure, etc. results in important conclusions concerning the nature and the character of intermolecular forces in the medium. The study of the kinetic characteristics of the medium in the critical region in which its physical properties change abruptly is of special importance.

Some results obtained using the rotating disc method to investigate kinetic processes in systems with variable physical properties will now be discussed.

Consider an isothermal case and assume that ρ, μ, and D depend only on concentration of the dissolved substance:

$$\rho = \rho^* R, \quad \mu = \mu^* N, \quad D = D^* \Delta . \tag{7.5}$$

The quantities ρ^*, μ^*, and D^* correspond to the values of density, viscosity, and diffusion coefficient, respectively, in the bulk solution, where the concentration is c^*. The functions R, N, and Δ depend only on the concentration at the given point.

In the presence of concentration gradients, the physical properties of the medium are different at different points [cf. Eq. (7.5)]. Equations (1.1) and (1.2) describing the motion of a viscous incompressible fluid are unsuitable for a medium with variable physical properties. They must be replaced by more general expressions which take into account variations of ρ and μ at different points of the liquid. Moreover, it can be easily demonstrated that when density and viscosity are concentration dependent, the equations of motion are connected with the equation

$$\rho v \text{ grad } c = \text{div } (\rho\ D \text{ grad } c) \qquad (7.6)$$

which describes the motion of dissolved substance, and they form a complete set of equations.

The latter problem was analyzed by Daguenet and Schuhmann [68], by Newman and Hsueh [69], and by Olander [70]. They showed that for mass transfer to an infinite disc uniformly rotating in a liquid whose physical properties depend on concentration of the transferred species [Eq. (7.5)], the general set of equations can be reduced to a system of relatively simple ordinary differential equations. This requires, by analogy with § 1.2, introduction of the dimensionless distance from the disc

$$\zeta = z \sqrt{\omega\rho^{\bullet}/\mu^{\bullet}} \qquad (7.7)$$

and of the functions F, G, H related with the radial (u), azimuthal (v), and axial (w) components of the liquid velocity:

$$u = r\omega F\ (\zeta), \qquad v = r\omega G\ (\zeta), \qquad w = \sqrt{\mu^{\bullet}\omega/\rho^{\bullet}}H\ (\zeta). \qquad (7.8)$$

With these expressions, the equations of liquid motion and the mass-transfer equation (7.6) can be transformed into ordinary differential equations (where the prime designates differentiation with respect to ζ)†:

$$2F + H' = -\frac{HC'}{R}\frac{dR}{dC},$$
$$F^2 + F'H - G^2 - \frac{NF''}{R} = \frac{F'C'}{R}\frac{dN}{dC},$$
$$2FG + HG' - \frac{NG''}{R} = \frac{G'C'}{R}\frac{dN}{dC},$$
$$HC' = \frac{D^{\bullet}\rho^{\bullet}}{\mu^{\bullet}}\left[\Delta C'' + \frac{C'^2}{R}\frac{d\ (R\Delta)}{dC}\right],$$

$$(7.9)$$

---

†For simplicity, the equation describing pressure distribution is omitted.

where

$$C = c/c^*. \tag{7.10}$$

It can be easily shown that for liquids with constant physical properties, Eqs. (7.9) reduce to the corresponding equations (1.12) obtained by von Kármán.

Far from the disc, the rotating motion disapears, and therefore in the bulk solution we have

$$F \to 0, \quad G \to 0, \quad C \to 1 \qquad \text{as } \zeta \to \infty . \tag{7.11}$$

The boundary conditions at the surface of the disc depend on the character of the heterogeneous process in which the transferred species participates. If, for example, the disc surface is impermeable to the substance supplied and the rotating disc entrains all the liquid, the following conditions obtain, by analogy with (1.13):

$$F = 0, \quad G = 1, \quad H = 0, \quad C = C_S \qquad \text{for } \zeta = 0. \tag{7.12}$$

The flux of the substance to the surface of the disc is given in this case by

$$j = \left(-\rho D \frac{\partial c}{\partial z} \right)_s = -\frac{D^* \rho^{*1/2}}{\mu^{*1/2}} \omega^{1/2} c^* (R_S \Delta_S C_S'), \tag{7.13}$$

where R_S, Δ_S, and C_S' denote the respective functions at the disc surface. It follows from Eq. (7.13) that the flux of the reactant is, in this case, also proportional to $\omega^{1/2}$.

Daguenet and Schuhmann [68], who derived Eqs. (7.9), did not attempt to solve the latter, nor did they study any actual dependence of the R, N, and Δ functions on the concentration C of the transferred species. Newman and Hsueh [69] calculated the rate of copper deposition from aqueous solutions of copperas at 25°C. The latter authors did not consider the analytical form of the dependence of ρ, μ, and D on c, but used directly in their calculations the experimental data concerning this dependence. Their results are shown in Fig. 7.11 as a plot of $i_d \omega^{-1/2}$ vs the concentration of the discharging Cu^{2+} ions (curve 1). Curve 2 in the same figure corresponds to calculations made according to Eq. (2.37), i.e.,

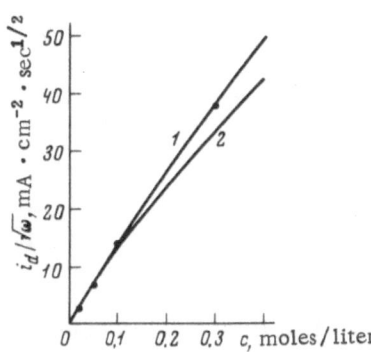

Fig. 7.11. Limiting current density for copper deposition from $CuSO_4$ solutions as a function of concentration [69]. The points are experimental data.

neglecting the variation of physical properties within the diffusion layer. The difference beween curves 1 and 2 is small in dilute solutions (1.6% in 0.02 M $CuSO_4$), but it increases considerably with concentration (up to 13% in 0.3 M solution). The experimental anc calculated results are in good agreement [69].

Olander [70], who calculated the mass transfer away from the rotating disc, took into account not only the variation of physical properties but also the possible effect of the rate of the transfer away from the solid surface. His calculations were first made for systems with high Schmidt numbers Sc_S (at the disc surface) using the perturbation method. Later, the calculations were extended to the gas phase as well (in particular, to the ternary system $I_2 - GeI_4 -$ inert gas). It was shown that the rate of mass transfer in concentrated mixtures can considerably exceed that in dilute systems.

The physical properties of the medium depend particularly strongly on composition in the critical region. The most affected are kinetic coefficients (viscosity, heat conductivity, diffusion coefficients).

According to the thermodynamics of irreversible processes, the rate of diffusional transfer of a solution component is proportional to the gradient of its chemical potential. Since at the critical point this gradient (at least for binary solutions) is equal to zero, the diffusion flux should also disappear. Experimental verification of this conclusion is rather important, both for elucidation of the structural details of the medium as well as from the point of view of verification of the basic principles of irreversible thermodynamics.

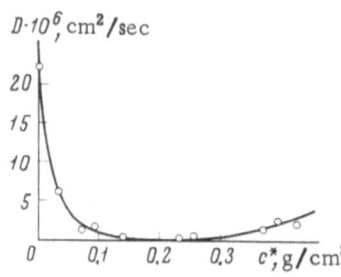

Fig. 7.12. Diffusion coefficient of hexa-
methylimine as a function of its concentration in mixtures with water at 340.6°K
[72].

The first study of molecular diffusion near the critical point
was carried out by Krichevskii et al. [71]. Diffusion in binary
liquid systems was first studied near the critical point of solution
stratification. This critical phenomenon is relatively simple and
can occur under isothermal conditions. The experimental study
of the diffusion coefficient (using the thin capillary method) demonstrated that the latter abruptly decreases when the concentration
approaches the point of stratification. The dependence of the diffusion coefficient on composition of the hexamethylenimine—water
system at 340.6°K is shown in Fig. 7.12 [72]. It can be seen that
already a small concentration increase results in decrease of the
diffusion coefficient by about 230 times.

Similar peculiarities are observed in studies of the effect of
the critical state on diffusion in ternary liquid systems. Krichevskii et al. [73] showed that in ternary systems, e.g., triethylamine—
water containing small additions of butylamine or phenol, the diffusion coefficient of triethylamine abruptly decreases when the
system approaches the critical point, independently of the presence
of the third component. No peculiarities in the behavior of the
third component, which forms a dilute solution, were observed.

The thin capillary method, extensively used in the earlier
work of Krichevskii et al., is experimentally rather difficult.
Therefore Krichevskii and Tsekhanskaya [74] applied the rotating
disc method in their later studies of critical phenomena. They
measured the rate of dissolution of organic acids (terephthalic,
sebacic, adipic, and salicylic) in aqueous solutions of triethylamine
and hexamethylenimine near the critical point.†

---

†The critical temperature of the triethylamine—water system is 18.3°C (26.1 wt.%
triethylamine); that of the hexamethylenimine—water system is 68.1°C (24.7 wt.%
hexamethylenimine).

The above acids are virtually insoluble, or sparingly soluble, in water, whereas their salts with triethylamine and hexamethylen-imine are very soluble. The acids were pressed into a disc shape and placed in a holder rotating with constant angular velocity. Laminar conditions obtained for discs 10 mm in diameter and for rotation speeds 300-1000 rpm. Turbulent conditions obtained for discs 32 mm (and larger) in diameter and 2000 rpm (Reynolds numbers were in the range $2 \cdot 10^4 - 2 \cdot 10^5$). The dissolution rate was limited by the supply of base (triethylamine or hexamethylen-imine) to the disc surface. It was determined experimentally by measuring the decrease in weight of the disc. Diffusion fluxes thus obtained are shown in Fig. 7.13.

The rate of dissolution of the disc under laminar conditions (Fig. 7.13a) shows that the experimentally measured diffusion fluxes are independent of the nature of the acid and coincide with values calculated according to Eq. (2.37) in the region far from the critical point. With increasing concentration of triethylamine or hexamethylenimine, i.e., upon closer approach to the critical point, diffusion fluxes become virtually independent of composition. Similar phenomena are observed under turbulent conditions (Fig. 7.13b). A quantitative theory of the processes involved was developed by Levich and Ivanov [44, 75]. First, the qualitative description of triethylamine transfer to the disc will be considered.

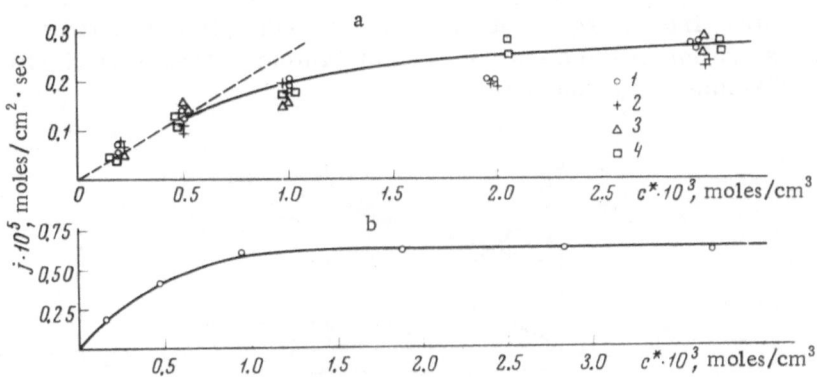

Fig. 7.13. Diffusion flux of triethylamine as a function of concentration [74]. Dissolution of terephthalic (1), sebacic (2), adipic (3), and salicylic (4) acids in water—triethylamine mixtures at 17°C. a) Laminar flow [dashed line is cal-culated according to Eq. (2.37)]; b) turbulent flow.

Let the concentration of triethylamine in the bulk solution be c^*. Convective diffusion supplies triethylamine to the disc surface, where it reacts with the organic acid (e.g., terephthalic acid). The reaction is fast, and the surface concentration of triethylamine can thus be considered equal to zero. Soluble reaction products are carried away into the bulk solution. It is assumed that the concentration of the soluble salt is low and does not significantly affect the kinetics of processes occurring in the system.

If the concentration of triethylamine is lower than the critical value c_{cr} at which stratification occurs ($c^* < c_{cr}$), no special features are observed in the mass transfer of triethylamine to the surface. Conversely, when $c^* > c_{cr}$, the mass transfer occurs in a different way. The dependence of the concentration, viscosity, and diffusion coefficient on distance from the disc is shown in Fig. 7.14. In the region where $c^* > c_{cr}$, the diffusion coefficient is very small, and the substance is almost exclusively transferred by convection. However, at the surface of the disc, $c_s = 0$; the change from c^* to c_s occurs, according to Levich and Ivanov, as follows.

In a thin layer of solution of thickness δ_1, the concentration decreases to the critical value (from c^* to c_{cr}). Simultaneously, the diffusion coefficient of triethylamine changes from its bulk value D^* to $D = 0$. Also, the viscosity coefficient μ changes within this layer from μ^* to a value μ_{cr} corresponding to the critical concentration of the solution. Triethylamine is transferred within the distance δ_1 by convection.

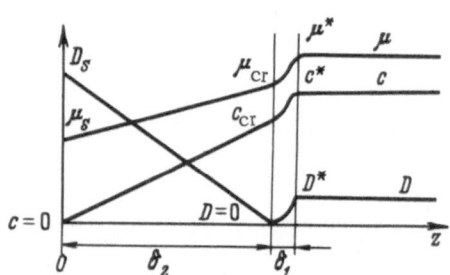

Fig. 7.14. Dependence of concentration c, viscosity μ, and diffusion coefficient D in the critical region on the distance from the surface of the disc [44].

The concentration of triethylamine decreases upon further approach to the disc surface. Simultaneously, the viscosity of the solution decreases. However, the experimental data show (cf. Fig. 7.14) that the diffusion coefficient increases to a value D_S corresponding to low triethylamine concentration. The viscosity attains a value μ_S. In this region of thickness δ_2, mass transfer consists in convective diffusion.

The relations described above are too complex for analytical solution of the problem. Therefore, Levich and Ivanov made some simplifying assumptions. Since the thickness δ_1 is small, and the mass transfer therein consists practically in convection only, the region can be neglected. In region δ_2, D and μ are assumed to depend linearly on triethylamine concentration. Then the functions introduced in expressions (7.5) can be approximated as follows:

$$N = 1, \qquad \Delta = 1 \qquad \text{when } z > \delta_2,$$
$$N = \mu_{cr}/\mu^* - [(\mu_{cr} - \mu_S)/\mu^*] \ (1 - c/c_{cr});$$
$$\Delta = D_S/D^* \ (1 - c/c_{cr}) \quad \text{when } z < \delta_2, \quad R = 1 \quad \text{when } z \geqslant 0.$$

Substitution of functions R, N, and Δ into Eq. (7.9) gives a set of equations describing the liquid motion and mass transfer to the rotating disc in the critical region.

Details of the derivations of the expressions given below can be found in the original papers of Levich and Ivanov [44, 75]. The diffusion flux density j to the disc surface is expressed by

$$j = A c_{cr} D_S \sqrt{\omega \rho^*/\mu_S}. \tag{7.14}$$

The dimensionless coefficient A was calculated with respect to the experiments of Krichevskii and Tsekhanskaya [74]. Its dependence on the triethylamine concentration c* in the bulk was found to be as follows:

c^*, mg/cm^2	100	200	300	400
A	6.06	6.61	6.74	6.96

Thus, the diffusion flux j depends little on the bulk concentration of triethylamine c* and is proportional to $\omega^{1/2}$.

The dependence of j on concentration c* is shown in Fig. 7.15. When c* < c_{cr} (i.e., in the precritical concentration range), the diffusion flux j is proportional to c*, whereas in the postcritical concentration range (c* > c_{cr}), the flux is proportional to c_{cr} and depends little on c*. Physically, this unusual dependence of j on c* is due to the fact that in the postcritical concentration range the rate of mass transfer is controlled mainly by diffusion in the δ_2 layer. The state of this layer depends only on the value of c_{cr} and depends little on c*.

§ 7.6. Thermal Diffusion near the Surface of a Rotating Disc

It was shown in § 2.1 that one of the possible reasons for the appearance of a flux can be a temperature difference between two points of the medium. This phenomenon is called thermal diffusion, or (in the case of condensed systems) the Ludwig—Soret effect.

Thermal diffusion was discovered in electrolyte solutions by Ludwig [76] in 1876. Later, thermal diffusion (particularly in the gas phase) was extensively applied as a method for separating different substances and isotopes [77, 78].

The design of a separation column requires a knowledge of the thermal diffusion coefficient D_T. Measurements of this quantity are extremely complex. Various methods for the experimental determination of D_T are described in the literature [78]. The majority of them, however, suffer from two considerable deficien-

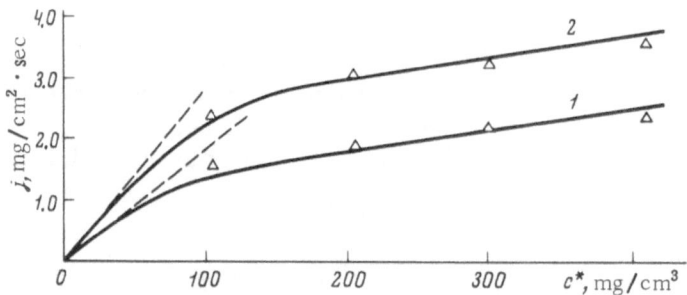

Fig. 7.15. Comparison of calculated [Eq. (7.14)] (triangles) and experimentally measured (solid lines) diffusion fluxes j in the critical region [74]. Dissolution of terephthalic acid in water—triethylamine mixtures at 17°C, laminar flow. 1) 500 rpm; 2) 1000 rpm. The dashed line is calculated according to (2.37).

cies: 1) significant effects of natural convection accompany the presence of a temperature gradient, and 2) the theoretical interpretation of the experimental data is complex.

The advantages of the rotating disc method allowed Levich and Baranovski to suggest, in 1957, an application of this method to studies of thermal diffusion. The theory of thermal diffusion processes near the surface of a rotating disc was discussed by Suponitskii [79], and for the case of electrolytic solutions, by Levich, Markin, and Chirkov [80].

Consider a disc electrode rotating with a uniform speed in a solution of concentration c* and temperature T*. The disc temperature T_S is assumed to differ from T*. It is also assumed that the dissolved reactant undergoes an electrode reaction at the disc at a rate depending on the surface concentration c_S. The simultaneous gradients of concentration and temperature near the surface of the disc result in mass and heat transfer.

When thermal diffusion is significant, the flux of the reactant is given by [81]

$$\mathbf{j}_c = vc - D \operatorname{grad} c - D_T c\,(1 - c)\operatorname{grad} T. \qquad (7.15)$$

The last term in Eq. (7.15) describes the influence of the Ludwig— Soret effect. The concentration gradient at the surface of the disc results in an additional heat effect (diffusion thermal effect, or Dufour effect). The heat flux to the disc surface is given by

$$\mathbf{j}_q = vT - \varkappa_T \operatorname{grad} T - \varkappa_D \operatorname{grad} c, \qquad (7.16)$$

where $\varkappa_T$ is the thermal conductivity and $\varkappa_D$ is a coefficient which characterizes the magnitude of the diffusion thermal effect. Numerical values of $\varkappa_D$ are little known at present; however, it can be assumed, for a number of reasons [81], that the diffusion thermal effect is extremely small in liquids. Therefore, it is assumed further that $\varkappa_D = 0$.

Substitution of Eqs. (7.15) and (7.16) into the corresponding conservation equations gives expressions describing the variations of concentration and temperature near the disc surface. Owing to the uniformity of the disc surface, both temperature and concentration depend only on the distance from the surface of the disc:

$$w \frac{dc}{dz} = D \frac{d^2c}{dz^2} + Ds \frac{d}{dz}\left[c\,(1-c)\frac{dT}{dz}\right], \qquad (7.17a)$$

$$w \frac{dT}{dz} = \varkappa_T \frac{d^2T}{dz^2} . \qquad (7.17b)$$

The thermal diffusion coefficient D_T is replaced here by the Soret coefficient s, defined by

$$D_T = Ds. \qquad (7.18)$$

The boundary conditions for Eqs. (7.17) are as follows:

for $z \to \infty$ $\qquad\qquad c \to c^*, \qquad T \to T^*, \qquad (7.19a)$

for $z = 0$ $\qquad D\left(\frac{dc}{dz}\right)_S + Dsc_S(1-c_S)\left(\frac{dT}{dz}\right)_S = k\,(T_S)\,f\,(c_S),$

$$T = T_S, \qquad (7.19b)$$

where $k(T_S)\,f\,(c_S)$ is the rate of the electrode process.

It can easily be seen that, within the framework of the assumption made, the problem of temperature distribution becomes independent of the mass-transfer problem.

The solution of the thermal problem is similar to that of the concentration distribution near the rotating disc, considered in § 2.3. The temperature distribution is therefore of the form

$$T\,(z) = (T^* - T_S)\frac{\displaystyle\int_0^z \exp\left[\frac{1}{\varkappa_T}\int_0^\xi wdz\right]d\xi}{\displaystyle\int_0^\infty \exp\left[\frac{1}{\varkappa_T}\int_0^\xi wdz\right]d\xi} + T_S. \qquad (7.20)$$

Substitution of Eq. (7.20) into Eq. (7.17a) gives a relation describing the concentration distribution of the transferred species. The latter equation has been solved for two special cases [79, 80]. Some of the results obtained are described below.

It should be mentioned first that, in aqueous solutions, the Soret coefficient s is usually small. Therefore, the last term on the right side of Eq. (7.17a) which characterizes the thermal diffusion effect can be considered to be small. This validates the use of the method of consecutive approximations used in the solution of Eq. (7.17a). Using this method, Suponitskii [79] obtained the concentration profile near an inert disc (disc surface impermeable for the transferred species, i.e., $k = 0$).

Levich, Markin, and Chirkov [80] considered a practically more important case of the transferred species undergoing a heterogeneous chemical or electrochemical reaction of the first order $[f(c_S) = c_S]$ at the disc surface. These authors restricted their case to dilute solutions and obtained the following expression describing the effect of thermal diffusion on the disc current:

$$i = i_k \left\{ 1 + s(T^* - T_S) \left[\frac{c_S}{c^*} \left(\frac{\delta_d}{\delta_T} - \frac{D}{\varkappa_T} \right) + \frac{c^* - c_S}{c^*} \left(\frac{v\gamma}{\omega \delta_T \delta_d} - \frac{D}{\varkappa_T} \right) \right] \right\}, \quad (7.21)$$

where δ_d and δ_T refer to the diffusion and thermal boundary layers, respectively, v is the viscosity of the solution, ω is the rotation speed, c_S is the surface concentration of the transferred species equal to [cf. Eq. (3.27)]

$$c_S = \frac{c^*}{1 + \frac{\delta_d}{D} k(T_S)}, \quad (7.22)$$

and $i_k = nFk(T_S)c_S$ is the current density at the rotating disc under isothermal conditions. The numerical coefficient γ in Eq. (7.21) is given by (according to [80]):

$$\gamma = 1.47 (D/v)^{1/3} + 0.96 (D/v). \quad (7.23)$$

When the electrode reaction is slow $[k(T_S) \ll D/\delta_d]$, the surface concentration differs little from that in the bulk. The current then is given by

$$i = nFk(T_S) c^* [1 + s(T^* - T_S) (\delta_d/\delta_T - D/\varkappa_T)]. \quad (7.24)$$

Conversely, when the electrode reaction is fast $[k(T_S) \gg D/\delta_d]$ and $c_S = 0$, the disc current is

$$i = i_d \left[1 + s(T^* - T_S) \left(\frac{v\gamma}{\omega \delta_T \delta_d} - \frac{D}{\varkappa_T} \right) \right]. \quad (7.25)$$

Equation (7.25) describing the change of the diffusion limiting current due to thermal diffusion can conveniently be used to determine the Soret coefficient s. Under normal conditions in aqueous solutions $\nu/\varkappa_T = 6.7$, $\nu/D \simeq 10^3$, $\delta_d = 0.19(\nu/\omega)^{1/2}$, $\delta_T = 1.64(\nu/\omega)^{1/2}$, and $\gamma = 0.016$. Therefore,

$$i = i_d [1 + s(T^* - T_s)0.3]. \qquad (7.26)$$

Since $T = T^* - T_S$ does not exceed a few tens of degrees and s is estimated to be of the order of 10^{-2} deg^{-1}, the contribution to the limiting current due to thermal diffusion is of the order only of a few percent. Thus, in spite of the simple theoretical interpretation of the Ludwig–Soret effect at a rotating disc, the measurement of the absolute value of this effect constitutes, even in the rotating disc system, a rather difficult, although realizable, experimental problem.

Two further comments should be given at the end of this section. All calculations described in the latter have been made with the assumption that diffusion coefficients, viscosity, thermal conductivity, etc. were independent of the concentration of the transferred species and temperature. However, such dependence (particularly the temperature dependence of the coefficients listed) may considerably influence cross effects similar to thermal diffusion discussed above.

The steady state is rather difficult to attain under conditions of measurements of the Soret coefficient by standard methods [78]. This is connected with the fact that concentration and temperature gradients extend over distances of the order of 10 cm or more. Therefore the time required to reach a stationary concentration distribution is 20-100 days, which is extremely inconvenient from the experimental point of view. The application of a rotating disc at which the concentration gradient exists in the diffusion boundary layer only ($\delta_d \approx 10^{-3}$ cm) considerably decreases the relaxation time of the system.

Measurements carried out in cells with a temperature gradient between the working electrode and solution necessitate consideration of the thermoelectromotive force. The theory and method of measurement of the latter at a rotating disc have been discussed by Sundheim and Sauerwein [82].

References

1. Yu. V. Pleskov, Dokl. Akad. Nauk SSSR, 117:645 (1957); Zh. Fiz. Khim., 34:623 (1960).
2. Yu. V. Pleskov and B. N. Kabanov, Dokl. Akad. Nauk SSSR, 123:884 (1958).
3. G. R. Johnson and D. R. Turner, J. Electrochem. Soc., 109:918 (1962).
4. L. I. Lyamina and K. M. Gorbunova, Élektrokhimiya, 1:41 (1965); Electro-chim. Acta, 11:457 (1966).
5. A. N. Doronin and O. L. Kabanova, Élektrokhimiya, 4:1460 (1968).
6. O. L. Kabanova, Zh. Fiz. Khim., 17:796 (1962).
7. Z. Zembura and W. Ziółkowska, Bull. Acad. Polon. Sci., Ser. Sci. Chim., 13:217 (1965); Z. Zembura and W. Głodzińska, Roczn. Chem., 40:715 (1966).
8. L. Müller, R. Wetzel, and H. Otto, J. Electroanal. Chem., 24:175 (1970).
9. R. M. Vishomirskis, Kinetics of Electrodeposition of Metals from Complex Electrolytes, Nauka, Moscow (1969); R. M. Vishomirskis et al., Tr. Akad. Nauk Litovsk. SSR, Ser. B, No. 4(16), p. 39 (1958); No. 4 (20), p. 103 (1959); No. 4 (27), p. 87 (1961); No. 2 (29), p. 19 (1962); No. 2 (33), p. 23 (1963); No. 4 (39), p. 3, 17 (1964); No. 1 (44), p. 37, 49, 83, 97 (1966); Zashchita Metallov, 1:703 (1965); O. K. Gal'dikene, I. V. Dagite, and Yu. Yu. Matulis, Studies in the Field of Electroplating, Novocherkassk (1965), p. 33.
10. L. A. Taran, S. I. Berezina, and G. S. Vozdvizhenskii, Tr. Kazansk. Khim. Technol. Inst., Khim. Nauki, 36:591 (1967); R. K. Dorsch, J. Electroanal. Chem., 21:495 (1969).
11. V. V. Gorodetskii, A. G. Alenina, and V. V. Losev, Élektrokhimiya, 5:227 (1969).
12. F. Hilbert, Naturwissenschaften, 56:215 (1969); Extended Abstracts, 20th CITCE Meeting, Strasbourg (1969), p. 161.
13. V. N. Boronenkov, O. A. Esin, and P. M. Shurygin, Dokl. Akad. Nauk SSSR, 151:872 (1963); Zh. Fiz. Khim., 38:1148 (1964); Izv. Vuzov. Tsvet. Metal-lurgiya, No. 3, p. 45 (1964).
14. M. S. Shapnik, K. A. Zinkicheva, and N. V. Gudin, Zaschchita Metallov, 5:647 (1969).
15. O. R. Brown and H. R. Thirsk, Electrochim. Acta, 10:383 (1965).
16. N. Ibl and K. Schadegg, J. Electrochem. Soc., 114:54 (1967).
17. S. S. Kruglikov et al., Dokl. Akad. Nauk SSSR, 149:911 (1963); Tr. MKhTI im D. I. Mendeleeva, 44:74 (1963); Electrochim. Acta, 10:253 (1965); Élektrokhimiya, 2:100 (1966); Zashchita Metallov, 3:92 (1967); Progress in Science. Electrochemistry, 1965, VINITI, Moscow (1967), p. 117.
18. G. T. Rogers and K. J. Taylor, Electrochim. Acta, 8:887 (1963); 13:109 (1968).
19. I. Epelboin, M. Froment, and R. Wiart, Compt. Rend., 260:3379 (1965).
20. P. Javet, N. Ibl, and H. E. Hintermann, Electrochim. Acta, 12:781 (1967); Schweiz. Archiv. Angew. Wiss. und Technik, No. 7, p. 3 (1969).
21. S. S. Kruglikov, Yu. D. Gamburg, and N. T. Kudryavtsev, Electrochim. Acta 1967, 12:1129 (1967); Tr. MKhTI im. D. I. Mendeleeva, 54:180 (1967); Zashchita Metallov, 5:81 (1969).

22.   S. S. Kruglikov, Yu. I. Sinyakov, and N. T. Kudryavtsev, Tr. MKhTI im.
       D. I. Mendeleeva, 49:130 (1965).

23.   Z. Zembura, Bull. Acad. Polon. Sci., Ser. Sci. Chim., 11:271 (1963).

24.   M. Daguenet, I. Epelboin, and M. Froment, Compt. Rend., 258:3694 (1964).

25.   L. Kiss and A. Elek, Magyar. Kem. Folyoirat, 72:159 (1966).

26.   R. D. Amstrong and G. M. Bulman, J. Electroanal. Chem., 25:121 (1970).

27.   E. S. Varenko, V. P. Galushka, V. N. Kovtun, P. M. Fedash, and Yu. M.
       Loshkarev, Zashchita Metallov, 6:103 (1970).

28.   V. N. Boronenkov, O. A. Esin, and P. M. Shurygin, Élektrokhimiya, 1:592
       (1965); V. N. Boronenkov, O. A. Esin, G. A. Toporishchev, and A. S.
       Churkin, Izv. Vuzov, Chernaya Metallurgiya, No. 3, p. 10 (1967).

29.   A. Caprani, I. Epelboin, and Ph. Morel, Extended Abstracts, 20th CITCE
       Meeting, Strasbourg (1969), p. 206; Compt. Rend., Ser. C, 269:1087 (1969).

30.   I. Epelboin, M. Keddam, and M. Froment, Compt. Rend., 259:137 (1964);
       260:5534 (1965).

31.   T. I. Popova, V. S. Bagotskii, and B. N. Kabanov, Dokl. Akad. Nauk SSSR,
       132:639 (1960).

32.   R. Memming, Surface Sci., 4:109 (1966).

33.   A. D. Davydov, V. D. Kashcheev, and B. N. Kabanov, Élektrokhimiya, 5:221
       (1969).

34.   U. E. Ebersbach, K. Schwabe, and P. König, Electochim. Acta, 14:773 (1969).

35.   P. M. Fedash, Izv. Kazansk. Filiala Akad. Nauk SSSR, Ser. Khim., No. 5,
       p. 91 (1959).

36.   G. P. Dezider'ev, S. I. Berezina, and G. A. Gorbachuk, Izv. Kazansk. Filiala
       Akad. Nauk SSSR, Ser. Khim., No. 5, p. 99 (1959).

37.   A. N. Frumkin and V. G. Levich, Zh. Fiz. Khim., 15:748 (1941); A. N. Frum-
       kin, Zh. Fiz. Khim., 23:1477 (1949); N. D. Tomashov, Corrosion of Metals
       with Oxygen Depolarization, Izd. AN SSSR, Moscow (1947).

38.   Ya. V. Durdin and Z. U. Dukhnyakova, Collection of Papers on General
       Chemistry, Vol. 1, Izd. AN SSSR, Moscow—Leningrad (1953), p. 157.

39.   Z. Zembura, Corrosion Sci., 8:703 (1968).

40.   Z. Zembura and A. Fulinski, Electrochim. Acta, 10:859 (1965).

41.   Z. Zembura and W. Glodzinska, Roczn. Chem., 40:911 (1966); 42:1525 (1968).

42.   J. Sedzimir, Z. Zembura, and W. Ziolkowska, Transactions of the 3rd Inter-
       national Congress on Metal Corrosion, Vol. 1, Mir, Moscow (1968), p. 207.

43.   D. P. Gregory and A. C. Riddiford, J. Chem. Soc., 1956:3756.

44.   V. G. Levich, Physicochemical Hydrodynamics, Prentice Hall, Inc. (1962).

45.   R. Landsberg, K. Kresse, D. Molch, and W. Geissler, Z. Phys. Chem., 227:401
       (1964).

46.   D. P. Gregory and A. C. Riddiford, J. Electrochem. Soc., 107:950 (1960).

47.   L. A. Poluboyartseva, P. I. Zarubin, and E. A. Sidel'nikova, Élektrokhimiya,
       3:878 (1967).

48.   I. I. Il'yashevich, I. A. Kosnareva, and Yu. B. Kholmanskikh, Tr. Uralsk.
       Nauchn.-Issled. i Proekt Inst. Mednoi Promyshlennosti, 7:393 (1963).

49.   E. Heitz, Electrochim. Acta, 10:49 (1965).

50. A. A. Ravdel' and G. N. Gorelik, Zh. Prikl. Khim., 37:65, 275, 522, 527(1964);
     A. A. Ravdel' and I. G. El'kin, Zh. Prikl. Khim., 40:966 (1967); A. A. Ravdel'
     and G. A. Chulanova, Zh. Prikl. Khim., 41:2451 (1968).

51. I. A. Kakovskii and Yu. B. Kholmanskikh, Izv. Akad. Nauk SSSR, Otd. Tekh.
     Nauk, Metallurgiya i Toplivo, No. 5, p. 97 (1959); No. 5, p. 207 (1960).

52. I. A. Kakovskii and A. N. Lebedev, Dokl. Akad. Nauk SSSR, 164:614 (1965);
     Tsvetnye Metally, 38(7):17 (1965); Élektrokhimiya, 2:1079 (1966).

53. I. A. Kakovskii and V. A. Svetlov, Izv. Akad. Nauk SSSR, Metally, No. 3,
     p. 50 (1965); Izv. Vuzov, Tsvetnaya Metallurgiya, No. 2, p. 68 (1967); I. A.
     Kakovskii, V. A. Svetlov, and B. D. Khalezov, Izv. Vuzov, Tsvetnaya Metal-
     lurgiya, No. 3, p. 51 (1969).

54. O. K. Shcherbakov and I. A. Kakovskii, Tsvetnye Metally, No. 6, p. 15
     (1966); Trudy Inst. Uralmekhanobr., 13:157, 167 (1967).

55. M. L. Episkoposyan and I. A. Kakovskii, Tsvetnye Metally, No. 10, p. 15
     (1965); Izv. Vuzov, Tsvetnaya Metallurgiya, No. 1, p. 34 (1966); I. A.
     Kakovskii and O. K. Scherbakov, Izv. Akad. Nauk SSSR, Metally, No. 1,
     p. 76 (1967).

56. H. R. Copson, Corrosion, 16:130 (1960).

57. I. V. Riskin and A. V. Turkovskaya, Zashchita Metallov, 5:443 (1969).

58. V. M. Novakovskii, E. P. Lapshina, and M. Sh. Blokh, Tr. Ural'sk. Khim.
     Inst., 9:101 (1961); Z. Phys. Chem., 230:313 (1966); L. A. Poluboyartseva
     and V. M. Novakovskii, Tr. Ural'sk. Khim. Inst., 9:93 (1961); L. A. Polu-
     boyartseva, P. I. Zarubin, and V. M. Novakovskii, Zh. Prikl. Khim., 36:1264
     (1963); P. I. Zarubin, L. A. Poluboyartseva, and V. M. Novakovskii, Zash-
     chita Metallov, 1:297 (1965); P. I. Zarubin, L. A. Poluboyartseva, L. N.
     Yurlova, and V. M. Novakovskii, Zashchita Metallov, 3:552 (1967).

59. V. M. Novakovskii and S. N. Fishman, Tr. Ural'sk. Khim. Inst., 9:71 (1961).

60. I. L. Rozenfel'd, Corrosion and Protection of Metals, Metallurgiya, Moscow
     (1969); I. I. Rozenfel'd and O. I. Vashkov, Zashchita Metallov, 1:70 (1965).

61. R. R. Dogonadze, V. G. Levich, and Yu. A. Chizmadzhev, Zh. Fiz. Khim.,
     34:2320 (1960).

62. A. A. Sharin and K. N. Shabalin, Zh. Fiz. Khim., 36:209 (1962).

63. F. Lohmann and W. Mehl, Ber. Bunsenges., 71:493 (1967); Electrochim.
     Acta, 13:1469 (1968).

64. V. N. Boronenkov, O. A. Esin, P. M. Shurygin, and B. A. Kukhtin, Élektrokhi-
     miya, 1:1245 (1965); Dokl. Akad. Nauk SSSR, 160:151 (1965); P. M.
     Shurygin, V. N. Boronenkov, V. I. Kryuk, and V. V. Revebtsov, Izv. Vuzov,
     Chernaya Metallurgiya, No. 2, p. 23 (1965).

65. I. A. Kakovskii and Yu. M. Potashnikov, Izv. Akad. Nauk SSSR, Otd. Tekh.
     Nauk, Metallurgiya i Toplivo, No. 3, p. 41 (1962); No. 5, p. 81 (1962);
     Dokl. Akad. Nauk SSSR, 145:1311 (1962); 158:714 (1964).

66. P. M. Shurygin, L. N. Barmin, and O. A. Esin, Izv. Vuzov, Chernaya Metal-
     lurgiya, No. 12, p. 5 (1962); P. M. Shurygin and V. N. Boronenkov, Ognechpory,
     No. 12, p. 561 (1963); V. N. Boronenkov, P. M. Shurygin, and V. D. Shantarin,
     Izv. Akad. Nauk SSSR, Metallurgiya i Gornoe Delo, No. 6, p. 97 (1964); V. D.

Shantarin, P. M. Shurygin, and V. V. Utochkin, Izv. Akad. Nauk SSSR, Metally, No. 3, p. 31 (1966).

67. A. A. Ravdel' and Yu. V. Sharikov, Zh. Prikl. Khim., 39:65, 70 (1966).

68. M. Daguenet and D. Schuhmann, Compt. Rend., 260:2811 (1965); M. Daguenet, Theses, Faculte des Sciences de l'Universite de Paris (1965).

69. J. Newman and L. Hsueh, Electrochim. Acta, 12:417, 429 (1967).

70. D. R. Olander, Int. J. Heat Mass Transfer, 5:765 (1962); Ind. Eng. Chem. Fundamentals, 6:188 (1967).

71. I. R. Krichevskii, N. E. Khazanova, and L. P. Linshits, Dokl. Akad. Nauk SSSR, 99:113 (1954); I. R. Krichevskii and Yu. V. Tsekhanskaya, Zh. Fiz. Khim., 30:2315 (1956); N. E. Khazanova and M. V. Kal'sina, Zh. Fiz. Khim., 4:43 (1961); I. R. Krichevskii, N. E. Khazanova, and Yu. V. Tsekhanskaya, Zh. Fiz. Khim., 34:1250 (1960).

72. N. E. Khazanova, Heat and Mass Transfer, Vol. 3, Nauka i Tekhnika, Minsk (1965), p. 21.

73. I. R. Krichevskii, Phase Equilibria in Solutions at High Pressures, Goskhimizdat, Moscow—Leningrad (1952); I. R. Krichevskii, N. E. Khazanova, and L. B. Linshits, Dokl. Akad. Nauk SSSR, 119:975 (1958); I. R. Krichevskii, N. E. Khazanova, and M. V. Kal'sina, Inzh. Fiz. Zh., 5:93 (1962).

74. I. R. Krichevskii and Yu. V. Tsekhanskaya, Zh. Fiz. Khim., 33:2331 (1959); I. R. Krichevskii and Yu. V. Tsekhanskaya, Trudy GIAP, 9:5 (1959).

75. Yu. B. Ivanov and V. G. Levich, Zh. Fiz. Khim., 32:592 (1958).

76. C. Ludwig, Sitzber. Akad. Wiss., Vienna, 20:539 (1856).

77. K. E. Grew and T. L. Ibbs, Thermal Diffusion in Gases, Cambridge University Press (1952).

78. H. J. C. Tyrell, Diffusion and Heat Flow in Liquids, Butterworths, London (1961).

79. A. M. Suponitskii, Zh. Prikl. Mekhan. i Tekh. Fiz., No. 2, p. 126 (1962); No. 5, p. 48 (1963).

80. V. G. Levich, V. S. Markin, and Yu. G. Chirkov, Élektrokhimiya, 1:1416 (1965).

81. S. R. De Groot, Thermodynamics of Irreversible Processes, GITTL, Moscow (1956).

82. B. R. Sundheim and W. Sauerwein, J. Phys. Chem., 69:4042 (1965); Rev. Scient. Instrum., 38:229 (1967).

Chapter 8

Rotating Ring-Disc Electrode

One of the most suitable methods of investigating the kinetics of multistep electrode reactions is that of the rotating ring–disc electrode. It was first described by Frumkin and Nekrasov and the quantitative theory was developed by Levich and Ivanov [1–3]. In recent years, a series of calculations connected with practical applications of a rotating ring–disc electrode were made by Bruckenstein, Albery, and others [4–14]. The method is intended for studies of multistep electrode processes (proceeding with formation of unstable intermediates) and of chemical (bulk) reactions of the products of electrode processes, for analytical determinations, as well as for investigation of adsorption processes accompanying electrode reactions.

The rotating ring–disc electrode (Fig. 8.1) consists of three concentric parts: the central disc electrode $\mathscr{D}$ (r_{10} in radius), the ring electrode $\mathscr{R}$ (internal radius r_{20}, external radius r_{30}), and a thin insulating separator, $\mathscr{G}$ between them (from $r_{10}-r_{20}$). The working surfaces of both electrodes and the surface of the insulator lie in the same plane, forming a single flat surface.

Rotation of the electrode results in convective motion of the electrolyte. The important components of this motion are the normal (to the surface) and radial fluxes. The general hydrodynamic picture does not differ from that discussed in § 1.2.

In order to understand the physical principles underlying the method of the rotating ring–disc electrode (RDR), the case of a

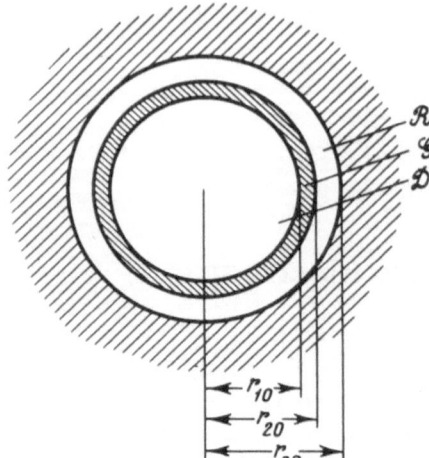

Fig. 8.1. Schematic diagram of a rotating ring-disc electrode. $\mathcal{D}$) disc; $\mathcal{I}$) insulating separator; $\mathcal{R}$) ring.

two-step electrode reaction proceeding at the disc,

$$A + n_2 e^- \xrightarrow{k_2} B + n_3 e^- \xrightarrow{k_3} P_2, \tag{8.1}$$

will first be discussed.

It is assumed for simplicity that the intermediate B is stable. It is obvious that if the rate constant of the second step (k_3) is not sufficiently high, the intermediate B formed in the first reaction step cannot be fully reduced at the disc in the second step. The nonreduced part of the intermediate is carried away by the radial flux to the disc edges. Simultaneously, B diffuses into the bulk solution. If the electrode (the ring) capable of detecting the presence of B is placed in its way, quantitative determination is possible of that fraction of B which did not enter the second step of the reaction (8.1). At the same time, the rate constant k_3 can be found. For this purpose, it is first necessary to calculate the fraction of B arriving at the ring, and also the dependence of this fraction on (a) the rates of electrochemical reactions at the disc and at the ring, and (b) the rotational speed of the electrode.

Evidently, the total amount of B carried away from the disc in the solution flux could be collected at the ring if the latter were sufficiently large and if the discharge rate of B on the ring were sufficiently high.

In practice, however, the surface area of the ring electrode is relatively small and, at best, some 40% of the intermediate B carried away from the disc can be detected.

Furthermore, the intermediate formed in the first step of reaction (8.1) is often unstable and undergoes chemical reactions in the bulk on its way to the ring. Thus, the fraction arriving at ring can be further decreased.

The optimum dimensions of the disc $\mathscr{D}$, the separator $\mathscr{G}$, and the ring $\mathscr{R}$ are determined by the individual experimental problem to which the method is applied. In particular, if the basic aim of the experiment is to study the nature of unstable intermediates, the disc electrode area should be increased and the width of the separator decreased [thus decreasing the time of the transfer of intermediates from the disc to the ring (the so-called transit time) and decreasing the probability of their decomposition on their way to the ring]. However, use of an oversize disc electrode can result in turbulence; under these conditions, application of the rotating ring-disc electrode becomes pointless. The above methodological considerations must be combined with the purely electrotechnical demands of sufficiently good insulation between the disc and ring electrodes.

§ 8.1. Calculation of the Current

at the Ring Electrode

Theoretical calculations of the ring current include the determination of current detected at the ring electrode for the given r_{10}, r_{20}, and r_{30} values, as well as determination of the dependence of this current on the kinetics of the disc electrode reaction.

The problem was solved first by Levich and Ivanov [3, 15]. Albery and Bruckenstein [4, 5] repeated the calculation in 1966 using a different method of solving the convective mass-transport equation. Subsequently, Albery, Bruckenstein, and others [7-12] took into account the possible chemical reactions of the intermediate in the solution and the possible nonstationary phenomena accompanying polarization of the disc electrode.

The calculation of the ring current, according to the method of Albery and Bruckenstein, is presented below for the simplest case, i.e., where a stable product is formed at the disc at a constant rate j_0.

The convective mass transport in the vicinity of a rotating disc (cf. § 2.1) is described by

$$u \frac{\partial c}{\partial r} + w \frac{\partial c}{\partial z} = D \frac{\partial^2 c}{\partial z^2}, \tag{8.2}$$

where c is the concentration of the product formed at the disc; u and w are the radial and axial components of the solution velocity, respectively. It was shown in § 1.2 that near the disc surface

$$u = a_0 r z \omega^{3/2} \nu^{-1/2}, \qquad w = -a_0 z^2 \omega^{3/2} \nu^{-1/2}, \tag{8.3}$$

where the factor $a_0 = 0.51$.

The velocity components u and w are related by the continuity equation

$$\frac{\partial u}{\partial r} + \frac{\partial w}{\partial z} = -\frac{u}{r}. \tag{8.4}$$

Introducing the function

$$\psi = \frac{a_0}{2} r^2 z^2 \omega^{3/2} \nu^{-1/2}, \tag{8.5}$$

we obtain the u vs w relation in the form

$$ru = \frac{\partial \psi}{\partial z}, \qquad rw = -\frac{\partial \psi}{\partial r}. \tag{8.6}$$

The variables r and z are now replaced by the new variables r and ψ.† Equation (8.2) is then transformed into

$$\frac{1}{r^2} \frac{\partial c}{\partial r} = D \frac{\partial}{\partial \psi}\left(u \frac{\partial c}{\partial \psi}\right). \tag{8.7}$$

---

†r and ψ coordinates are usually called "Mises coordinates." The standard method of substitution of Mises variables is described in a general way in [15] and in connection with electrochemical problems in [16].

The radial velocity component in Eq. (8.7) depends only on the variable ψ. It follows from Eqs. (8.3) and (8.5) that

$$u = (2a_0\omega^{3/2}\nu^{-1/2})\,\psi^{1/2}. \tag{8.8}$$

Replacing again the variables r and ψ by the variables η and ξ defined by

$$\eta = \left(\frac{2}{9}\,a_0\omega^{3/2}\nu^{-1/2}\right)^{1/2} r^3, \tag{8.9a}$$

$$\xi = \psi^{1/2}, \tag{8.9b}$$

we can write Eq. (8.7) in the form

$$\frac{\partial c}{\partial \eta} = \frac{D}{4\xi}\frac{\partial^2 c}{\partial \xi^2} \tag{8.10}$$

The above equation is solved below.

The boundary conditions for Eq. (8.2) are formulated as follows.

At the surface of the uniformly accessible central disc a stable product is formed at a constant rate, j_0, i.e.,

$(\mathscr{D})$ $$D\left(\frac{\partial c}{\partial z}\right)_S = j_0 \qquad \text{for } z = 0, \quad 0 < r < r_{10}. \tag{8.11}$$

No discharge occurs at the insulated part of the electrode, i.e.,

$(\mathscr{G})$ $$D\left(\frac{\partial c}{\partial z}\right)_S = 0 \qquad \text{for } z = 0, \quad r_{10} < r < r_{20}. \tag{8.12}$$

The boundary conditions (8.11) and (8.12) can be combined into one boundary condition

$$D\left(\frac{\partial c}{\partial z}\right)_S = f(r) \qquad \text{for } z = 0, \quad 0 < r < r_{20}, \tag{8.13a}$$

where $f(r)$ is a step function:

$$f(r) = \begin{cases} j_0 & \text{for } 0 < r < r_{10} \tag{8.13b} \\ 0 & \text{for } r_{10} < r < r_{20}. \tag{8.13c} \end{cases}$$

It is assumed for simplicity that the species formed at the disc undergoes a fast discharge at the ring electrode,† i.e.,

$$c = 0 \quad \text{for } z = 0, \quad r_{20} < r < r_{30}. \tag{8.14}$$

The product is absent in the bulk; therefore,

$$c \to 0 \quad \text{as } z \to \infty, \quad r \geqslant 0. \tag{8.15}$$

In addition to the above conditions, a further condition will be required later; it is

$$c = 0 \quad \text{for } r = 0, \quad z > 0. \tag{8.16}$$

The relation (8.16) corresponds to the condition that the solution arriving at the rotating electrode is not depleted.

The new variables η and ξ are also substituted in the boundary conditions, utilizing the fact that the points at the disc surface with coordinates $r = 0$, $r = r_{20}$, $r = r_{30}$ correspond to the points

$$\eta_0 = 0, \quad \eta_{10} = \left(\frac{2}{9} a_0 \omega^{3/2} \nu^{-1/2}\right)^{1/2} r_{10}^3, \quad \eta_{20} = \left(\frac{2}{9} a_0 \omega^{3/2} \nu^{-1/2}\right)^{1/2} r_{20}^3,$$

$$\eta_{30} = \left(\frac{2}{9} a_0 \omega^{3/2} \nu^{-1/2}\right)^{1/2} r_{30}^3. \tag{8.17}$$

The plane $z = 0$ is represented in the new coordinates by the plane $\xi = 0$; also, when $z \to \infty$, $\xi \to \infty$.

It can be easily shown that the mass flux at the electrode surface in the axial direction is given in terms of the new variables by

$$j = D\left(\frac{\partial c}{\partial z}\right)_S = \frac{3D}{2}\left(\frac{2}{9} a_0 \omega^{3/2} \nu^{-1/2}\right)^{1/3} \eta^{1/3} \left(\frac{\partial c}{\partial \xi}\right)_S. \tag{8.18}$$

---

†It must be mentioned that neglect of the kinetics of the heterogeneous ring reaction is an approximation. In fact, as will be shown in the text following (and as was demonstrated in §2.8), the internal edge of the ring (at $r = r_{20}$) is exposed to a continuously arriving nondepleted solution. Therefore, the current density at $r = r_{20}$ turns out to be infinite. This indicates the necessity of taking into account the finite rate of the heterogeneous ring reaction. However, if this rate is very high, the error arising from neglect of the kinetics of the ring reaction is insignificant.

The boundary conditions now have the following form:

$(\mathscr{D})$ $\qquad \left(\dfrac{\partial c}{\partial \xi}\right)_s = \dfrac{2j_0}{3D}\left[\dfrac{9\nu^{1/2}}{2a_0\omega^{3/2}}\right]^{1/3}\eta^{-1/3}$ $\quad$ for $\xi = 0$, $\quad 0 < \eta < \eta_{10}$, $\qquad$ (8.19)

$(\mathscr{G})$ $\qquad\qquad\qquad \left(\dfrac{\partial c}{\partial \xi}\right)_s = 0$ $\quad$ for $\xi = 0$, $\quad \eta_{10} < \eta < \eta_{20}$, $\qquad$ (8.20)

$(\mathscr{R})$ $\qquad\qquad\qquad\quad c = 0$ $\quad$ for $\xi = 0$, $\quad \eta_{20} < \eta < \eta_{30}$, $\qquad$ (8.21)

$\qquad\qquad\qquad\qquad c \to 0$ $\quad$ as $\xi \to \infty$, $\quad \eta > 0$, $\qquad$ (8.22)

$\qquad\qquad\qquad\qquad c = 0$ $\quad$ for $\xi > 0$, $\quad \eta = 0$. $\qquad$ (8.23)

Conditions (8.19) and (8.20) can again be combined into the condition

$$\left(\dfrac{\partial c}{\partial \xi}\right)_s = \begin{cases} \dfrac{2j_0}{3D}\left[\dfrac{9\nu^{1/2}}{2a_0\omega^{3/2}}\right]^{1/3}\eta^{-1/3} & \text{for } \xi = 0, \quad 0 < \eta < \eta_{10}, \quad (8.24\text{a}) \\[2mm] 0 & \text{for } \xi = 0, \quad \eta_{10} < \eta < \eta_{20}. \quad (8.24\text{b}) \end{cases}$$

Thus, the calculation of the ring current is reduced to the solution of Eq. (8.10) with the boundary conditions (8.19)–(8.23). It is convenient to solve the problem in two stages. Initially, the product distribution in regions $\mathscr{D}$ and $\mathscr{G}$ is found. It is convenient to use here the boundary condition (8.24a)–(8.24b). Thereafter, using the solution obtained in the first stage, we find the concentration distribution in region $\mathscr{R}$ (ring electrode), or, what is much simpler, the ring current.

Solution in Regions $\mathscr{D}$ and $\mathscr{G}$. The Laplace transform (with respect to the variable η) of the function $\bar{c}(p, \xi)$ is given by

$$\bar{c}(p, \xi) = p \int_0^\infty \exp(-p\eta)\, c(\eta, \xi)\, d\eta.$$

Equation (8.10) (with the boundary condition (8.23) now becomes

$$p\bar{c} = \dfrac{D}{4\xi}\dfrac{d^2\bar{c}}{d\xi^2}. \qquad (8.25)$$

This equation must be supplemented by the boundary conditions $\bar{c} \to 0$ as $\xi \to \infty$ and $(d\bar{c}/d\xi)_S = \bar{f}(p)$, which correspond to the conditions (8.19), (8.20), and (8.22): $\bar{f}(p)$ is a Laplace transform of the step function (8.24a)–(8.24b) and is equal to [17]

$$\bar{f}(p) = \frac{2j_0}{3D}\left(\frac{9v^{1/2}}{2a_0\omega^{3/2}}\right)^{1/3} p^{1/3}\gamma\left(^2/_3,\ \eta_{10}p\right), \qquad (8.26)$$

where $\gamma\left(2/3,\ \eta_{10}p\right)$ is an incomplete gamma function.

The general solution of Eq. (8.25) can be written in terms of Airy functions of the first and second kind [18]:

$$\bar{c}(p,\ \xi) = A_1 Ai\left[\left(\frac{4p}{D}\right)^{1/3}\xi\right] + A_2 Bi\left[\left(\frac{4p}{D}\right)^{1/3}\xi\right]. \qquad (8.27)$$

The function $Bi[(4p/D)^{1/3}\xi]$ tends to infinity as $\xi \to \infty$ and does not satisfy the required behavior of $\bar{c}(p,\ \xi)$ as $\xi \to \infty$. Therefore $A_2 = 0$. Using the boundary condition at $\xi = 0$ and Eq. (8.26) for $\bar{f}(p)$, A_1 can be found. The final result is

$$\bar{c}(p,\ \xi) = \frac{2j_0}{3D}\left(\frac{9v^{1/2}}{2a_0\omega^{3/2}}\right)^{1/3}\left(\frac{D}{4}\right)^{1/3}\frac{Ai\ [(4p/D)^{1/3}\xi]}{Ai'\ (0)}\ \gamma\left(^2/_3,\ \eta_{10}p\right). \qquad (8.28)$$

The distribution of the product concentration in regions $\mathcal{D}$ and $\mathcal{G}$ can be found by performing the inverse transformation of the Laplace expression (8.28). The recovery of the original function is a complex mathematical problem. Albery and Bruckenstein [5] present the original function in the form of a double infinite series. It will be assumed for the moment that the problem is solved and that an analytical solution has been found expressing the product concentration $c(\xi,\ \eta)$.

At the disc surface, for $\xi = 0$, Eq. (8.28) simplifies to

$$\bar{c}(p,\ \xi = 0) = \frac{2j_0}{3D}\left(\frac{9v^{1/2}}{2a_0\omega^{3/2}}\right)^{1/3}\left(\frac{D}{4}\right)^{1/3}\frac{Ai\ (0)}{Ai'\ (0)}\gamma\left(^2/_3,\ \eta_{10}p\right). \qquad (8.29)$$

The original function is given by [17], viz.,

$$c(\xi = 0,\ \eta) = -\frac{2j_0}{3D}\left(\frac{9v^{1/2}}{2a_0\omega^{3/2}}\right)^{1/3}\left(\frac{D}{4}\right)^{1/3}\times$$

$$\times \frac{1}{3^{1/3}\Gamma\,(2/3)} \begin{cases} \displaystyle\int_0^{\eta} \frac{d\zeta}{\zeta^{1/3}\,(\eta - \zeta)^{2/3}}, & \eta < \eta_{10} \qquad (8.30a) \\[3em] \displaystyle\int_0^{\eta_{10}} \frac{d\zeta}{\zeta^{1/3}\,(\eta - \zeta)^{2/3}}, & \eta > \eta_{10}. \qquad (8.30b) \end{cases}$$

Expression (8.30a) describes the distribution of the product concentration at the surface of the disc, $\mathscr{D}$. The integral in Eq. (8.30a) can be expressed by Γ-functions [19]. After suitable substitution, the product concentration in region $\mathscr{D}$ is expressed by

$$c\,(\xi = 0,\, \eta) = -\frac{2j_0}{3D}\left(\frac{9\nu^{1/2}}{2a_0\omega^{3/2}}\right)^{1/3}\left(\frac{D}{4}\right)^{1/3}\frac{\Gamma\,(1/3)}{3^{1/3}}. \qquad (8.31)$$

It can easily be shown that Eq. (8.31) has a form corresponding to the expression for a uniform rotating disc:

$$c_S = \frac{j_0\delta_d}{D}, \qquad (8.32)$$

i.e., the surface concentration of the intermediate product is uniform in the whole $\mathscr{D}$ region (the surface $\mathscr{D}$ is uniformly accessible), and the diffusion-layer thickness δ_d is as described by Eq. (2.20).

Equation (8.30b) describes the concentration distribution at the surface of the insulating separator, $\mathscr{G}$. In particular, the surface concentration of the intermediate product at the external edge of the separator (for $\eta = \eta_{20}$) is given by

$$c\,(\xi = 0,\, \eta = \eta_{20}) = -\frac{2j_0}{3D}\left(\frac{9\nu^{1/2}}{2a_0\omega^{3/2}}\right)^{1/3}\left(\frac{D}{4}\right)^{1/3}\frac{1}{3^{1/3}\Gamma\,(2/3)}\int_0^{\eta_{10}}\frac{d\zeta}{\zeta^{1/3}(\eta_{20} - \zeta)^{2/3}}. \qquad (8.33)$$

The diffusional problem in the region of the ring electrode will now be solved.

Solution in Region $\mathscr{R}$. In order to find the product distribution at the ring electrode, it is necessary to solve Eq. (8.10) with the boundary conditions (8.21) and (8.22). However, the initial condition (8.23) should be replaced by a new one. In fact, the liquid flux arrives at the ring electrode from region $\mathscr{G}$, i.e., from the insulating separator. The product distribution in the arriving liquid is determined by $c(\xi, \eta)$ at the external edge of the insulating separator. Therefore, Eq. (8.23) should be replaced by the condition

$$c(\xi, \eta = \eta_{20}) = c_{\mathscr{g}}(\xi, \eta_{20}),\qquad(8.34)$$

where $c_{\mathscr{g}}(\xi, \eta_{20})$ is the concentration distribution found by solution of the first part of the problem.

Applying the Laplace transformation method to Eq. (8.10), we obtain, instead of Eq. (8.25), an inhomogeneous equation

$$p\,[\bar{c}_2 - c_{\mathscr{g}}(\xi, \eta_{20})] = \frac{D}{4\xi}\frac{d^2\bar{c}_2}{d\xi^2},\qquad(8.35)$$

with boundary conditions

$$\bar{c}_2 \to 0 \quad \text{as } \xi \to \infty, \quad \bar{c}_2 = 0 \quad \text{for } \xi = 0.\qquad(8.36)$$

The solution of Eq. (8.35) which satisfies the boundary conditions (8.36) is in the form

$$\bar{c}_2(\xi, p) = \frac{Ai\,[(4p/D)^{1/3}\xi]}{w}\left\{\int\limits_0^{(4p/D)^{1/3}\xi} sBi(s)\,c_{\mathscr{g}}\,[(D/4p)^{1/3}s,\ \eta_{20}]\,ds - \right.$$

$$\left. - \frac{Bi(0)}{Ai(0)}\int\limits_0^{\infty} sAi(s)\,c_{\mathscr{g}}\,[(D/4p)^{1/3}s,\ \eta_{20}]\,ds\right\} + \frac{Bi\,[(4p/D)^{1/3}\xi]}{w}\times$$

$$\times \int\limits_{(4p/D)^{1/3}\xi}^{\infty} sAi(s)\,c_{\mathscr{g}}\,[(D/4p)^{1/3}s,\ \eta_{20}]\,ds,\qquad(8.37)$$

where Ai(s) and Bi(s) are Airy functions of the first and second kind, and $w = Ai(s)Bi'(s) - Bi(s)Ai'(s) = 1/\pi$ is the Wronskian.

The derivative at the ring surface can be easily found from Eq. (8.37):

$$\left(\frac{d\bar{c}_2}{d\xi}\right)_S = \frac{(4p/D)^{1/3}}{Ai(0)}\int\limits_0^{\infty} sAi(s)\,c_{\mathscr{g}}\,[(D/4p)^{1/3}s,\ \eta_{20}]\,ds.\qquad(8.38)$$

The first two of the series of papers by Albery and Bruckenstein [4, 5] concern the calculation of the integral in Eq. (8.38) and the recovery of the original expression. The equations obtained by the latter authors are very cumbersome. Therefore, further discussion will be restricted to one special case. General results are tabulated in Table 8.1.

Consider the case of a very thin ring. In this situation the magnitude of flux is of interest near the boundary between the separator and the ring only, i.e., for $\eta \approx \eta_{20}$. These values of the variable correspond to high values of the parameter p. It can be easily shown that in this case the function $c_{\#}[(D/4p)^{1/3}s, \eta_{20}]$ in Eqs. (8.37) and (8.38) can be replaced by

$$c_{\#}[(D/4p)^{1/3}s, \eta_{20}] \approx c_{\#}(0, \eta_{20}). \qquad (8.39)$$

TABLE 8.1. Values of the Collection Efficiency
Coefficient N

r_{30}/r_{20}	r_{20}/r_{10}								
	1.02	1.03	1.04	1.05	1.06	1.07	1.08	1.09	1.10
1.02	0.1013	0.0976	0.0947	0.0922	0.0902	0.0884	0.0869	0.0855	0.0843
1.03	0.1293	0.1250	0.1215	0.1186	0.1162	0.1140	0.1121	0.1104	0.1089
1.04	0.1529	0.1483	0.1444	0.1412	0.1385	0.1360	0.1339	0.1320	0.1302
1.05	0.1737	0.1687	0.1647	0.1612	0.1582	0.1556	0.1533	0.1512	0.1483
1.06	0.1923	0.1872	0.1829	0.1793	0.1761	0.1733	0.1708	0.1686	0.1665
1.07	0.2092	0.2039	0.1996	0.1958	0.1925	0.1896	0.1869	0.1846	0.1824
1.08	0.2247	0.2194	0.2149	0.2110	0.2076	0.2046	0.2019	0.1994	0.1972
1.09	0.2392	0.2338	0.2292	0.2252	0.2217	0.2186	0.2158	0.2133	0.2110
1.10	0.2526	0.2412	0.2426	0.2385	0.2350	0.2318	0.2289	0.2263	0.2240
1.12	0.2772	0.2717	0.2670	0.2629	0.2593	0.2560	0.2530	0.2503	0.2479
1.14	0.2992	0.2938	0.2890	0.2849	0.2812	0.2778	0.2748	0.2720	0.2695
1.16	0.3192	0.3138	0.3090	0.3048	0.3011	0.2977	0.2947	0.2919	0.2893
1.18	0.3375	0.3321	0.3274	0.3232	0.3154	0.3161	0.3130	0.3101	0.3075
1.20	0.3544	0.3490	0.3443	0.3402	0.3364	0.3330	0.3290	0.3271	0.3245
1.22	0.3701	0.3648	0.3601	0.3560	0.3523	0.3489	0.3458	0.3429	0.3403
1.24	0.3848	0.3795	0.3749	0.3708	0.3671	0.3637	0.3606	0.3577	0.3551
1.26	0.3985	0.3933	0.3887	0.3847	0.3810	0.3776	0.3745	0.3717	0.3691
1.28	0.4115	0.4063	0.4018	0.3977	0.3941	0.3907	0.3877	0.3849	0.3822
1.30	0.4237	0.4186	0.4141	0.4101	0.4065	0.4032	0.4001	0.3973	0.3947
1.32	0.4353	0.4302	0.4258	0.4218	0.4183	0.4150	0.4119	0.4092	0.4066
1.34	0.4463	0.4413	0.4369	0.4330	0.4294	0.4262	0.4232	0.4204	0.4278
1.36	0.4567	0.4518	0.4475	0.4436	0.4402	0.4369	0.4339	0.4311	0.4286
1.38	0.4667	0.4619	0.4576	0.4538	0.4503	0.4471	0.4441	0.4414	0.4389
1.40	0.4762	0.4715	0.4673	0.4635	0.4600	0.4566	0.4539	0.4512	0.4487

Equation (8.38) is replaced now by

$$\left(\frac{d\bar{c}_2}{d\xi}\right)_s = \frac{(4p/D)^{1/3}}{Ai\,(0)}\,c_{\mathscr{D}}\,(0,\,\eta_{20}) \int_0^\infty sAi\,(s)\,ds = -\,c_{\mathscr{D}}\,(0,\,\eta_{20})\left(\frac{4p}{D}\right)^{1/3}\frac{Ai'\,(0)}{Ai\,(0)}\,,$$

(8.40)

where $c_{\mathscr{D}}\,(0,\,\eta_{20})$ is described by Eq. (8.33).

By performing the inverse transformation in Eq. (8.40) [17], it is found for the $\eta > \eta_{20}$ region that

$$\left(\frac{\partial c_2}{\partial \xi}\right)_s = -\,c_{\mathscr{D}}\,(0,\,\eta_{20})\frac{Ai'\,(0)}{Ai\,(0)}\left(\frac{4p}{D}\right)^{1/3}\frac{1}{\Gamma\,(^2/_3)\,(\eta - \eta_{20})^{1/3}}\,.$$

(8.41)

Combining Eqs. (8.18), (8.33), and (8.41), we easily obtain the following expression for the density of the diffusional flux to the ring:

$$j_{\mathscr{R}} = -\,\frac{j_0}{\Gamma\,(^1/_3)\,\Gamma\,(^2/_3)}\frac{\eta^{1/3}}{(\eta - \eta_{20})^{1/3}}\int_0^{\eta_{10}}\frac{d\zeta}{\zeta^{1/3}\,(\eta_{20} - \zeta)^{2/3}}\,.$$

(8.42)

Experimentally, only the total flux to the ring electrode can be measured. The latter can be found by integration of Eq. (8.42) between the limits r_{20} and r_{30}:

$$\bar{j}_{\mathscr{R}} = 2\pi \int_{r_{20}}^{r_{30}} j_{\mathscr{R}}\,rdr.$$

(8.43)

Substituting the value of $j_{\mathscr{R}}$ from Eq. (8.41) and taking into account the relation between the variables η and r [Eq. (8.9a)], we obtain the flux of the product to the ring electrode:

$$\bar{j}_{\mathscr{R}} = -\,\frac{j_0\pi}{\Gamma\,(^1/_3)\,\Gamma\,(^2/_3)}\int_0^{\eta_{10}}\frac{d\zeta}{\zeta^{1/3}\,(\eta_{20} - \zeta)^{2/3}}\,(r_{30}^3 - r_{20}^3)^{2/3}.$$

(8.44)

The minus sign indicates that the flux j_0 of the product directed toward the bulk solution at the disc has an opposite direction at the ring electrode, i.e., from the bulk to the ring.

A coefficient N, important in the further discussion, will now be introduced. The coefficient N is defined as follows:

$$N = -\,\bar{j}_{\mathscr{R}}/\bar{j}_{\mathscr{D}} = -\,\bar{j}_{\mathscr{R}}/\pi r_{10}^2 j_0\,,$$

(8.45)

where $\bar{j}_\mathscr{D} = \pi r_{10}^2 j_0$ is the amount of the product forming per unit time at the disc. The coefficient N, usually called the collection efficiency, indicates the fraction of the product formed at the disc which can be collected by the ring.

Combining Eqs. (8.44) and (8.45) gives

$$N = \frac{1}{\Gamma\,(1/3)\,\Gamma\,(2/3)} \int_0^{\eta_{10}} \frac{d\zeta}{\zeta^{1/3}\,(\eta_{20} - \zeta)^{2/3}}\ [(r_{30}/r_{10})^3 - (r_{20}/r_{10})^3]^{2/3}. \qquad (8.46)$$

Assuming, for simplicity, that the width of the insulating separator is small, i.e., $\eta_{10} \approx \eta_{20}$, we can reduce Eq. (8.46) to

$$N \approx [(r_{30}/r_{10})^3 - (r_{20}/r_{10})^3]^{2/3}. \qquad (8.47)$$

Equation (8.47) coincides with that of Albery and Bruckenstein for the case discussed where both the ring and separator are very narrow (cf. Eq. (8.1) in [5]).

It was mentioned already that the above calculation has a qualitative character only. In fact, calculation of the integral (8.38) involves a rather crude approximation of the integrand; replacement of Eq. (8.46) by (8.47) also results in a considerable error. Therefore, Eq. (8.47) is of a qualitative nature and is an approximation only of the character of the dependence of N on the geometric properties of the electrode. However, an important feature should be stressed: it follows from Eq. (8.47) that the collection efficiency does not depend on the rotational velocity of the electrode and is solely dependent on the dimensions of the disc, separator, and ring.

The calculation of N was first made by Levich and Ivanov [3][†] who treated the case of a narrow separator and a narrow ring electrode $(r_{20} - r_{10} \ll r_{20},\ r_{30} - r_{20} \ll r_{20})$. Their final expression is close to Eq. (8.47). However, the authors themselves stressed the low degree of accuracy of their calculations.

Albery and Bruckenstein calculated N for various r_{10}, r_{20}, and r_{30} values with high accuracy. Their N values are shown in Table 8.1. The columns in the table differ in the ratio of the external

---

[†]The integral in the equation describing N in [3] was calculated numerically. Subsequently, Bruckenstein [20] obtained an analytical expression for N.

separator radius (r_{20}) to the radius of the central disc electrode (r_{10}). Horizontal rows correspond to various dimensions of the ring electrode (r_{30}/r_{20}). Using Table 8.1, we can easily find the magnitude of N for various ring-disc systems:

§ 8.2. Quantitative Verification of the Theory

The coefficient N was first determined experimentally by Nekrasov [1, 3, 21] who studied the electrochemical reduction of quinone, of oxygen in alkaline solution, and of divalent copper in chloride solution. He measured oxidation currents for respective products at the ring electrode as a function of the disc potential. All the above reactions result in stable products which do not decompose in the solution.

The value of N calculated for platinum and gold electrodes used was 0.39.

In Fig. 8.2, a polarization curve for oxygen reduction at an amalgamated gold disc electrode is shown. In alkaline solution, the reaction proceeds in two stages:

$$O_2 + H_2O + 2e^- \rightarrow HO_2^- + OH^-,$$

$$HO_2^- + H_2O + 2e^- \rightarrow 3OH^-, \qquad\qquad (8.48)$$

represented on the polarogram by two waves. In the potential

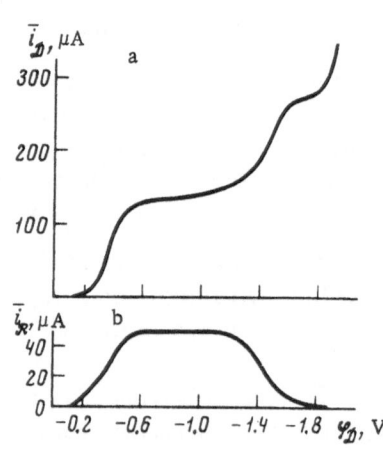

Fig. 8.2. Cathodic reduction of oxygen at a ring-disc electrode. a) Polarization curve at a gold disc in 0.1 N NaOH at 2150 rpm; b) limiting current of HO_2^- oxidation at a gold ring as a function of disc potential (vs normal calomel electrode) [1].

range of the first wave at a mercury electrode, hydrogen peroxide
is completely stable and can accumulate in the solution; in the
potential range of the second wave hydrogen perioxide can be reduced electrochemically and thus is an unstable intermediate.
Simultaneously, the limiting anodic current at the gold ring was
measured as a function of the disc potential (Fig. 8.2b). In the
absence of the cathodic disc current, and upon reaching the limiting current of the second wave, the ring current is virtually equal
to the residual current. In the region of the first wave for O_2 reduction the ring current reaches a considerable magnitude, demon-
strating the presence of hydrogen peroxide at the electrode; N
was found to be 0.37.

In the case of quinone, ~38% of hydroquinone, the stable product
of quinone reduction at the disc, was detected by oxidation at the
ring electrode.

The reduction of Cu^{2+} ions in KCl solutions proceeds in two
steps. The univalent cation, the product of the first step, can be
reoxidized at the ring. The measured N value was 0.41-0.42.

Albery and Bruckenstein [5] carried out systematic measurements of N at platinum and gold electrodes with varying r_{20}/r_{10}
and r_{30}/r_{20} ratios:

I. Reduction of Cu^{2+} at the disc (in 10^{-4} M $CuCl_2$ + 0.5 M KCl
    solution) with oxidation of Cu^+ at the ring (similar to the
    work of Nekrasov).

II. Oxidation of Br^- at the disc (in 0.1 M KBr + 1 M H_2SO_4)
    with Br_2 reduction at the ring.

III. Anodic dissolution of silver plated at the disc in 0.1 M
     $HClO_4$ with Ag^+ reduction at the ring.

TABLE 8.2. Calculated and Measured N Values

Electrode	A	B	C	D	E
r_{10}, cm	0.3480	0.3888	0.3672	0.3723	0.2500
r_{20}, cm	0.3860	0.3988	0.3763	0.3817	0.2738
r_{30}, cm	0.4325	0.4445	0.4369	0.4920	0.3592
N_{calc}	0.261	0.268	0.318	0.414	0.401
N_{exp}	0.264±0.003 (I)	0.266±0.003 (I)	0.328±0.004(III)	0.430±0.002 (I)	0.38
	0.262±0.02 (II)	0.268±0.002 (II)		0.424±0.005 (II)	

The characteristic electrode dimensions, together with experimental and calculated N values, are shown in Table 8.2. Roman numbers pertain to the three systems described. The last column contains Bruckenstein's results [20] obtained for the reduction of $K_3Fe(CN)_6$ in 0.5 M Na_2SO_4 at the gold disc with oxidation of $K_4Fe(CN)_6$ at the ring.

The good agreement of experiment with theory allows the ring-disc system to be used for quantitative determinations of the kinetic parameters of electrode reactions.

§ 8.3.   Application of the Rotating

Ring-Disc Electrode to Mechanistic

and Kinetic Studies of Multistep

Electrode Reactions (Theory)

The fraction N of particles formed at the disc which reach the ring electrode and can be detected there was calculated in § 8.1. It was shown that the flux $\bar{j}_{\mathscr{R}}$, arriving at the ring electrode is †

$$\bar{j}_{\mathscr{R}} = N\bar{j}_{\mathscr{D}} = Nsj_0, \tag{8.49}$$

where j_0 is the rate of formation of the product per unit area of the disc, and s is the surface area of the disc.

If the formation of the product at the disc and its disappearance at the ring electrode are accompaned by transfers of $n_{\mathscr{D}}$ and $n_{\mathscr{R}}$ electrons, respectively, Eq. (8.49) can be rewritten in terms of electric currents measured at the disc $(i_{\mathscr{D}})$ and ring $(i_{\mathscr{R}})$ electrodes:

$$i_{\mathscr{R}} = N\frac{n_{\mathscr{R}}}{n_{\mathscr{D}}} i_{\mathscr{D}}. \tag{8.50}$$

The magnitude of $i_{\mathscr{D}}$ depends on the kinetics of the disc electrode process. If the latter consists of several steps, or if it involves parallel electrochemical reactions, knowledge of the total current $i_{\mathscr{D}}$, is not sufficient for mechanistic analysis. However,

---

† In the text following, only the absolute magnitude of the current will be discussed, and the minus sign in Eq. (8.45) will be omitted.

additional information concerning separate reaction steps can be
obtained by measurements of the ring current $i_{\mathscr{R}}$.

The rate constants for the separate stages of two-stage oxygen
reduction were first measured using the ring-disc system by
Muller and Nekrasov [22]. Subsequently, Bockris et al. [23] proposed an original approach to the analysis of complex reactions
with parallel and consecutive steps (using oxygen reduction as an
example) by a study of the dependence of the disc-to-ring-current
ratio $i_{\mathscr{D}}/i_{\mathscr{R}}$ on the angular velocity of the electrode ω. More recently, Bagotskii, Tarasevich, Shumilova, and one of the present
authors [24-27] developed criteria for measurements of the most
complete set (up to five) of kinetic parameters of this complex
reaction.

The application of the rotating ring-disc electrode is demonstrated below for several examples.

1. The first example concerns the analysis of a two-stage
process (8.1):

$$A + n_2 e^- \xrightarrow{k_2} B + n_3 e^- \xrightarrow{k_3} P_2.$$

A part of the intermediate B is transferred to the bulk and subsequently arrives at the ring electrode. The ring reaction

$$B \pm n_{\mathscr{R}} e^- \rightarrow P_{\mathscr{R}} \tag{8.51}$$

is assumed to be sufficiently fast. The ring is polarized to a value
of the potential at which the discharge of the reactant A is prevented.

The reactions considered are of the first order, and the
adsorption equilibrium (if the reactant adsorbs at the electrode
before discharge) is assumed to be established fast and not to in-
terfere with the reaction kinetics.

The material balance describing the concentration charges of
A and B at the disc is expressed by

$$\frac{D_A}{\delta_{dA}} (c_A^* - c_{AS}) = k_2 c_{AS}, \tag{8.52a}$$

$$k_2 c_{AS} = \frac{D_B \, c_{BS}}{\delta_{dB}} + k_3 c_{BS}, \tag{8.52b}$$

where c_{AS} and c_{BS} are the concentrations of A and B at the disc surface, respectively, c_A^* is the bulk concentration of A, D and δ_d are the diffusion coefficient and the thickness of the diffusion layer of the respective species, and k_2 and k_3 are the rate constants for the first and second steps of the electrochemical reaction.

Solution of Eqs. (8.52a)-(8.52b) gives

$$c_{AS} = \frac{c_A^*}{1 + k_2 \delta_{dA}/D_A},$$ (8.53)

$$c_{BS} = \frac{c_{AS}k_2/k_3}{1 + D_B/k_3\delta_{dB}}.$$ (8.54)

The total current $i_{\mathscr{D}}$, measured at the disc electrode consists of the sum of the currents of both steps of reaction (8.1):

$$i_{\mathscr{D}} = i_{1\mathscr{D}} + i_{2\mathscr{D}} = sF\,(n_2 k_2 c_{AS} + n_3 k_3 c_{BS}).$$ (8.55)

Substituting Eq. (8.54) into Eq. (8.55) gives

$$i_{\mathscr{D}} = sF\left(n_2 + \frac{n_3}{1 + D_B/k_3\delta_{dB}}\right)k_2 c_{AS}.$$ (8.56a)

Using Eqs. (8.53) and (8.56a), we can express the current at the disc electrode by

$$i_{\mathscr{D}} = \frac{i_{dA}}{(n_2 + n_3)(1 + D_A/k_2\delta_{dA})}\left(n_2 + \frac{n_3}{1 + D_B/k_3\delta_{dB}}\right),$$ (8.56b)

where $\bar{i}_{dA} = (n_2 + n_3)sFD_A c_A^*/\delta_{dA}$ is the limiting diffusion current of the reduction of the reactant A at the disc accompanied by transfer of $(n_2 + n_3)$ electrons.

According to Eq. (8.49), the current at the ring electrode, $i_{\mathscr{R}}$, reflects a fraction N of the diffusional flux of B removed from the disc surface, i.e.,

$$i_{\mathscr{R}} = n_{\mathscr{R}}sFND_B c_{BS}/\delta_{dB}.$$ (8.57a)

Using Eqs. (8.53), (8.54), and (8.57a), we obtain the expression for the ring current:

$$i_{\mathscr{R}} = \frac{n_{\mathscr{R}}N i_{dA}}{(n_2 + n_3)(1 + D_A/k_2\delta_{dA})}\frac{D_B/k_3\delta_{dB}}{(1 + D_B/k_3\delta_{dB})}.$$ (8.57b)

Equations (8.56b) and (8.57b) can be used to determine the rate constants of the separate steps of reaction (8.1) from the measured $i_{\mathscr{D}}$ and $i_{\mathscr{R}}$ currents. However, since the functional dependences in Eqs. (8.56b) and (8.57b) are rather complex, k_2 and k_3 are usually determined by special graphical constructions.

From Eqs. (8.56b) and (8.57b), we obtain the ratio $i_{\mathscr{D}}/i_{\mathscr{R}}$ as

$$i_{\mathscr{D}}/i_{\mathscr{R}} = [n_2 + (n_2 + n_3)\,k_3\delta_{dB}/D_B\,]\,/Nn_{\mathscr{R}}. \qquad (8.58a)$$

Since the thickness of the diffusion layer δ_d is inversely proportional to $\omega^{1/2}$ ($\delta_d = \gamma/\sqrt{\omega}$), Eq. (8.58a) can be rewritten as follows:

$$i_{\mathscr{D}}/i_{\mathscr{R}} = [n_2 + (n_2 + n_3)\,k_3\gamma_B/D_B\,\sqrt{\omega}]\,/Nn_{\mathscr{R}}, \qquad (8.58b)$$

where $\gamma_B = 1.61 D_B^{1/3}\nu^{-1/6}$.

The disc-to-ring-current ratio $i_{\mathscr{D}}/i_{\mathscr{R}}$ depends linearly on $1/\sqrt{\omega}$. The intercept on the ordinate axis represents the number of electrons n_2 transferred in the first reaction step. The slope of the line is a direct measure of the rate constant k_3.

Determination of the rate constant k_2 requires, according to [24], the construction of an expression for $(i_{dA} - i_{\mathscr{D}})/i_{\mathscr{R}}$. Using (8.56b) and (8.57b), we obtain

$$(i_{dA} - i_{\mathscr{D}})/i_{\mathscr{R}} = \left[n_3 + (n_2 + n_3)\frac{D_A\delta_{dA}k_3}{D_B\delta_{dB}k_2} + (n_2 + n_3)\frac{D_A}{k_2\delta_{dA}} \right] /n_{\mathscr{R}}N. \qquad (8.59a)$$

Introducing again explicitly the dependence on the rotational velocity of the electrode ($\delta_d = \gamma/\sqrt{\omega}$), we can rewrite Eq. (8.59) in the form

$$(i_{dA} - i_{\mathscr{D}})/i_{\mathscr{R}} = \left[n_3 + (n_2 + n_3)\frac{D_A\gamma_A k_3}{D_B\gamma_B k_2} + \frac{(n_2 + n_3)\,D_A\,\sqrt{\omega}}{k_2\gamma_A} \right] /n_{\mathscr{R}}N. \qquad (8.59b)$$

The straight line obtained in the coordinates $(i_{dA} - i_{\mathscr{D}})/i_{\mathscr{R}}$ vs $\sqrt{\omega}$ intercepts the ordinate at a point which allows the magnitude of n_3 to be recovered. The slope of the line is a measure of the rate constant k_2.

Construction of the plots described above requires measurements of $i_{\mathscr{D}}$ and $i_{\mathscr{R}}$ at at least two rotational speeds of the electrode. This set of four experimental data is sufficient for determination of four parameters: n_2, n_3, k_2, and k_3. If the numbers of electrons transferred n_2 and n_3 are known, calculation of k_2 and k_3 is considerably simplified.

2. As a second example, the case of two parallel electrode reactions proceeding simultaneously at the disc with formation of stable products may be considered:

$$
A
\begin{array}{c}
\xrightarrow[\;k_1\;]{n_1e^-} P_1 \\
\xrightarrow[\;k_2\;]{n_2e^-} B
\end{array}
\tag{8.60}
$$

Let the product of one of the parallel reactions (species B) react at the ring electrode, according to reaction (8.51). The material balance for A and B at the disc electrode is described by

$$
\frac{D_A}{\delta_{dA}} \left(c_A^* - c_{AS} \right) = (k_1 + k_2) c_{AS}, \tag{8.61a}
$$

$$
k_2 c_{AS} = \frac{D_B}{\delta_{dB}} c_{BS}. \tag{8.61b}
$$

The current $i_{\mathscr{D}}$ is equal to

$$
i_{\mathscr{D}} = sF \left(n_1 k_1 + n_2 k_2 \right) c_{AS}. \tag{8.62}
$$

The ring current is described as previously by Eq. (8.57a):

$$
i_{\mathscr{R}} = sF n_{\mathscr{R}} N D_B c_{BS} / \delta_{dB}.
$$

Combining Eqs. (8.61a), (8.61b), (8.62), and (8.57a) gives

$$
i_{\mathscr{D}}/i_{\mathscr{R}} = \frac{n_1 k_1 + n k_2}{N n_{\mathscr{R}} k_2}, \tag{8.63}
$$

$$
\left(i_{dA} - i_{\mathscr{D}} \right) / i_{\mathscr{R}} = \frac{n_1 - n_2}{n_{\mathscr{R}} N} + \frac{n_1 D_A}{n_{\mathscr{R}} N k_2 \gamma_A} \sqrt{\omega}. \tag{8.64}
$$

The limiting diffusion current $\bar{I}_{dA}$ in the above expressions corresponds to the transfer of n_1 electrons.

Equations (8.64) and (8.63) allow measurements of current at different rotational velocities to be utilized for separation of the behavior of the two reaction steps in (8.60) and for determination of k_1 and k_2.

McIntyre [28] used Bockris' results [23] to calculate the $i_{\mathscr{D}}/i_{\mathscr{R}}$ vs ω relation for combined reactions (8.1) and (8.60), i.e., for the case

$$A \begin{array}{c} \xrightarrow{n_1 e^-} P_1 \\ \xrightarrow{n_2 e^-} B \xrightarrow{n_3 e^-} P_2. \end{array}$$

The same author discussed the catalytic reaction

$$A + n_2 e^- \rightarrow B, \quad B + n_3 e^- \rightarrow P_2,$$
$$\nu B \xrightarrow{k} A + P_3,$$

where k is the rate constant of a surface (chemical) reaction which results in regeneration of the reactant A.

3. The most significant results obtained by the rotating ring-disc method pertain to studies of oxygen reduction at various metals.

Oxygen ionization is a complex electrochemical process including both parallel and consecutive steps. According to modern views [29], the overall reaction can be reduced to the following steps. Generally, oxygen ionization proceeds in two parallel paths:

a) direct four-electron reduction to water (the rate constant is designated by k_1)

$$O_2 + 4H^+ + 4e^- \xrightarrow{k_1} 2H_2O; \tag{8.65}$$

b) the (previously discussed) O_2 reduction with formation of hydrogen peroxide as an intermediate. In acid solution, the reaction proceeds as follows:

$$O_2 + 2H^+ + 2e^- \xrightarrow{k_2} H_2O_2. \tag{8.66}$$

Hydrogen peroxide is partly reduced within a two-electron process:

$$H_2O_2 + 2H^+ + 2e^- \xrightarrow{k_3} 2H_2O, \tag{8.67}$$

and partly diffuses into the bulk solution (and thus can be detected at the ring electrode). Furthermore, catalytic H_2O_2 decomposition occurs at a series of metals with formation of oxygen and water:

$$H_2O_2 \xrightarrow{k_4} \frac{1}{2} O_2 + H_2O. \tag{8.68}$$

At potentials close to the reversible potential of reaction (8.66), the reverse reaction of H_2O_2 oxidation must also be taken into account:

$$H_2O_2 \xrightarrow{k_2} O_2 + 2H^+ + 2e^-. \tag{8.69}$$

The process listed above can be summarized in terms of the following scheme for oxygen ionization:

$$\tag{8.70}$$

The dashed arrow indicates the removal of H_2O_2 by diffusion into the bulk solution.

Scheme (8.70) can be treated as the "optimum" one. On the one hand, it is sufficiently general to include the majority of investigated cases. On the other hand, it is not too complex and allows, as shown below, rate constants of all five separate steps to be determined.

The experimental results obtained in studies of oxygen reduction are briefly reviewed in the next section. At present we are only concerned with the possibilities offered by the ring-disc method for investigation of such a complex electrochemical process.

The material-balance equations describing the changes of oxygen concentration (c_A) and hydrogen peroxide concentration

(c_B) at the disc electrode are as follows:

$$D_A(c_A^{\cdot} - c_{AS})/\delta_{dA} = (k_1 + k_2)c_{AS} - (k_2' + k_4/2)c_{BS}, \quad (8.71a)$$

$$k_2 c_{AS} = (k_3 + k_2' + k_4)c_{BS} + (D_B c_{BS})/\delta_{dB}. \quad (8.71b)$$

The total current measured at the disc is described by

$$i_{\mathscr{D}} = sF\,[(4k_1 + 2k_2)c_{AS} + (2k_3 - 2k_2')c_{BS}]. \quad (8.72)$$

The ring current is equal to

$$i_{\mathscr{R}} = 2sFND_B c_{BS}/\delta_{dB}. \quad (8.73)$$

As in paragraphs 1 and 2 of the present section, the expressions $i_{\mathscr{D}}/i_{\mathscr{R}}$ and $(i_{\,dO_2} - i_{\mathscr{D}})/i_{\mathscr{R}}$ can now be calculated using the above material-balance equations (8.71a) and (8.71b). After tedious but not difficult algebraic transformations, we obtain

$$i_{\mathscr{D}}/i_{\mathscr{R}} = \frac{1}{N}\left\{ \frac{2k_1 + k_2}{k_2} + \left[\frac{(2k_1 + k_2)(k_3 + k_2' + k_4)}{k_2} + (k_3 - k_2') \right] \frac{\Upsilon_B}{D_B\sqrt{\omega}} \right\},$$
$$(8.74)$$

$$(i_{dO_2} - i_{\mathscr{D}})/i_{\mathscr{R}} = \frac{1}{N}\left\{ 1 + 2\,\frac{(k_3 + k_2' + k_4)}{k_2}\frac{D_A \Upsilon_B}{D_B \Upsilon_A} + \frac{2D_A\sqrt{\omega}}{k_2\Upsilon_A} \right\}.$$
$$(8.75)$$

Subscripts A and B designate the respective quantities for oxygen and hydrogen peroxide.

The limiting diffusion current of oxygen reduction, i_{dO_2}, corresponds to the maximum number (four) of electrons transferred.

Relations (8.74) and (8.75) plotted in $i_{\mathscr{D}}/i_{\mathscr{R}}$ vs $1/\sqrt{\omega}$ and $(i_{dO_2} - i_{\mathscr{D}})/i_{\mathscr{R}}$ vs $\sqrt{\omega}$ coordinates result in straight lines. The intercepts at the ordinate axes and the slopes can be utilized to obtain four rate constants k_1, k_2, k_3, k_4, and k_2'. It is obvious, however, that the four equations (characteric of the straight lines) are insufficient to determine all five rate constants. One of the present authors together with Bagotskii and Tarasevich [25] used (in the paper containing the above calculation) additional thermodynamic considerations to obtain the required relation between the rate constants; namely, the rate constant of H_2O_2 formation, k_2, and that of H_2O_2

oxidation, k_2', are related by

$$k_2' = k_2 \exp\left[\frac{2F}{RT}(\varphi - \varphi_r)\right],\tag{8.76}$$

where φ_e is the reversible potential of the O_2/HO_2^- couple.

For the practical purpose of studying oxygen ionization, the general scheme (8.70) is simplified by neglecting a part of the reaction scheme included in (8.70) on the basis of these or other physicochemical considerations. Thus, the calculations of rate constants in the first paper [30] describing application of the rotating ring-disc assembly to oxygen ionization were carried out neglecting oxidation of hydrogen peroxide)8.69). Subsequently Bockris et al. [23] neglected both oxidation and catalytic decomposition of hydrogen peroxide. The experimental data were interpreted in terms of the $i_{\mathscr{D}}/i_{\mathscr{R}}$ ratio. The equations obtained by the latter authors can be easily derived from Eq. (8.74) by equating $k_4 = k_2' = 0$. This allowed the rate constants k_1, k_2, and k_3 to be determined. Figure 8.3 illustrates the extrapolation of the $i_{\mathscr{D}}/i_{\mathscr{R}}$ vs $1/\sqrt{\omega}$ plots for various special cases of scheme (8.70) when $k_4 = k_2' = 0$.

A paper of one of the authors with Bagotskii and Shumilova [24] demonstrated that the introduction of the second quantity, $(i_{dO_i} - i_{\mathscr{D}})/i_{\mathscr{R}}$, allows k_1, k_2, k_3, and k_4 to be determined. (It was assumed that no oxidation of hydrogen peroxide occurs, i.e., $k_2' = 0$.) In fact, Eqs. (8.74) and (8.75) become considerably simplified where $k_2' = 0$.

On the other hand, sometimes scheme (8.70) must be made more complex. Thus, Tarasevich [27] took into account the finite desorption rate of hydrogen peroxide formed in reaction (8.66) at the disc. All previous considerations assumed adsorption equilibrium, whereas slow adsorption and desorption kinetics can considerably affect the disc-to-ring-current ratio. The self-consistency of scheme (8.70) was indirectly confirmed in a study of oxygen ionization at platinum [31]. Assuming, on the basis of general considerations, that the rate constant of the catalytic H_2O_2 decomposition depends little on potential, Tarasevich et al. [31] determined all five rate constants without using Eq. (8.76). Thereafter, on the basis of Eq. (8.76) they determined the reversible potential φ_e of the O_2/HO_2^- couple and compared the latter with the experi-

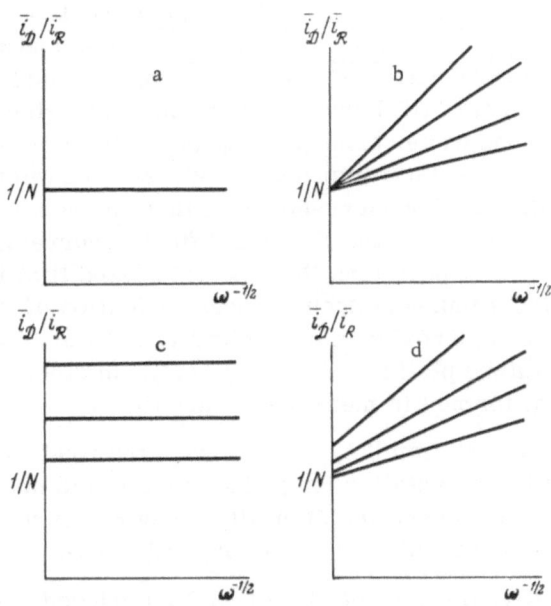

Fig. 8.3. Schematic representation of the $i_\mathscr{D}/i_\mathscr{R}$ vs $1/\sqrt{\omega}$ plots for various mechanisms of oxygen reduction [23]. a) H_2O_2 stable at the disc ($k_3 = 0$), reaction (8.63) absent ($k_1 = 0$); b) H_2O_2 reduces at the disc [$k_3 = k_3 (\varphi_\mathscr{D})$], reaction (8.65) absent ($k_1 = 0$); c) H_2O_2 stable at the disc ($k_3 = 0$), reaction (8.65) proceeds (k_1/k_2 depends on $\varphi_\mathscr{D}$); d) more general case: rates of reactions (8.65) and (8.67) depend on $\varphi_\mathscr{D}$.

mental data. Close agreement of φ_e values obtained by independent methods supports the self-consistency of the scheme (8.70).

§ 8.4. Experimental Studies of Oxygen

Reduction

This reaction, which has important applications, has been studied in great detail, especially at platinum electrodes. Without attempting a full review of the present state of knowledge of oxygen ionization, this reaction will be used as an example illustrating the wide range of possibilities for determination of kinetic parameters using the rotating ring-disc.

a. Separate Measurements of the Direct and Two–State
Reduction of Oxygen at Platinum Electrode. Polarization curves
obtained at a smooth active platinum electrode in alkaline solutions
exhibit (as opposed to the behavior at the mercury electrode; cf.
Fig. 8.2a) only one wave corresponding (at sufficiently low rotational speeds) to a four-electron process. When more intensive stirring is used, the current increases with increasing rotational
velocity slower than does the limiting diffusion current. When the
measured disc current is less than the calculated limiting diffusion current for a four-electron process, oxidation of the intermediate, hydrogen peroxide, can be observed at the anodically
polarized platinized platinum ring (Fig. 8.4). In other cases,
hydrogen is not formed in measurable quantities.

Nekrasov and Müller [22, 30] who were the first to study oxygen ionization by the rotating ring–disc system, did not observe
the four-electron process (8.65) at all. They measured the rate
constants of the consecutive steps (8.66) and (8.67).

Subsequently, Bockris et al. [23, 32] introduced a graphical
treatment of experimental data involving extrapolation of the
$i_{\mathcal{D}}/i_{\mathcal{R}}$ vs $1/\sqrt{\omega}$ line.

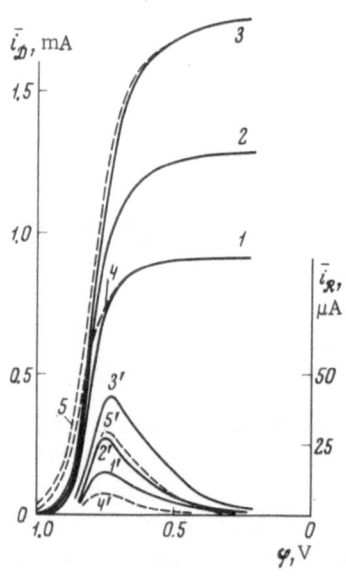

Fig. 8.4. Polarization curves for oxygen
ionization at a platinum disc electrode in
0.1 N KOH (1-5) and the dependence of the
limiting current of H_2O_2 oxidation at the
ring on potential (1'-5') [31]. Curves 4,
5 and 4', 5' are obtained using an electrode with prereduced surface. Rotation
speed (rpm): 1, 1', 4, 4') 1080; 2, 2')
1960; 3, 3', 5, 5') 3700.

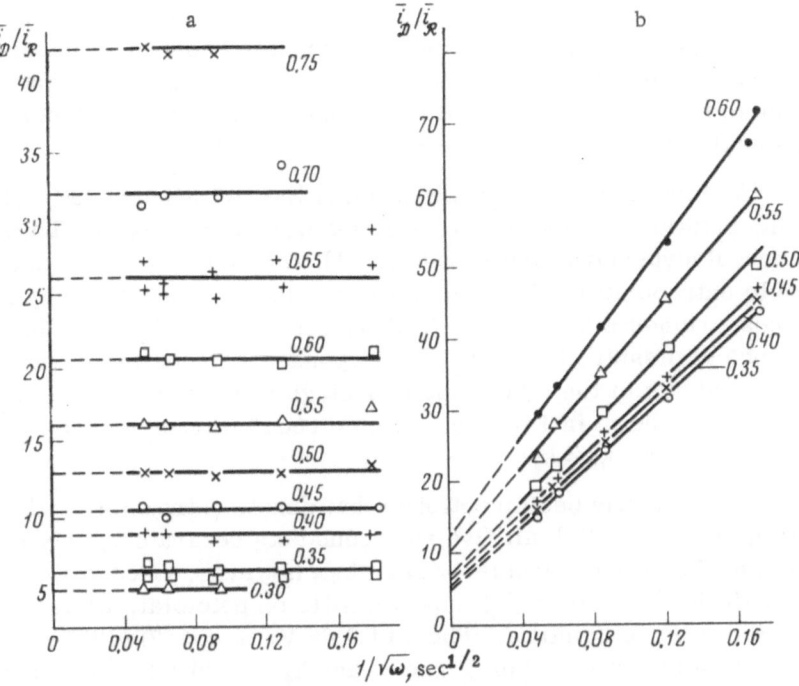

Fig. 8.5. Dependence of $i_{\mathscr{D}}/i_{\mathscr{R}}$ on $1/\sqrt{\omega}$ for oxygen reduction at various potentials of the platinum disc electrode [32]. a) 1 N H_2SO_4 (not highly purified); b) 1 N KOH. Disc potentials (V) are indicated on the curves.

In highly purified H_2SO_4 solutions, the ring current was found to be zero over a wide range of disc potentials, indicating direct oxygen reduction to water [reaction (8.65)]. In insufficiently purified acid solutions, as well as in alkaline solutions, ring current is detected, indicating the formation of hydrogen peroxide on the disc. Linear $i_{\mathscr{D}}/i_{\mathscr{R}}$ vs $1/\sqrt{\omega}$ plots are shown in Fig. 8.5. Comparison of Figs. 8.5a and 8.3b shows that in nonpurified acid solutions direct reduction to water (8.65) is accompanied by the parallel formation of hydrogen peroxide (8.66). The ratio of the two reduction currents decreases with decreasing disc potential; the rate constant of reaction (8.67), k_3, is equal to zero; i.e., hydrogen peroxide produced at the disc is not further reduced. In alkaline solutions the two reactions always proceed simultaneously, and both the k_1/k_2 ratio and k_3 change with the disc potential (cf. Figs. 8.3d and 8.5b).

The results of Bockris' et al., most systematically described in [33], were to a certain extent confirmed in subsequent papers of other authors [34-36]. It must be mentioned, however, that in much of this work the catalytic decomposition of hydrogen peroxide was neglected.

The most complete investigation of the O_2/HO_2^- couple in alkaline solutions at platinum electrodes was carried out by Tarasevich, Radyushkina, and Burstein [31]. The latter authors used double extrapolations [according to Eqs. (8.74) and (8.75)] applied to measurements carried out in O_2 and in H_2O_2 solutions in the absence of dissolved oxygen. The dependence of the measured rate constants of the direct and two-stage reduction on electrode potential is shown in Fig. 8.6. At electrodes whose surfaces were not prereduced, $k_1 \approx k_2$.

It has already been mentioned that double extrapolation is not sufficient to establish all five rate constants separately, only combinations being measurable. Assuming, however, that the rate of catalytic decomposition (k_4) depends little on potential, all the constants can be determined. Thus, at 0.85 V, $k_3 = (6-7) \cdot 10^{-3}$ cm · $\sec^{-1}$, $k_2' = (9-10) \cdot 10^{-3}$ cm · $\sec^{-1}$, and $k_4 = (5-7) \cdot 10^{-3}$ cm · $\sec^{-1}$; i.e., the rates of the catalytic decomposition of hydrogen peroxide (k_4) and of its electrochemical decomposition (sum of the oxidation k_2 and reduction k_3 rates) are approximately the same at this potential.

At a prereduced electrode surface, $k_1/k_2 \approx 7$, i.e., direct reduction (8.65) contributes most to the current.

It should be mentioned that measurements of the rate constants of separate stages and of their ratios (as was previously

Fig. 8.6. Logarithm of the rate constants k_1 and k_2 as a function of potential. Oxygen ionization at platinum in 0.1 N KOH [31].

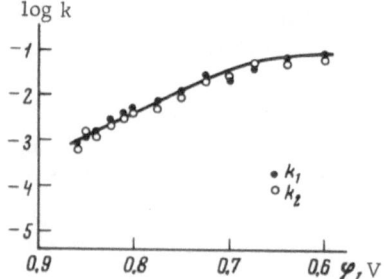

mentioned and will be discussed later in more detail) strongly depend on the experimental conditions (degree of surface oxidation, presence of contaminants in solution, etc.). This, apparently, is the reason for the discrepancies between the results of earlier [22, 30] and more recent work, which are of a purely qualitative character.

b. The Effect of Platinum Surface Oxides on the Rates of Separate Reaction Stages. Platinum surface oxides have specific effects on various stages of oxygen reduction [37]. In alkaline solution, reduction of O_2 to H_2O_2 [Eq. (8.66)] is slowed down at preoxidized surfaces (i.e., the half-wave potential becomes more negative), whereas further reduction of HO_2^- becomes faster: the current density in the potential range 0.5-0.1 V increases.

The latter effect on HO_2^- reduction was explained in terms of catalytic decomposition of HO_2^- at oxidized platinum surfaces. Oxygen produced in the latter reaction is reduced faster than HO_2^-.

The amount of oxygen produced by catalytic decomposition of HO_2^- cannot be determined by measurements of the cathodic current at the ring only. The latter includes a contribution of the reduction of the initial product HO_2 as well. Nekrasov and Müller [37] separated the two reactions by measuring the difference between the cathodic and anodic currents at the ring† (anodic polarization results in oxidation of HO_2^- only).

A new method of investigation of surface oxides effects on the rates of separate stages of oxygen reduction was proposed in [38, 39]: rotating ring-disc electrode subjected to a triangular potential sweep at the disc. For the same purposes, the platinum disc is prereduced in hydrogen and subsequently degassed [40], or its surface is treated in a programmed way (anodic-cathodic cyclic polarization [41].

c. Effect of Ionic Adsorption on Platinum. Divalent cations (Ca^{2+}, Sr^{2+}, Ba^{2+}) affect oxygen reduction in alkaline solution in a similar way to chemisorbed oxygen: they decrease the rate of the first stage and increase the rate of the second stage of the reaction. This effect (as well as the adsorbability) increases in

---

†It is convenient to use, for this purpose, a disc electrode with a sectioned ring (cf. §8.8).

the series $Ca^{2+} < Sr^{2+} < Ba^{2+}$. Müller and Sobol [42] explained the effect of alkaline earth cations in terms of an increased Pt$-$O bond strength in the presence of adsorbed cations.

The strongly adsorbed CNS^- anion inhibits formation of HO_2^- on platinum and, even more, its further reduction. Therefore, the ring current strongly increases in the presence of CNS^- [43].

The effect of anions is much more pronounced in acid solutions. With increasing adsorption, smooth platinum electrodes loose their activity with time. The anionic effect increases in the series $ClO_4^- < SO_4^{2-} < Cl^- < Br^-$ [22, 39]. Alkaline earth cations do not affect the reaction rates in acid solutions, as opposed to their behavior in alkaline solution. Cd^{2+} inhibits reduction of H_2O_2 at platinized platinum and the formation of the latter, as well, at smooth platinum. Tl^+ ion inhibits mainly the first reaction stage (H_2O_2 formation [43]). Ionization of oxygen is especially strongly inhibited in the presence of Cu^{2+} and Ag^+ ions [44].†

d. Oxygen Ionization at Other Metals. Other metals of the platinum group $-$ palladium [45], rhodium [46-49], and ruthenium [50] $-$ behave similarly to platinum. Rhodium has been studied in a more detailed manner [48] than the others.

Reduction of oxygen at Rh in alkaline solution proceeds simultaneously by two parallel paths (8.65) and (8.66), the first of which is faster (Fig. 8.7). The dominating role of the four-electron process, as compared with platinum, can apparenty be explained in terms of a greater affinity of rhodium for oxygen and, consequently, a faster dissociative chemisorption of oxygen on rhodium than on platinum.

The rate constants measured by the double extrapolation procedure indicate that the further disappearance of HO_2^- formed in reaction (8.66) occurs at potentials > 0.7 V mainly via catalytic decomposition, and at potentials < 0.7 V via electrochemical reduction of HO_2^-.

At gold electrodes [51] in acid solutions the reaction proceeds by two parallel paths, (8.65) and (8.66); the direct path predominates on preoxidized electrodes.

---

†Depending on potential, these metal ions may be "underpotential" deposited at Pt. $-$ Editor.

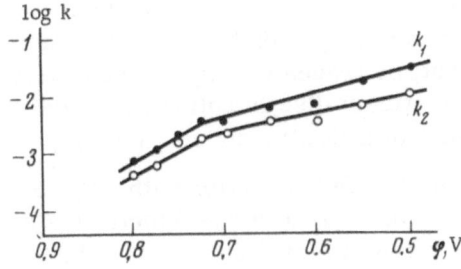

Fig. 8.7. Logarithm of rate constants k_1 and k_2 as a function of potential. Oxygen ionization at rhodium electrode in 0.1 N KOH [48].

Fig. 8.8. Logarithm of rate constants of oxygen reduction at silver electrode in 0.1 N KOH as a function of potential [55].

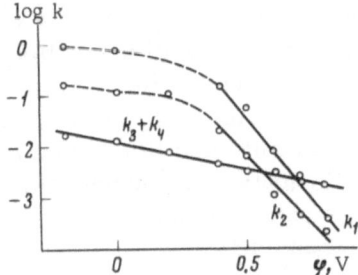

In alkaline solutions, both the direct and two-stage reduction processes are observed ($k_1 \approx k_2$). Within the potential range 0.4–0.8 V the HO_2^- formed is stable; at more negative potentials, it is further reduced at the electrode [52, 53].[†]

Oxygen reduction on silver [54, 55] in alkaline solutions proceeds through the intermediate formation of HO_2^-, the latter being reduced at a faster rate at a prereduced silver surface than at an oxidized silver surface. Direct four-electron reduction to OH^- ions proceeds in parallel to the two-stage reaction, the former predominating at not too negative potentials. The rate constants k_1, k_2, and ($k_3 + k_4$) were found by the extrapolation procedure [Eqs. (8.74) and (8.75)] (Fig. 8.8).

[†]Another interpretation of the observed relations can be found in [53]. Direct reduction to water (8.65) does not proceed at all. HO_2^- ions formed in reaction (8.66) are partly desorbed and partly catalytically decomposed in the adsorbed state into products which subsequently desorb. Although the graphical treatment of experimental data according to Eqs. (8.74) and (8.75) results in $k_1 \neq 0$, the existence of the four-electron process (8.65) is only apparent.

As opposed to dissolved molecular oxygen, oxygen chemisorbed on platinum and silver reduces upon cathodic polarization directly to hydroxyl ions. In fact, no traces of HO_2^- can be detected at the ring when a preoxidized platinum disc is polarized by triangular cathodic potential pulses (in a N_2 atmosphere) [38, 54].

Both processes (direct reduction and reduction with HO_2^- formation) proceed with commensurate rates at nickel electrodes in alkaline solutions. Surface oxidation of nickel strongly inhibits HO_2^- reduction. This is demonstrated in Fig. 8.9 in which the ratio $i_{\mathscr{D}}/i_{\mathscr{R}}$ is plotted vs $1/\sqrt{\omega}$ (compare with Fig. 8.3). The rate of catalytic HO_2^- decomposition was measured separately in oxygen-free, alkaline solutions of H_2O_2 by comparison of the cathodic and anodic ring currents. A combination of several methods has led to the determination of the rate constants k_1, k_2, and $(k_3 + k_4)$. Some results are illustrated in Fig. 8.10. It can be seen that $k_3 > k_4$ (catalytic decomposition is rather slow) [55, 60]. The sum of the rate constants, $k_3 + k_4$, has been measured at a thallium electrode [61].

The two-stage process is only observed at pyrographite and glassy carbon electrodes in KOH and H_2SO_4 solutions. At low polarizations, H_2O_2 is formed with a yield close to 100% (i.e., $k_3 = 0$). At higher cathodic polarization, hydrogen peroxide starts being reduced according to reaction (8.67). Catalytic decomposition (8.68) is practically absent [62-64].

Finally, the rotating ring-disc has been used to demonstrate [65] that ozone is reduced on gold and platinum in acid solutions to

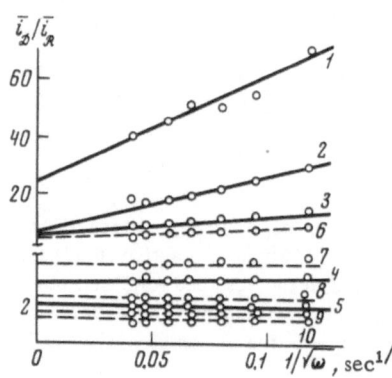

Fig. 8.9. Dependence of $i_{\mathscr{D}}/i_{\mathscr{R}}$ on $1/\sqrt{\omega}$ for oxygen ionization at a reduced (solid lines) and an oxidized (dashed lines) nickel electrode in 0.1 N KOH [59]. Disc potential: 1, 6) 0.1 V; 2, 7) 0.2 V; 3, 8) 0.3 V; 4, 9) 0.4 V; 5, 10) 0.5 V.

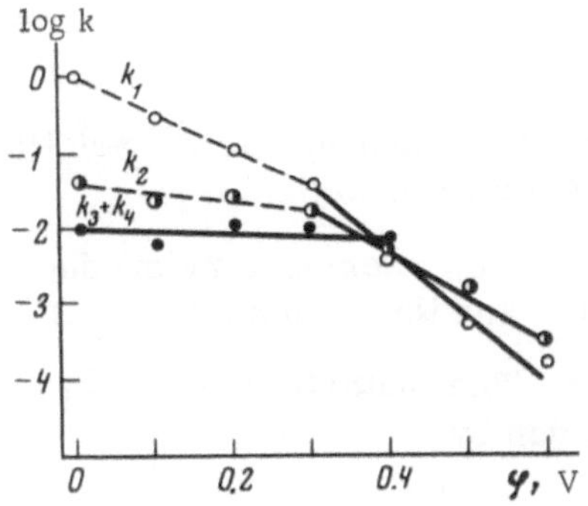

Fig. 8.10. Logarithm of the rate constants of separate stages of oxygen ionization at a nickel electrode in 0.1 N KOH as a function of potential [58].

O_2, according to the scheme

$$O_3 + 2H^+ + 2e^- \rightarrow O_2 + H_2O.$$

Hydrogen peroxide is not formed in this reaction (although formerly the reaction products were thought to be $H_2O_2 + \frac{1}{2}O_2$).

§ 8.5. Determination of Reaction Orders of Separate Stages of an Electrochemical Process

It was assumed until now that all electrode reactions proceeding at the disc are of the first order. However, each stage accompanied by charge transfer can in turn consist of several consecutive or parallel steps. If these steps include adsorption or chemical surface reactions, the resulting kinetics of the discharge stage can be nonlinear. Furthermore, the nature of the intermediates is very often completely unknown. Additional information concerning the reaction mechanism may be obtained in these cases by determination of reaction orders.

The following considerations involve methods of utilizing the measured disc $i_\mathcal{D}$ and ring $i_\mathcal{R}$ currents to determine the reaction orders in simple cases [66].

a. Let the mechanism of a two-stage disc electrode reaction be

$$\mu_2 A + n_2 e^- \xrightarrow{k_2} B, \qquad (8.77a)$$

$$\mu_3 B + n_3 e^- \xrightarrow{k_3} P, \qquad (8.77b)$$

where μ_2 and μ_3 are the reaction orders of the first and second stages, respectively.

The intermediary product B is partly removed from the solution and thus can be detected at the ring electrode [reaction (8.51)].

The material balance of species A and B at the disc surface is given by

$$\frac{D_A}{\delta_{dA}} (c_A^* - c_{AS}) = k_2 c_{AS}^{\mu_2}, \qquad (8.78a)$$

$$k_2 c_{AS}^{\mu_2} = \frac{D_B}{\delta_{dB}} c_{BS} + k_3 c_{BS}^{\mu_3}. \qquad (8.78b)$$

The disc current equals

$$i_{\mathscr{D}} = sF (n_2 k_2 c_{AS}^{\mu_2} + n_3 k_3 c_{BS}^{\mu_3}). \qquad (8.79)$$

The ring current, as previously, is proportional to the amount of intermediate B removed from the disc by convective diffusion into the solution bulk:

$$i_{\mathscr{R}} = sF n_{\mathscr{R}} N D_B c_{BS}/\delta_{dB}. \qquad (8.57a)$$

The expression for the reaction order μ_2 can be derived by elimination of the surface concentration c_{BS} from Eqs. (8.79) and (8.57a) using Eq. (8.78b):

$$i_{\mathscr{D}} = sF (n_2 k_2 c_{AS}^{\mu_2} + n_3 k_2 c_{AS}^{\mu_2} - n_3 D_B c_{BS}/\delta_{dB}),$$

or

$$i_{\mathscr{D}} = sF (n_2 + n_3) k_2 c_{AS}^{\mu_2} - n_3 i_{\mathscr{R}}/n_{\mathscr{R}} N,$$

from which

$$sF (n_2 + n_3) k_2 c_{AS}^{\mu_2} = i_{\mathscr{D}} + n_3 i_{\mathscr{R}} N. \qquad (8.80)$$

The surface concentration c_{AS} of species A is thus given by

$$c_{AS} = \left[\frac{i_{\mathscr{D}} + n_3 i_{\mathscr{R}}/n_{\mathscr{R}}N}{(n_2 + n_3)\, F s k_2} \right]^{1/\mu_2} \tag{8.81}$$

The material balance equation for A [Eq. (8.78a)] can be re-written in the form

$$\frac{D_A \overset{\bullet}{c_A}}{\delta_{dA}} - k_2 c_{AS}^{\mu_2} = \frac{D_A\, c_{AS}}{\delta_{dA}}.$$

Substitution of Eq. (8.81) into the above expression gives

$$\frac{D_A \overset{\bullet}{c_A}}{\delta_{dA}} - \frac{i_{\mathscr{D}} + n_3 i_{\mathscr{R}}/n_{\mathscr{R}}N}{(n_2 + n_3)\, F s} = \frac{D_A}{\delta_{dA}} \left[\frac{i_{\mathscr{D}} + n_3 i_{\mathscr{R}}/n_{\mathscr{R}}N}{(n_2 + n_3)\, F s} \right]^{1/\mu_2}. \tag{8.82}$$

Following the general method of determination of reaction orders of heterogeneous processes described in § 3.3, we construct the ratio of expressions (8.82) written for two different rotational speeds of the disc:

$$\frac{1 - (i_{\mathscr{D}}^{(1)} + n_3 i_{\mathscr{R}}^{(1)}/n_{\mathscr{R}}N)\,(i_{dA}^{(1)})^{-1}}{1 - (i_{\mathscr{D}}^{(2)} + n_3 i_{\mathscr{R}}^{(2)}/n_{\mathscr{R}}N)\,(i_{dA}^{(2)})^{-1}} = \left[\frac{i_{\mathscr{D}}^{(1)} + n_3 i_{\mathscr{R}}^{(1)}/n_{\mathscr{R}}N}{i_{\mathscr{D}}^{(2)} + n_3 i_{\mathscr{R}}^{(2)}/n_{\mathscr{R}}N} \right]^{1/\mu_2},$$

where $\bar{i}_{dA} = s F(n_2 + n_3) D_A c_A^{*}/\delta_{dA}$ is the total diffusion-limiting disc current and the superscripts (1) and (2) designate quantities which change with changing rotation speed.

Taking logarithms, we can rewrite the above expression as follows:

$$\ln \left[\frac{1 - (i_{\mathscr{D}}^{(1)} + n_3 i_{\mathscr{R}}^{(1)}/n_{\mathscr{R}}N)\,(i_{dA}^{(1)})^{-1}}{1 - (i_{\mathscr{D}}^{(2)} + n_3 i_{\mathscr{R}}^{(2)}/n_{\mathscr{R}}N)\,(i_{dA}^{(2)})^{-1}} \right] = \frac{1}{\mu_2} \ln \left[\frac{i_{\mathscr{D}}^{(1)} + n_3 i_{\mathscr{R}}^{(1)}/n_{\mathscr{R}}N}{i_{\mathscr{D}}^{(2)} + n_3 i_{\mathscr{R}}^{(2)}/n_{\mathscr{R}}N} \right]. \tag{8.83}$$

The only unknown in Eq. (8.83) is μ_2 since $i_{\mathscr{D}}$ and $i_{\mathscr{R}}$ are measured experimentally, and the limiting diffusion current, $\bar{i}_{dA}$, can also be measured or theoretically calculated.

The reaction order of the second stage μ_3, is found by elimination of the surface concentration c_{AS} from Eq. (8.79) using

Eq. (8.78b):

$$i_{\mathscr{D}} - n_2 i_{\mathscr{R}}/n_{\mathscr{R}}N = sF(n_2 + n_3)\,k_3 c_{\mathrm{BS}}^{\mu_3}.$$

Repeating the procedure used in the derivation of Eq. (8.83) from Eq. (8.80), we obtain the following expression:

$$\ln \frac{i_{\mathscr{R}}^{(1)}\delta_{\mathrm{dB}}^{(2)}}{i_{\mathscr{R}}^{(2)}\delta_{\mathrm{dB}}^{(1)}} = \frac{1}{\mu_3}\ln \frac{i_{\mathscr{D}}^{(1)} - n_2 i_{\mathscr{R}}^{(1)}/n_{\mathscr{R}}N}{i_{\mathscr{D}}^{(2)} - n_2 i_{\mathscr{R}}^{(2)}/n_{\mathscr{R}}N}, \qquad (8.84)$$

where the superscripts (1) and (2) designate quantities changing with the rotation speed.

The diffusion layer thickness of species B at the rotating disc can be calculated from Eq. (2.20).

The quantity μ_3 can be determined by substitution of the values of currents $i_{\mathscr{D}}$ and $i_{\mathscr{R}}$ measured at ω_1 and ω_2 into Eq. (8.84).

   b. Consider now the case of two parallel reactions consuming species A at the disc:

$$\mu_2 A + n_1 e^- \xrightarrow{k_1} P_1, \qquad (8.85a)$$

$$\mu_2 A + n_2 e^- \xrightarrow{k_2} B. \qquad (8.85b)$$

The product of one of the reactions, species B, is further reduced at the disc electrode to the final product P_2:

$$\mu_3 B + n_3 e^- \xrightarrow{k_3} P_2. \qquad (8.85c)$$

A part of the intermediate B diffuses into the solution and undergoes reduction at the ring [(8.51)].

   In addition to the simplifying assumption of the same reaction orders μ_2 of both parallel reactions (8.85a) and (8.85b), it is assumed that the number of electrons transferred in the reduction to P_1 is equal to the number of electrons transferred in the two-stage reduction, i.e.,

$$n_1 = n_2 + n_3. \qquad (8.86)$$

The above assumptions (especially the first one) obviously rule out further discussions of generality. However, it must be realized that the determination of reaction orders of two parallel electrode reactions is extremely complex in the general case.

The material balance (8.78a) must now be replaced by the condition

$$\frac{D_A}{\delta_{dA}}(c_A^* - c_{AS}) = (k_1 + k_2) c_{AS}^{\mu_2}. \tag{8.87}$$

The second equation, (8.78b), remains unchanged.

The disc current is described by

$$i_{\mathscr{D}} = sF\left[(n_1 k_1 + n_2 k_2)\, c_{AS}^{\mu_2} + n_3 k_3 c_{BS}^{\mu_3}\right]. \tag{8.88}$$

The total kinetic current for species A is obtained from Eqs. (8.57a) and (8.78b):

$$sF\,(n_2 + n_3)(k_2 + k_3)\, c_{AS}^{\mu_2} = i_{\mathscr{D}} + n_3 i_{\mathscr{R}}/n_{\mathscr{R}} N.$$

The above expression is analogous to Eq. (8.80) for the case of two parallel processes.

The surface concentration of A at the disc is equal to

$$c_{AS} = \left[\frac{i_{\mathscr{D}} + n_3 i_{\mathscr{R}}/n_{\mathscr{R}} N}{sF\,(n_2 + n_3)(k_2 + k_3)}\right]^{1/\mu_2}. \tag{8.89}$$

By replacing Eq. (8.81) by Eq. (8.89), one can easily arrive at Eq. (8.83) which can be used, as previously, to determine the reaction order μ_2. The reaction order of the second stage, μ_3, however, cannot be simply determined [using Eq. (8.84)] in this case.

Nekrasov et al. [67-69] studied experimentally both schemes, a and b, with consecutive and parallel reactions using the ring-disc system. The first scheme was represented by disproportionation of U(V) ions produced by the cathodic reduction of uranyl at the disc. Analysis of the dependence of the yield of U(V) ions at the ring on the rotation speed of the electrode showed that the disproportionation of U(V) ions at the disc in moderately acid solutions is a reaction of the first order. This is confirmed by the fact that the U(V) yield is independent of the concentration of

U(VI) in the solution. In strongly acidic solutions, disproportionation occurs in the bulk solution as well, the reaction having an
order higher than the first with respect to U(V).

The first scheme was also studied by reference to dimerization of radical anions formed at an amalgamated gold electrode
during reduction of aromatic carbonyl compounds ($\mu_2 = 2$).

The second scheme, with two parallel reactions, was represented by oxygen reduction, which, as previously discussed
[Eq. (8.70)] results in two simultaneous products: H_2O_2 and H_2O
(or OH^-). Both reactions were found to be of the first order at
smooth platinum (in agreement with the results obtained using
another method, described in § 3.3).

§ 8.6. Voltammetric Curve at the Ring

The voltammetric characteristics of a uniform rotating disc
were derived in § 4.9. The derivation was based on the assumed
uniform accessibility of the disc surface. Consequently, the surface concentration of each reaction component was assumed to be
constant over the entire disc surface.

The surface of a nonhomogeneous rotating disc electrode, including that of a disc with a ring, is not uniformly accessible to
the reacting species. In particular, as shown in studies of the
limiting diffusion current at a partially "coated" disc (cf. § 2.8),
the thickness of the diffusion layer δ_d and the limiting current
density vary along the electrode radius.

In general, the concentrations of electrochemically active
species vary along the surface of a potentiostated ring. Therefore,
the voltammetric curve on the ring can differ from an analogous
curve at a uniformly accessible disc.

The polarization curve of a ring electrode depends considerably on the processes occurring at the central disc electrode. In-
deed, depending on the disc potential $\varphi_{\mathscr{D}}$, solution arriving at the
ring is enriched or depleted in one or another species.

Finally, as was already shown in § 8.1, the electrode reaction
proceeds under activation control in the region close to the internal boundary of the ring electrode (close to r_{20}).

In order to obtain the voltammetric characteristics of the ring reaction, it is useful to consider the case of a simple redox reaction

$$Ox + n\,e^- \rightleftarrows Red \qquad\qquad (8.90)$$

occurring at the disc and ring electrodes.

Transport equations have the same form for both species†:

$$u\,\frac{\partial c_O}{\partial r} + w\,\frac{\partial c_O}{\partial z} = D\,\frac{\partial^2 c_O}{\partial z^2}, \qquad\qquad (8.91a)$$

$$u\,\frac{\partial c_R}{\partial r} + w\,\frac{\partial c_R}{\partial z} = D\,\frac{\partial^2 c_R}{\partial z^2}. \qquad\qquad (8.91b)$$

Let the potential of the ring electrode, $\varphi_{\mathscr{R}}$, be close to the reversible potential of the system (8.90). Then the concentrations of the oxidized and reduced forms at the ring can be assumed to obey the Nernst equation. Polarization of the disc is arbitrary.

The system of boundary conditions for Eqs. (8.91a) and (8.91b) is as follows:

at the disc $\mathscr{D}$ $(0 < r < r_{10})$

$$D\left(\frac{\partial c_O}{\partial z}\right)_S = -D\left(\frac{\partial c_R}{\partial z}\right)_S = \frac{i_{\mathscr{D}}}{nFs} \qquad \text{for } z = 0; \qquad (8.92a)$$

at the seperator $\mathscr{G}$ $(r_{10} < r < r_{20})$

$$D\left(\frac{\partial c_O}{\partial z}\right)_S = -D\left(\frac{\partial c_R}{\partial z}\right)_S = 0 \qquad \text{for } z = 0; \qquad (8.92b)$$

at the ring $\mathscr{R}$ $(r_{20} < r < r_{30})$

$$D\left(\frac{\partial c_O}{\partial z}\right)_S = -D\left(\frac{\partial c_R}{\partial z}\right)_S \qquad \text{for } z = 0, \qquad (8.92c)$$

$$c_{OS} = c_{RS}\theta_{\mathscr{R}} \qquad \text{for } z = 0, \qquad (8.92d)$$

---

†For simplicity, it is assumed that $D_O = D_R = D$.

where $\theta_{\mathscr{R}} = \exp\ [nF\ (\varphi_{\mathscr{R}} - \varphi_0)/RT]$, $i_{\mathscr{D}}$ is the disc current, and φ_0 is the standard potential of the system.

In the bulk solution the concentrations of both species are constant:

$$c_O \to \overset{\bullet}{c}_O, \quad c_R \to \overset{\bullet}{c}_R \qquad \text{as } z \to \infty, \quad r > 0. \qquad (8.93)$$

The same concentrations are maintained in the flux of arriving species:

$$c_O = \overset{\bullet}{c}_O, \quad c_R = \overset{\bullet}{c}_R \qquad \text{for } z > 0, \quad r = 0. \qquad (8.94)$$

Instead of solving the mathematical problem formulated above, let the latter be divided into two parts. It will become clear below that each of the two problems thus obtained has been solved earlier. The final result will be found using the already existing relations.

Two new functions are now introduced:

$$\Phi_1 = c_O + c_R, \qquad (8.95a)$$

$$\Phi_2 = c_O - \theta_{\mathscr{R}} c_R. \qquad (8.95b)$$

It can be easily shown that the equations satisfied by functions Φ_1 and Φ_2 retain their previous form (8.91a)-(8.91b). However, the boundary conditions are changed:

at the disc $(0 < r < r_{10})$

$$D\left(\frac{\partial \Phi_1}{\partial z}\right)_S = 0, \quad D\left(\frac{\partial \Phi_2}{\partial z}\right)_S = \frac{i_{\mathscr{D}}(1 + \theta_{\mathscr{R}})}{nFs} \qquad \text{for } z = 0; \quad (8.96a)$$

at the separator $(r_{10} < r < r_{20})$

$$D\left(\frac{\partial \Phi_1}{\partial z}\right)_S = 0, \quad D\left(\frac{\partial \Phi_2}{\partial z}\right)_S = 0 \qquad \text{for } z = 0; \qquad (8.96b)$$

at the ring $(r_{20} < r < r_{30})$

$$D\left(\frac{\partial \Phi_1}{\partial z}\right)_S = 0, \quad \Phi_{2S} = 0 \qquad \text{for } z = 0. \qquad (8.96c)$$

In the solution bulk

$$\Phi_1 \to \overset{*}{\Phi}_1 = \overset{*}{c}_O + \overset{*}{c}_R, \quad \Phi_2 \to \overset{*}{\Phi}_2 = \overset{*}{c}_O - \theta_{\mathscr{R}}\overset{*}{c}_R \qquad \text{as } z \to \infty, \ r > 0.$$

$$(8.96d)$$

The same conditions exist at the disc axis, i.e., at $r = 0$, $z \geq 0$.

It is easy to establish from conditions imposed on function Φ_1 that $\Phi_1 \equiv \Phi_1^* = c_0^* + c_R^*$. This indicates that the total solution concentration remains constant. It is the ratio of concentrations of Ox and Red forms only that varies in the process of convective diffusion.

The function Φ_2 is sought in the form of a sum of two functions†

$$\Phi_2 = \Phi_{21} + \Phi_{22}. \qquad (8.97)$$

Each of the functions Φ_{21} and Φ_{22} is chosen so that they satisfy transport equations of the type (8.91a)-(8.91b) and the following boundary conditions:

$$D\left(\frac{\partial \Phi_{21}}{\partial z}\right)_S = \frac{i_{\mathscr{D}}(1 + \theta_{\mathscr{R}})}{nFs}; \quad D\left(\frac{\partial \Phi_{22}}{\partial z}\right)_S = 0 \quad \text{for } 0 < r < r_{10}, z = 0;$$

$$(8.98a)$$

$$D\left(\frac{\partial \Phi_{21}}{\partial z}\right)_S = 0; \quad D\left(\frac{\partial \Phi_{22}}{\partial z}\right)_S = 0 \quad \text{for } r_{10} < r < r_{20} \quad z = 0; \quad (8.98b)$$

$$\Phi_{21S} = 0, \quad \Phi_{22S} = 0 \qquad \text{for } r_{20} < r < r_{30}, \quad z = 0; \quad (8.98c)$$

$$\Phi_{21} \to 0; \quad \Phi_{22} \to \overset{*}{\Phi}_2 = \overset{*}{c}_O - \theta_{\mathscr{R}}\overset{*}{c}_R \quad \text{for } r > 0, \quad z \to \infty. \quad (8.98d)$$

The same boundary conditions are obtained here as in § 8.1 when seeking the ring current. The only exception is condition (8.97) for regions far removed from the electrode: previously the concentration of the transferred species in the bulk solution was equal to zero.

It can be seen that the above choice of functions Φ_{21} and Φ_{22} ensures that the boundary conditions (8.96a)-(8.96d) are satisfied by Φ_2. Also, the boundary problems for functions Φ_{21} and Φ_{22} have been previously discussed, for function Φ_{21} in §8.1 and for Φ_{22} in §2.8.

---

†A similar method was used by Bruckenstein [20].

The total ring current consists, in the present case, of the sum of the previously found quantities:

$$i_{\mathscr{R}} = nF2\pi D \int_{r_{20}}^{r_{30}} \left(\frac{\partial c_O}{\partial z}\right)_S r\,dr = \frac{nF \cdot 2\pi D}{1 + \theta_{\mathscr{R}}} \int_{r_{20}}^{r_{30}} \left[\left(\frac{\partial \Phi_1}{\partial z}\right)_S + \left(\frac{\partial \Phi_{21}}{\partial z}\right)_S + \left(\frac{\partial \Phi_{22}}{\partial z}\right)_S\right] r\,dr.$$

Since $\Phi_1 \equiv \Phi_1^*$, integration of the first term gives

$$\int_{r_{20}}^{r_{30}} \left(\frac{\partial \Phi_1}{\partial z}\right)_S r\,dr = 0. \qquad (8.99a)$$

Using the results of § 8.1 to integrate the second term, we obtain

$$\frac{nF \cdot 2\pi D}{1 + \theta_{\mathscr{R}}} \int_{r_{20}}^{r_{30}} \left(\frac{\partial \Phi_{21}}{\partial z}\right)_S r\,dr = -Ni_{\mathscr{D}}. \qquad (8.99b)$$

The relations derived in § 2.8 allow the results of integration of the third term to be presented in the form

$$\frac{nF \cdot 2\pi D}{1 + \theta_{\mathscr{R}}} \int_{r_{20}}^{r_{30}} \left(\frac{\partial \Phi_{22}}{\partial z}\right)_S r\,dr = \frac{(1 - \theta_{\mathscr{R}} c_R^*/c_O^*)}{1 + \theta_{\mathscr{R}}} i_{dA} [(r_{30}/r_{10})^3 - (r_{20}/r_{10})^3]^{1/3},$$

$$(8.99c)$$

where $i_{dO} = nFsDc_O^*/\delta_d$ is the total diffusion-limiting current of the Ox species to the disc.

Substituting Eqs. (8.99a)–(8.99c) into the expression for current finally gives

$$i_{\mathscr{R}} = -Ni_{\mathscr{D}} + \frac{(1 - \theta_{\mathscr{R}} c_R^*/c_O^*)}{1 + \theta_{\mathscr{R}}} i_{dO}\beta^{1/3}, \qquad (8.100)$$

where $\beta = (r_{30}/r_{10})^3 - (r_{20}/r_{10})^3$.

From Eq. (8.100) and the Nernst equation describing the dependence of $\theta_{\mathscr{R}}$ on the ring potential $\varphi_{\mathscr{R}}$, the voltammetric characteristics of the ring electrode are found in the form

$$\varphi_{\mathscr{R}} = \varphi_0 + \frac{RT}{nF} \ln \frac{i_{dO}\beta^{1/3} - i_{\mathscr{R}} - Ni_{\mathscr{D}}}{i_{dR}\beta^{1/3} - i_{\mathscr{R}} + Ni_{\mathscr{D}}}, \qquad (8.101)$$

where $i_{dR} = nFsDc_R^*/\delta_d$ is the total diffusion-limiting current for oxidation of the Red species.

The voltammetric characteristics of the disc electrode can be presented in the case of a reversible redox reaction in the form

$$\varphi_{\mathcal{D}} = \varphi_0 + \frac{RT}{nF} \ln \frac{i_{d0} - i_{\mathcal{D}}}{i_{dR} + i_{\mathcal{D}}} . \tag{8.102}$$

Equation (8.102) can be easily derived from the general voltammetric characteristics (4.104) for high exchange current densities.

Comparison of the voltammetric characteristics of the ring [(8.101)] and disc [(8.102)] electrodes shows that the difference between the two expressions is due to the effects of the disc electrode processes on the behavior of the ring electrode.

A more detailed analysis of the disc effects on the electrode processes at the ring requires substitution in Eq. (8.100) of the explicit expression for the dependence of $i_{\mathcal{R}}$ on $\varphi_{\mathcal{D}}$ and $\varphi_{\mathcal{R}}$,

$$i_{\mathcal{R}} = i_{d0} \left\{ - N \, \frac{\exp\left[-nF\left(\varphi_{\mathcal{D}} - \varphi_0\right)/RT\right] - c_R^*/c_O^*}{\exp\left[-nF\left(\varphi_{\mathcal{D}} - \varphi_0\right)/RT\right] + 1} + \right. $$
$$\left. + \beta^{2/3} \frac{\exp\left[-nF\left(\varphi_{\mathcal{R}} - \varphi_0\right)/RT\right] - c_R^*/c_O^*}{\exp\left[-nF\left(\varphi_{\mathcal{R}} - \varphi_0\right)/RT\right] + 1} \right\} . \tag{8.103}$$

Equation (8.103) is considerably simplified when the disc potential is equal to its reversible value, $\varphi_{\mathcal{D}e}$, and $\exp\left[-nF\left(\varphi_{\mathcal{D}e} - \varphi_0\right)/RT\right] = c_R^*/c_O^*$,

$$i_{\mathcal{R}} = i_{d0} \beta^{2/3} \frac{\exp\left[-nF\left(\varphi_{\mathcal{R}} - \varphi_0\right)/RT\right] - c_R^*/c_O^*}{\exp\left[-nF\left(\varphi_{\mathcal{R}} - \varphi_0\right)/RT\right] + 1} . \tag{8.104}$$

Relation (8.104) is plotted in Fig. 8.11 (curve 2).

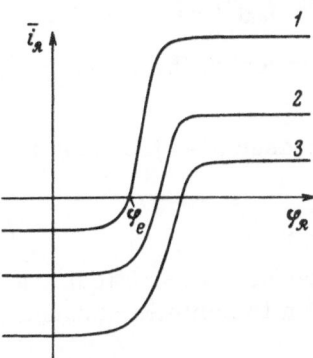

Fig. 8.11. Polarization curves of the ring electrode at various magnitudes of the disc current. 1) Anodic polarization of the disc; 2) disc potentiostated at the reversible potential; 3) cathodic polarization of the disc.

At high cathodic polarizations of the disc electrode, $|\varphi_{\mathscr{D}} - \varphi_0|$ $\gg RT/nF$, the ring current equals

$$i_{\mathscr{R}} = i_{dO} \left\{ -N + \beta^{2/3} \frac{\exp\left[-nF\left(\varphi_{\mathscr{R}} - \varphi_0\right)/RT\right] - c_R^*/c_O^*}{\exp\left[-nF\left(\varphi_{\mathscr{R}} - \varphi_0\right)/RT\right] + 1} \right\} \quad (8.105a)$$

(cf. Fig. 8.11, curve 3).

The limiting diffusion current at the ring (for $|\varphi_{\mathscr{R}} - \varphi_0|$ $\gg RT/nF$) is given by

$$i_{\mathscr{R} \text{ lim}} = i_{dO} \{-N + \beta^{2/3}\}. \quad (8.105b)$$

The decrease of the ring limiting current is due to the enrichment in the Red form of the solution arriving at the ring.

Conversely, anodic polarization of the disc results in an increased concentration of the Ox species at the ring. The ring current is then equal to

$$i_{\mathscr{R}} = i_{dR}N + i_{dO}\beta^{2/3} \frac{\exp\left[-nF\left(\varphi_{\mathscr{R}} - \varphi_0\right)/RT\right] - c_R^*/c_O^*}{\exp\left[-nF\left(\varphi_{\mathscr{R}} - \varphi_0\right)/RT\right] + 1} \quad (8.106a)$$

(Fig. 8.11, curve 1).

The limiting cathodic ring current is equal to

$$i_{\mathscr{R},\text{lim}}'' = i_{dR}N + i_{dO}\beta^{2/3}. \quad (8.106b)$$

The calculations of Albery and Bruckenstein [6] were made without assuming equality of diffusion coefficients. Their voltammetric characteristics differ somewhat from those corresponding to Eq. (8.101):

$$\varphi_{\mathscr{R}} = \varphi_0 + \frac{RT}{nF} \left\{ \ln\left(\frac{D_R}{D_O}\right)^{2/3} + \ln \frac{i_{dO}\beta^{2/3} - i_{\mathscr{R}} - Ni_{\mathscr{D}}}{i_{dR}\beta^{2/3} + i_{\mathscr{R}} + Ni_{\mathscr{D}}} \right\}. \quad (8.107)$$

This difference does not affect, however, the remaining part of the above discussion.

Albery and Bruckenstein [6] discussed also the case of a more complex electrode reaction

$$\mu\,Ox \pm n\,e^- \rightleftarrows Red, \quad (8.108)$$

proceeding at the ring and disc electrodes. Theoretical analysis of the general case is impossible owing to nonlinear boundary conditions.

It can be assumed that the Ox form is present in large excess in the solution and that changes in its concentration at the ring, due to the disc reaction, can be neglected. Under such conditions, the kinetics of the reversible redox system (8.108) is controlled by diffusion of the Red form.

It should be stressed that, as opposed to the previously discussed case, the surface concentration of the Red species on the ring electrode remains constant. In fact, since the ring potential is equal to $\varphi_{\mathscr{R}}$, and the Ox concentration c_O^* is assumed constant everywhere, the constancy of $c_{\mathscr{R}S}$ immediately follows from the Nernst equation for the system (8.108). A similar condition can be introduced in the case of ring dissolution.

Repeating the above derivation,[†] we can derive the following expression describing the ring current:

$$i_{\mathscr{R}} = -i_{dR}\{-N(1 - \exp[-nF(\varphi_{\mathscr{D}} - \varphi_e)/RT] +$$

$$+ \beta^{2/3}(1 - \exp[-nF(\varphi_{\mathscr{R}} - \varphi_e)/RT]\}, \qquad (8.109)$$

where $\varphi_{\mathscr{D}}$, and $\varphi_{\mathscr{R}}$ are the potentials of the disc and of the ring, respectively, and φ_e is the reversible value.

It is interesting to note that in the present case the ring current drops to zero at a potential $\varphi_{\mathscr{R}e}$, different from the reversible potential φ_e of the system (8.108).

It follows from Eq. (8.109) that the potential

$$\varphi_{\mathscr{R}e} = \varphi_e + \frac{RT}{nF}\ln\left(1 - \frac{N}{\beta^{2/3}}\left\{1 - \exp\left[-\frac{nF}{RT}(\varphi_{\mathscr{D}} - \varphi_e)\right]\right\}\right) , \qquad (8.110)$$

depends on polarization of the disc electrode.

In the absence of the Red form in the bulk solution ($c_R^* = 0$), the ring current is equal[‡] to

$$i_{\mathscr{R}} = i_{\mathscr{D}}\{-N + \beta^{2/3}\exp[-nF(\varphi_{\mathscr{R}} - \varphi_{\mathscr{D}})/RT\}, \qquad (8.111)$$

where $i_{\mathscr{D}}$ is the disc current.

[†]Equation (8.109) can be obtained directly from Eq. (8.103) assuming $\exp[nF(\varphi_{\mathscr{D}} - \varphi_0)/RT] \gg 1$ and $\exp[nF(\varphi_{\mathscr{R}} - \varphi_0)/RT] \gg 1$.

[‡]Equation (8.111) can be also easily obtained from Eq. (8.103) by adding to the conditions $\exp[nF(\varphi_{\mathscr{D}} - \varphi_0)/RT] \gg 1$ and $\exp[nF(\varphi_{\mathscr{R}} - \varphi_0)/RT] \gg 1$ a third condition, $c_R^* = 0$.

Choosing the potential difference between the disc and ring electrodes as

$$\varphi_\mathcal{R} - \varphi_\mathcal{D} = -\frac{RT}{nF} \ln \frac{N}{\beta^{2/3}}, \qquad (8.112)$$

we can obtain conditions such that the ring current is zero at any arbitrary values of the disc current.

Albery, Bruckenstein, and Napp [6] verified Eq. (8.112) experimentally using silver disc and ring electrodes. The potential difference between the ring and disc ($\varphi_\mathcal{R} - \varphi_\mathcal{D}$) was measured during the anodic dissolution of the silver disc in 0.1 M $HClO_4$. It can be seen from Table 8.3 that the experimental results are in good agreement with theory.

§ 8.7. Kinetic Currents and
Nonstationary Processes

It has been previously assumed that all the components of electrochemical processes at the disc are stable, i.e., that they do not enter chemical reactions in the bulk. This assumption precludes utilization of the relations derived above in the analysis of electrode reactions with unstable products.

Taking into account chemical reactions of the reactants makes the already complex problem of calculation of the ring current even more difficult. The theoretical analysis of various types of chemical reactions involving intermediates carried out by Albery,

TABLE 8.3. Difference of the Disc
and Ring Potentials during Anode
Dissolution of a Silver Disc Electrode [6]

$\bar{i}_\mathcal{D}$	$\varphi_\mathcal{R} - \varphi_\mathcal{D}$
9	21.5
10	21.0
19.3	21.0
23	22.95
28.4	21.0
Mean	21.5±0.7
Calculated from (8.112)	20.9

Bruckenstein, and others [7-13] required a series of simplifying assumptions. Below, basic results are described; for details of the calculations, the reader is referred to the original papers.

a. Bulk Reaction of the First Order. Consider first the case where decomposition of the disc electrode reaction product (species B) occurs in the bulk, proceeding with a rate $\rho_c \sigma$.† A fraction of B which did not decompose in the solution arrives at the ring electrode and is detected thereon. The general scheme of the processes in question can be represented as follows:

$$\text{at the disc} \qquad A + n_{\mathscr{D}} e^- \rightarrow B, \qquad (8.113a)$$

$$\text{in solution} \qquad \mu_B B \xrightarrow{\rho_c \sigma} P_1, \qquad (8.113b)$$

$$\text{at the ring} \qquad B + n_{\mathscr{R}} e^- \rightarrow P_2, \qquad (8.113c)$$

where μ_B is the order of the reaction in which B disappears.

The convective diffusion equation describing the behavior of the species B is given in the present case by

$$u \frac{\partial c_B}{\partial r} + w \frac{\partial c_B}{\partial z} = D \frac{\partial^2 c_B}{\partial z^2} - \rho_c \sigma F(c_B). \qquad (8.114)$$

The boundary conditions for Eq. (8.114) retain their previous form (8.11)-(8.16).

The term $\rho_c \sigma F(c_B)$, describing the chemical transformation of the species B, results in serious mathematical complications.

The problem consists in finding the ring current:

$$i_{\mathscr{R}} = 2\pi n_{\mathscr{R}} F D \int_{r_{20}}^{r_{30}} \left(\frac{\partial c_B}{\partial z} \right)_S r \, dr.$$

The collection efficiency coefficient N_c which characterizes the fraction of B collected at the ring when the chemical reaction of B is taken into account is defined by

$$N_0 = - \frac{n_{\mathscr{D}} i_{\mathscr{R}}}{n_{\mathscr{R}} i_{\mathscr{D}}} = - \frac{2 n_{\mathscr{D}} F D}{r_{10}^2 i_{\mathscr{D}}} \int_{r_{20}}^{r_{30}} \left(\frac{\partial c_B}{\partial z} \right)_S r \, dr. \qquad (8.115)$$

---

†In analogy with Chap. 6, we retain this designation of the rate constant of a chemical reaction involving an electroactive species.

where $i_{\mathscr{D}}$ is the current density at the disc electrode. The magnitude of $(\partial c_B / \partial z)_S$ at the surface of the ring electrode is determined in the process of solution of Eq. (8.114) and depends considerably on the rate of the chemical reaction $\rho_c \sigma$. Thus, in addition to the dependence on the geometric parameters (r_{20}/r_{10} and r_{30}/r_{10}), the magnitude of N_c depends also on the parameter $\varepsilon_c = \mu_c/\delta_d$. The latter quantity was shown in Chap. 6 to characterize the ratio of the convective transport rate (proportional to $1/\delta_d$) to the rate of disappearance of B in reaction (8.113b) (inversely proportional to the thickness of the kinetic reaction layer, μ_c).

The aim of the theoretical calculations is to examine the dependence of N_c on the above parameters. In the limiting case of slow chemical reactions involving B, when the rate of reaction (8.113b) is low ($\varepsilon_c \ll 1$), the presence of an additional term on the right side of Eq. (8.114) should significantly affect the distribution of c_B. The decrease of B concentration due to the chemical reaction is small and can be taken into account by considering the corresponding term in Eq. (8.114) as a small perturbation.

It is clear that the collection efficiency should be somewhat decreased by the occurrence of the chemical reaction, as compared with the N value obtained for a stable product (§ 8.1). This decrease should disappear when reaction (8.113b) ceases, and consequently it should be, to a first approximation, proportional to the rate of reaction (8.113b). Furthermore, upon the reversal of reaction (8.113b) with slow generation of B in the bulk solution, N_c should increase.

Thus, in the first approximation,

$$N_c = N(1 - b_1 \delta_d^2/\mu_c^2 + \ldots). \tag{8.116}$$

The form of the b coefficient and its dependence on the geometric parameters (r_{20}/r_{10} and r_{30}/r_{10}) can be obtained only by successive approximations.

If B disappears in a reaction of the first order, the thickness of the kinetic reaction layer $\mu_c = (D/\rho_c \sigma)^{1/2}$, and Eq. (8.116) can be transformed into

$$N_c = N(1 - b_1 \delta_d^2 \rho_c \sigma/D + \ldots). \tag{8.116a}$$

Albery and Bruckenstein calculated the b_1 coefficient first by a rough approximation [8] and subsequently with higher accuracy

[11]. The dependence of b_1 on the geometry of electrodes is discussed in [11]. These authors calculated also the next terms in Eq. (8.116a) (proportional to δ_d^4/μ_c^4). They concluded† that Eq. (8.116a) satisfactorily describes the effects of the first-order chemical reaction on N_c for reaction rates $\delta_c\sigma < 10^2$ sec^{-1}. Recently Prater and Bard [70] made a computer calculation of the kinetics of a first-order bulk reaction.

Experimental measurements of rate constants of bulk reactions were carried out [12, 71] in the following systems:

1. Disc: Bromine is regenerated by oxidation of bromide ions:

$$Br^- \rightarrow {}^1/_2 \, Br_2 + e^-;$$

   solution: bromine disappears in bromination of anisole or m-fluoroanisole;

   ring: excess of bromine becomes reduced to bromide ions.

2. Disc: reduction of ferric to ferrous ions:

$$Fe^{3+} + e^- \rightarrow Fe^{2+};$$

   solution:

$$Fe^{2+} + VO_2^+ + 2H^+ \rightarrow Fe^{3+} + VO^{2+} + H_2O;$$

   ring: excess Fe^{2+} ions are detected by oxidation to Fe^{3+} ions.

Using plots constructed for the above systems according to Eq. (8.116a) (dependence of N_c on the rate constant of the bulk reaction, $\rho_c\sigma$, with appropriate geometric parameters taken into account), we can easily calculate $\rho_c\sigma$ for the given measured $i_\mathcal{R}/i_\mathcal{D}$ values.

The rate constants of the reaction between Fe^{2+} and VO_2^+ ions measured at the rotating ring-disc system are shown in Fig. 8.12 as a function of pH of the solution. They are in good agreement with the literature data.

Adams et al. [72] proposed an empirical method of evaluation of the bulk reaction rates. Instead of theoretical calculations of N_c, an experimental dependence of the latter on $\rho_c\sigma$ is studied using a set of systems with known, independently measured, rate

---

† It was shown experimentally [12] that this calculation [11] is not very accurate either. The errors are mainly due to convective transport in the direction normal to the disc surface, which was neglected.

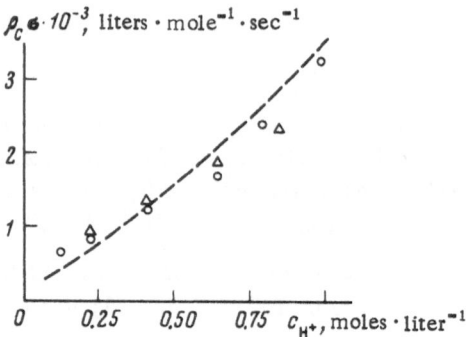

Fig. 8.12. Dependence of the rate constant of the $Fe^{2+} + VO_2^+$ reaction on hydrogen ion concentration in the solution [12]. $\triangle$, $\bigcirc$) Measured at the ring-disc system; dashed curve corresponds to literature data.

constants of the bulk reactions, $\rho_c \sigma$. The following reactions were chosen as standard systems: hydrolysis of quinoneimines of p-aminophenol and 3-methyl-p-aminophenol in 0.005 M H_2SO_4, for which the $\rho_c \sigma$ values were measured chronopotentiometrically. The rate constants thus obtained were used to construct a calibrated plot of $N_c = i_{\mathscr{R}}/i_{\mathscr{D}}$ vs $(\rho_c \sigma / \omega)^{1/2}$ (Fig. 8.13). Using the latter diagram, we found $\rho_c \sigma$ values for the same reactions at various sulfuric acid concentrations [72].

The ring-disc system was also used to evaluate the lifetime of univalent beryllium ions produced by dissolution of metallic beryllium in aqueous solutions [73].

The general expression (8.116) can also be used to estimate the effect on N_c of the disappearance of an intermediate in reactions of order higher than one (e.g., dimerization). It must be

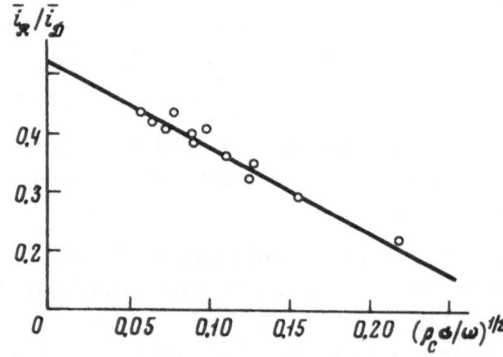

Fig. 8.13. Calibrated $i_{\mathscr{R}}/i_{\mathscr{D}}$ vs $(\rho_c \sigma / \omega)^{1/2}$ diagram obtained using the ring-disc system with a carbon paste disc. Hydrolysis of quinoneimines of p-aminophenol and 3-methyl-p-aminophenol in 0.05 M H_2SO_4 [72].

remembered, however, that Eq. (8.116) is valid only in the case
of relatively slow bulk reactions of intermediates, where reaction
(8.113b) affects the distribution of the B species very little.

In the case of fast chemical reactions, N_c depends in a more
complex manner on the parameters listed above. However, if
the chemical reaction rate is very high ($\varepsilon_c \ll 1$), B never reaches
the ring electrode and $N_c = 0$. Disappearance of the ring current
depends obviously on the dimensions of the disc and ring electrodes.

The case of a fast chemical reaction (8.113b) of the first order
was discussed in detail by Albery and Bruckenstein [8]. The
authors solved Eq. (8.114) with a linear function $F(c_B)$ separately
for each of the three regions — in the vicinity of the disc, of the
separator, and of the ring. The distribution of c_B at the disc was
calculated assuming convective terms in (8.114) to be negligible
at high $\rho_c \sigma$ values. Convective transport was taken into account
in the radial direction only in the separator and ring regions; convection in the axial direction was neglected.

The magnitude of N_c at rates $\rho_c \sigma \sim 10^3\ \mathrm{sec}^{-1}$ was found to
obey the relation

$$N_c \simeq 1.75 K^{-3} \exp\left[-4K^3 (r_{20} - r_{10})/r_{10}\right], \qquad (8.117)$$

where the parameter K used in [8] practically coincides with the
parameter $1/\varepsilon_c$ ($\varepsilon_c^3 \approx {}^1/_2\ K^3$).

Subsequent, more accurate calculations by Albery [11] also
included a series of intermediate cases (fast reaction using a narrow separator or a narrow ring, etc.).

The appearance at the ring of kinetic currents due to a fast
preceding or following chemical reaction of the first order was
investigated by Albery and Bruckenstein [10]. They calculated the
kinetic current at a partially "coated" disc; i.e., kinetic complications in a system described in §2.8 were analyzed. The effect
of the disc on the kinetic ring current was studied for a fast chemical reaction. Kinetic currents in the rotating ring-disc system
due to first-order chemical reactions were also discussed by Kiss
and Farkas [74].

b. Bulk Reaction of the Second Order. In their studies of
second-order chemical reactions at a ring-disc system, Albery
and Bruckenstein [7, 9, 13] paid special attention to titration pro-

cesses. Reactions involved in this process will be discussed using as a concrete example the titration of As^{3+} ions with bromine.

The ring-disc system is immersed in a solution of As^{3+} ions containing an excess of Br^- ions. The unknown As^{3+} concentration is to be determined by titration. The value of the disc potential $\varphi_{\mathscr{D}}$ ensures oxidation of Br^- ions to bromine and lack of electro-chemical activity of As^{3+} ions. A ring potential $\varphi_{\mathscr{R}}$ is chosen so that neither Br^- nor As^{3+} can undergo discharge, but molecular bromine can be reduced. Bromine produced at the disc undergoes a fast chemical reaction with As^{3+} ions:

$$Br_2 + As^{3+} \xrightarrow{p_c \sigma} As^{5+} + 2Br^-.$$

It will be assumed that the rate of the above reaction is high and the reaction zone (cf. § 6.5) very narrow. The location of the reaction zone with respect to the disc electrode is determined by the rate of bromine formation at the disc and the rate of supply of As^{3+} ions from the bulk solution (cf. § 6.5).

If the amount of bromine produced is small, the reaction zone lies practically at the surface of the disc. With increasing rate of Br_2 formation, the reaction zone extends more into the bulk solution and beyond the boundaries of the disc. One of the possible locations of the reaction zone boundary is shown in Fig. 8.14 (dot-dash curve). Here region I represents part of the space where bromine predominates and As^{3+} ions are practically absent. Conversely, practically no bromine penetrates into region II.

At a certain value of the disc current ($i_{\mathscr{D}}$), the reaction zone reaches the internal edge of the ring electrode. If the ring potential $\varphi_{\mathscr{R}}$ is chosen so that Br_2 becomes rapidly reduced to Br^-, the moment of bromine appearance at the ring coincides with the appearance of $i_{\mathscr{R}}$. Further increase of $i_{\mathscr{D}}$ results in increasing $i_{\mathscr{R}}$. It can be easily shown that the appearance of the ring current depends not only on $i_{\mathscr{D}}$, but on the bulk concentration of As^{3+} ions.

Integration in Eq. (8.115) depends obviously on the location of the reaction zone r_c at the disc surface. At $r_c \ll r_{20}$, the integrand is equal to zero and $N_c = 0$. When $r_{20} \leq r_c < r_{30}$, integration in Eq. (8.115) is carried out between r_{20} and r_c. As soon as the reaction zone extends beyond the ring region, the collection efficiency obeys expression (8.3).

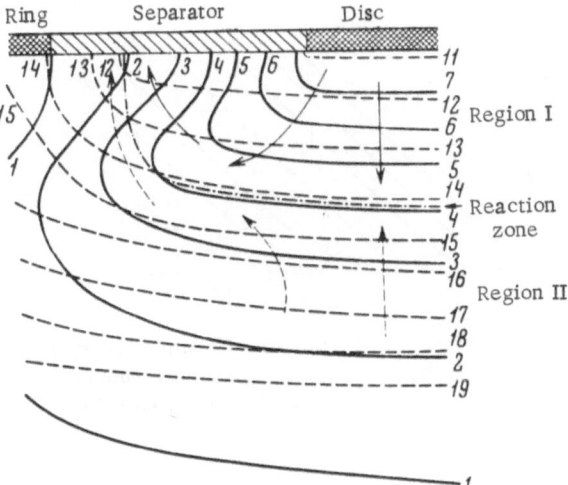

Fig. 8.14. Concentration profile of bromine (solid lines) and As^{3+} (dotted lines) in the vicinity of the rotating ring-disc electrode [9]. The reaction zone is shown by the dot-dash curve. As^{3+} concentration increases from 11 to 19, and bromine concentration, from 1 to 7.

In the latter case, the ring current can be obtained as follows. The total amount of bromine produced at the disc can be regarded as distributed in two parts: a) the part carried to the disc and reduced thereon, and b) the remaining part which enters the reaction with As^{3+} ions. The first part is proportional to $i_{\mathscr{D}}$ and depends on the collection efficiency N; the second part is proportional to the limiting diffusion flux which supplies As^{3+} ions to the disc surface. The calculations of Albery, Bruckenstein, and Johnson [7] gave the following expression for $i_{\mathscr{R}}$:

$$i_{\mathscr{R}} = -Ni_{\mathscr{D}} - i_{d,\,As}\beta^{3/2}, \qquad (8.118)$$

where $i_{d,\,As}$ is the limiting diffusion current of As^{3+} ions at the disc; $\beta = (r_{30}/r_{10})^3 - (r_{20}/r_{10})^3$.

The above method of "titration in the diffusion layer" using the ring-disc assembly was checked experimentally by Bruckenstein et al. [7, 13, 75-77]. In addition to the $As^{3+} + Br_2$ reaction, chemical reactions between As^{3+} and I_3^- and between allyl alcohol and Br_2 were studied (I_3^- and Br_2 are regenerated electrochemically at the disc, and their excess is detected at the ring).

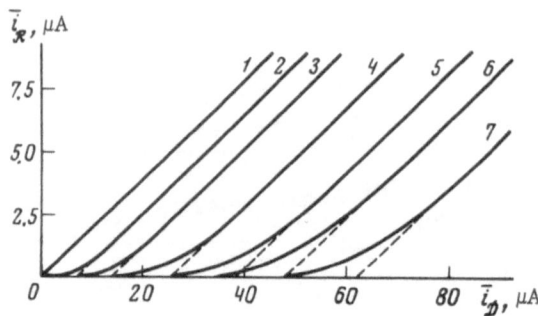

Fig. 8.15. Curves of titration of As^{3+} with electrogenerated bromine using the ring-disc electrode [75]. Bulk concentration of As^{3+} ($\times 10^5$ moles/liter): 1) 0; 2) 1.17; 3) 2.34; 4) 4.70; 5) 7.05; 6) 9.40; 7) 11.7.

A typical set of "titration curves," i.e., $i_{\mathscr{R}} - i_{\mathscr{D}}$ relations, is shown in Fig. 8.15. In agreement with Eq. (8.118), the ring current is proportional to the disc current at high $i_{\mathscr{D}}$ values only. Extrapolation to $i_{\mathscr{R}} \to 0$ results, for each As^{3+} concentration, in a value of disc current proportional to the limiting diffusion current of As^{3+} ions, which in turn is proportional to the unknown bulk concentration of As^{3+}. Plotting the extrapolated $i_{\mathscr{D}}$ values ($i_{\mathscr{R}} \to 0$) as a function of $c^*_{As^{3+}}$, we obtained a calibrated straight line permitting the determination of $c^*_{As^{3+}}$ from a single experimental titration curve. The method allows determination to be made of trivalent arsenic in the concentration range 10^{-8}-10^{-2} mole^{-1} with a relative error of $\pm 2\%$.

Utilization of the experimental dependence of $i_{\mathscr{R}}$ on the disc current $i_{\mathscr{D}}$ demonstrates, however, that the ring current appears somewhat earlier than expected according to the condition $r_c = r_{20}$. Albery and Bruckenstein [9] interpret this fact in terms of a reaction zone extending over a wider region than that assumed in the simple theory. In fact, the assumption of the instantaneous character of the bulk reaction and consequently the representation of the reaction zone as an extremely thin layer is rather crude. Thus, bromine penetrates to some extent into region II, and As^{3+} ions into region I (cf. Fig. 8.14). In Fig. 8.14 the concentration profile of bromine is shown by solid lines, and that of As^{3+} ions by dotted lines. The location of the reaction zone for the given concentration profiles is determined by the points of cross section of the Br_2 and As^{3+} curves corresponding to equal concentrations of the two substances. According to this picture, the appearance of the ring current can be expected even if the edge of the reaction zone is still in the separator region. This ring current is caused by the "leakage" of Br_2 beyond the reaction zone. Since the thickness of the reaction zone, and consequently $i_{\mathscr{R}}$, at the instant of its appearance is determined by the rate constant of the bulk reaction, a study of $i_{\mathscr{R}}$ in this region allows the rate constant of the chemical reaction to be found.

The calculations presented in [9] and subsequently improved by Albery, Hitchman, and Ulstrup [13] resulted in the following relation:

$$i_{\mathscr{R}} = 0.21 \pi r_{20}^2 n F D \omega^{3/2} \nu^{-1/2} / \rho_c \mathfrak{d}. \qquad (8.119)$$

The ring-disc was also used in a study of the dimerization of the unstable reduction product of manganate ions in alkaline solution [78].

c. Nonstationary Processes. Studies of nonstationary electrochemical processes by means of the ring-disc assembly are a very promising method (as was shown in § 8.8) for the investigation of the nature of intermediates in electrode reactions and of adsorption processes. Their application is limited at present by the absence of a more or less satisfactory theory of transient processes at a ring-disc. The theoretical description of transients, which results in an equation of the type of relation (8.114), is connected with the solution of a complex mathematical problem.

Albery [11] analyzed the dependence of the ring current on a varying disc current. His equations, however, are valid only after a sufficiently long time has elapsed from the beginning of the transient. The time lag between the signals at the disc and the ring was not discussed at all in [11].

Bruckenstein and Feldman [79], who first attempted an evaluation of the time lag $\tau_{\mathscr{R}}$, used rather rough approximations. The authors assumed that the product formed near the disc edge (r_{10}) reached the internal edge of the ring (r_{20}) by a simultaneous transfer in two directions: by diffusion away from the disc and by radial convective transport together with the solution.

Neglecting the convective transport along the rotation axis, Bruckenstein and Feldman [79] discussed the trajectory of motion of the particle and obtained the following estimate of the time lag:

$$m\tau_{\mathscr{R}} = 43.1 \, (\nu/D)^{1/3} \, [\log(r_{20}/r_{10})]^{2/3}, \tag{8.120}$$

where m is the rotation speed (rpm).

For the experimental verification of Eq. (8.120), Bruckenstein and Napp [14] studied the reduction of Ag^+ ions at a platinum electrode. The ring electrode was potentiostated at a potential in the region of the limiting current of this reaction. The disc potential was varied in a stepwise manner from an initial positive value $\varphi_{\mathscr{D}}^{(1)} = 0.5$ V (at which Ag^+ is not reduced) to the final value $\varphi_{\mathscr{D}}^{(2)} = 0.1$ in the limiting current region. Switching on the cathodic disc current results in a depletion of silver ions in the vicinity of the electrode and a progressive drop of the ring current. The ring current starts decreasing after a time $\tau_{\mathscr{R}}$ from the moment of switching on the disc current. The time lag $\tau_{\mathscr{R}}$ is necessary to let the depleted solution reach the internal edge of the ring. The measured $\tau_{\mathscr{R}}$ values are shown in Table 8.4. The experimental value of $m\tau_{\mathscr{R}}$ (17.5 sec·rpm) is in satisfactory agreement with the value 15.0 sec·rpm calculated according to Eq. (8.120) for the experimental conditions employed. Similar results were obtained [71] in studies of the oxidation of Br^- to Br_2 and $Fe(CN)_6^{3-}$ to $Fe(CN)_6^{4-}$.

Prater and Bard [80] calculated by means of a computer the transient ring current using a method of numerical modeling of

TABLE 8.4. Time Lag τ_R in the Rotating Ring-
Disc System as a Function of the
Rotational Speed [14]

m, rpm	$\tau_{\mathscr{R}} \cdot 10^3$, sec	$m\tau_{\mathscr{R}}$, sec · rpm
400	40	16.0
900	18	16.2
1600	10	16.0
2500	7	17.5
3600	5	18.0
4900	4	19.6
6400	3	19.2
8100	2	16.2
10000	1.8	18.0
	Mean	17.5±2.1

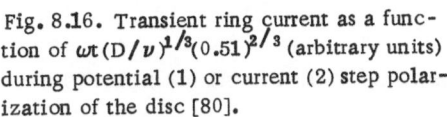

Fig. 8.16. Transient ring current as a function of $\omega t (D/\nu)^{1/3}(0.51)^{2/3}$ (arbitrary units) during potential (1) or current (2) step polarization of the disc [80].

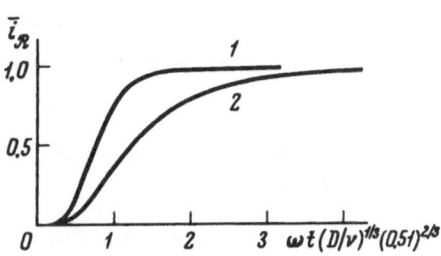

disc electrode processes developed by Feldberg [81]. Their results are close to those of Bruckenstein and Feldman [79]. The calculated ring-current transients resulting from stepwise variations of the disc potential (curve 1) or current (curve 2) are shown in Fig. 8.16. The faster rise of the ring current in the first case is due to the abrupt surge of the disc current at the instant of switching on the potential step on the disc.

§ 8.8. Application of the Ring-Disc Assembly to Studies of Anodic Dissolution, Corrosion and Passivation of Metals, and Electrode Reactions Involving Organic and Inorganic Compounds

Owing to the possibility of detecting unstable products or intermediates and measuring the kinetic characteristics of separate reaction steps, the ring-disc electrode has been widely used in recent years for various electrochemical purposes.

Anodic Dissolution, Corrosion, and Passivation of Metals. Indium (or indium amalgam) dissolves anodically with formation of In^+ ions [82]. The latter can be reduced to In^0 or oxidized to In^{3+} at the ring [83, 84]. Over the potential range which prevents In^{3+} reduction, the ratio of anodic to cathodic ring currents is equal to 2 but decreases with increasing disc current, although the total ring current remains constant at a value corresponding to a three-electron transfer. Evidently, at high dissolution rates, a considerable part of the current is used in the formation of intermediate In^{2+} ions in addition to In^+. The observed dependence of

the reversible ring potential on the indium disc dissolution current
led to the conclusion that the first dissolution step

$$In \to In^+ + e^-$$

is reversible: In^+ undergoes no chemical bulk reactions nor disproportionation.

A convenient method for studying intermediates capable of
both oxidation and reduction at the ring was proposed by Miller
and Visco [83, 85] who developed a rotating disc electrode with a
sectioned ring (the so-called split ring). The ring is cut along
the diameter in two insulated halves with independent current
leads.† The sections are potentiostated in the regions of the negative limiting reduction and oxidation currents of the intermediate.
Thus, three quantities can be simultaneously measured: the disc
current, the oxidation current of the intermediate at the half-ring
I, and the reduction current of the intermediate at the half-ring
II.‡ This arrangement is particularly useful in studies of the
catalytic H_2O_2 decomposition (cf. § 8.4). Obviously, the cathodic
and anodic polarization curves can be obtained in the usual ring-
disc system; however, the latter method fails if fast changes of
the electrode surface occur with time.

Miller studied anodic copper dissolution in alkaline medium
[86] using a sectioned ring. He was able to detect formation of
uni-, di-, and trivalent copper ions. Cu(I) and Cu(III) partly adsorb
at the copper anode surface and passivate the latter, whereas
Cu(II) fully passes into the solution. The same method was used
in studies of intermediates produced in anodic dissolution of silver
[87] and vanadium [88].

Attempts were made to use the ring-disc system for the
establishment of the mechanism of corrosion and anodic dissolution of certain alloys. The question was whether the more noble
metal passes from the alloy into the solution and subsequently
redeposits at the surface, or whether the process involves dissolution of the less noble component only. It was found that no gold

†Half-rings can also be made of different metals.
‡The latter two quantities are somewhat distorted, since the two halves are not fully
independent of each other. At small widths of the separating insulator, the spiral
current lines (cf. §2.8) from the end of one half pass through the beginning of the
other. However, the error thus introduced does not exceed 1% [85].

ions could be detected at the ring upon anodic dissolution of a
90 : 10 Cu–Au alloy in sulfate ion solution, i.e., corrosion proceeds
by leaching of copper only. Conversely, during dissolution of
copper-rich brasses (for copper contents > 70%) reduction of Cu^{2+}
ions can be detected at the ring. Consequently, both copper and
zinc dissolve, and copper redeposits at the electrode [89, 90].
Using a split ring in the same system [85], we can separately
measure Cu^+ and Cu^{2+} ions formed in brass corrosion. In acid
solutions, the fraction of Cu^+ ions is small, increasing sharply at
pH > 7.4.

An iron rotating disc electrode with a platinum ring was used
in studies of iron corrosion in an ammonia buffer [91].

Oshe and Kabanov [92, 93] used a zinc rotating disc electrode
with a zinc ring to define regions in which passivation of zinc in
alkaline solutions is due to adsorption or phase oxide formation.
The disc was subjected to an anodic potential step (0.05-0.1 V),
and the ring was potentiostated in the region of the limiting current
for zincate ion reduction. The $\bar{i}$ vs t curves measured simultaneously at the disc and ring electrodes are shown in Fig. 8.17. For
the sake of comparison of the zinc dissolution current $i_{\mathscr{D}}$ with the
zinc-ion reduction current $i_{\mathscr{R}}$ the latter is multiplied by $1/N$.

In a fraction of a second after imposing the potential step on
the disc, a quasi-stationary regime establishes itself at the ring
(cf. § 8.7). From this moment on, in the potential region corre-

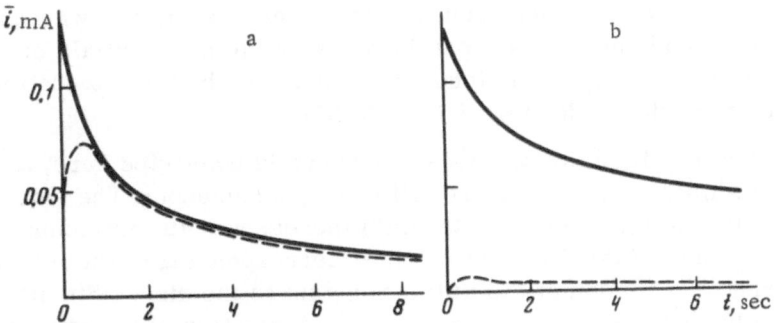

Fig. 8.17. Passivation of zinc in 0.1 N KOH [93]. Current transients on
the disc ($\bar{i}_{\mathscr{D}}$ — solid lines) and on the ring ($\bar{i}_{\mathscr{R}}/N$ — dashed lines) upon
imposing a potential step on the disc: a) from −0.9 to −0.8 V; b) from
1.0 to 1.1 V.

sponding to dissolution of active, or weakly passivated zinc, we have $i_{\mathscr{D}} \approx \bar{i}_{\mathscr{R}}/N$ (with an accuracy to 5%). This indicates that virtually the total disc current is used in zinc dissolution and no formation of a phase surface oxide occurs. Passivation in this region can be due only to chemisorbed oxygen species (cf. Fig. 8.17a).

Conversely, in the region of strong passivation, $i_{\mathscr{R}}/N$ is considerably lower than $i_{\mathscr{D}}$; i.e., a significant fraction of the disc current is used in formation of an oxide layer, which evidently causes zinc passivation (cf. Fig. 8.17b). Similar results were obtained in a study of anodic dissolution of passivated nickel: the whole current is used not in metal dissolution but in formation of an oxide film [94].

An anodic potential step applied to passivated iron in sulfuric acid solutions results initially in the formation of a thicker passivating oxide layer; after 1-2 min, a stationary regime is established, and the total current is used in iron dissolution [95].

<u>Multistep Reactions of Ion Reduction by Metals.</u> The polarization curve for Cu^{2+} ion reduction in sulfate solutions shows a single wave corresponding to the transfer of two electrons:

$$Cu^{2+} + 2e^- \to Cu^0.$$

Nekrasov and Berezina [21] demonstrated that the reaction proceeds in two steps. The product of the first step (Cu^+ ion) is unstable at the disc over the entire potential range (as opposed to the case of complexing solutions, for which two separate waves are observed and univalent copper ions are stable at potentials of the first wave [21, 76]). Cu^+ ions can be detected by their oxidation at the ring electrode (Fig. 8.18) [96, 97].

The maximum on the $i_{\mathscr{R}} - \varphi_{\mathscr{D}}$ curve is connected here, as in other similar cases, with the following phenomenon: The rate of formation of the intermediate (Cu^+) increases with increasing polarization of the disc up to a limit corresponding to the limiting current of diffusion of the initial reactant to the disc. Simultaneously, the rate of the next step — transformation of the intermediate (k_3) — increases as well. Superposition of the two effects results in a characteristic maximum on the $i_{\mathscr{R}} - \varphi_{\mathscr{D}}$ curve.

The reduction of Cu^{2+} (and anodic dissolution of metallic copper) in sulfuric acid solutions was also studied by Bruckenstein

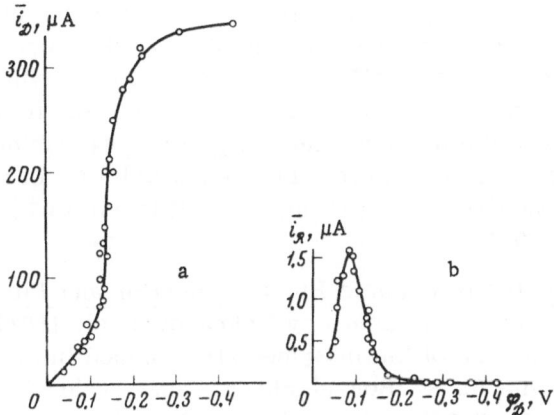

Fig. 8.18. Cathodic reduction of Cu^{2+} ions at a disc electrode with ring [21]. a) Polarization curve at a Pt disc in a solution of $3 \cdot 10^{-2}$ N $CuSO_4 + 1$ N Na_2SO_4; b) the limiting current of Cu^+ oxidation at a gold ring as a function of disc potential.

et. al. [98, 99]. The authors used cyclic voltammetry with linearly varying potential at the ring-disc electrode. They found, like others have, that the first monolayer of copper deposited on platinum differs considerably in its properties from a compact copper deposit.

Cu$^+$ ions are also formed in the absence of current according to

$$Cu + Cu^{2+} \rightleftarrows 2Cu^+.$$

The surface concentration of Cu^+ ions can be determined by ring current measurements. In this way, Tindall and Bruckenstein [100] determined the heterogeneous equilibrium constant in 0.2 M H_2SO_4:

$$\sigma = c_{Cu^+}^2 / c_{Cu^{2+}} = (5.6 \pm 0.4) \cdot 10^{-7} \text{ mole/liter } (25° C).$$

A very close value, $5.68 \cdot 10^{-7}$ mole/liter, was obtained by an independent method (coulometry at constant current).

The ring-disc assembly has also been used for polarographic determination of copper and silver in solutions containing both ions. The analytical method does not differ in principle from that usually used (preliminary deposition of copper and silver at the disc and subsequent anodic dissolution). Current measurements, however, during anodic dissolution of metals accumulated at the disc were carried out at the ring potentiostated at a potential ensuring Cu^{2+} (or Ag^+) reduction. Using a split ring with one half potentiostated in the potential region corresponding to the reduction of Cu^{2+}, and the other at that of Ag^+ ions, we can determine both elements from one curve. The reproducibility of analytical results for concentrations 10^{-8} and 10^{-5} M is within 15 and 5%, respectively [101].

Intermediates in Organic Electroreduction and Electroxidation Reactions. A series of papers by Nekrasov et al. [102] involve a study of the nature of the intermediates formed in the reduction of carbonyl compounds. Benzaldehyde, benzophenone, acetophenone, fluorenone, benzoylferrocene, and ferrocenylaldehyde when reduced at an amalgamated gold disc electrode form unstable products oxidizable at the ring. The yield of some intermediates reaches 100%; in other cases it is considerably less (1%).

In addition to current measurements at the disc and ring electrodes at various values of concentrations of the initial reactant, pH, and rotation speeds, Nekrasov and Korsun used UV and ESR spectroscopy to determine the nature of intermediates. ESR spectroscopy required measurements to be carried out in a frozen solution which contained a considerable concentration of the intermediate produced by preliminary electrolysis. Potentials of the individual anodic waves were compared with potentials of oxidation waves of stable substances which might be produced as intermediate products in the reactions studied.

Three types of intermediates were detected: anion-radicals, hydrodimers (e.g., Benzopinacol), and complexes formed by charge transfer between the anion-radical (free radical) and a molecule of the original carbonyl compound.

Anion-radicals of benzaldehyde and acetophenone, as well as those of benzoylferrocene and ferrocenylaldehyde, in aqueous solutions are detected first. Their lifetime is of the order of 10^{-3}-10^{-2} sec, and therefore they cannot be detected in the solution by ESR. Anion-radicals disappear in a bulk reaction of the second order (dimerization), and at sufficiently negative disc potentials they are reduced to the respective carbonyl compound.

Of special interest are free radicals (anion-radicals) stabilized by complex formation involving charge transfer with a molecule (protonated or not) of the carbonyl compound. Their lifetime is of the order of 10-100 sec.

Nitrobenzene is reduced in two stages to the anion-radical and phenylhydroxylamine (both species oxidize at the ring with two separate waves). The rate constant of the second stage was calculated using the theory described in § 8.3.

Oxidation of dimethylaniline at a carbon paste electrode proceeds with intermediate formation of tetramethylbenzidine [103].

An interesting application of the ring-disc system is reported in [104], namely, in studies of electrochemiluminescence. Usually, both the anion and cation radicals which react in solution with emission of light are produced at the same electrode polarized with ac. It was found that the process can be more conveniently studied under conditions of constant current in a two-electrode, ring-disc system. The two forms are produced by oxidation at one electrode and reduction at the other electrode. Electrochemiluminescence can be excited in this way in a solution of pyrene and tetramethyl-p-phenylenediamine in dimethylformamide (containing tetrabutylammonium perchlorate as inert electrolyte). Electrochemiluminescence accompanies the reaction of the pyrene anion with the tetramethyl-p-phenylenediamine cation (Wurster's Blue). The intensity of luminescence is proportional to the rotation speed.

Adsorption Processes. With the exception of [27, 48], the work described in this chapter has neglected adsorption of electroactive species at the electrode. In fact, however, establishment of the adsorption equilibrium can be inhibited, and the kinetics of adsorption or desorption processes can control the overall kinetics. Measurements of the extent of adsorption are also of primary interest. The methods of investigation of adsorption at the ring-disc system proposed by Bruckenstein and Napp [14] are described below.

Method I (quasistationary) consists in the following. The disc electrode is potentiostated at $\varphi_{\mathcal{D}}^{(1)}$, a potential at which species A adsorbs on the disc without undergoing electrode reaction. Upon establishment of the adsorption equilibrium, the disc potential is varied linearly with time from an initial value $\varphi_{\mathcal{D}}^{(1)}$ to a certain final value $\varphi_{\mathcal{D}}^{(2)}$. At the latter potential, A oxidizes to B:

$$A \to B + ne^-,\qquad\qquad (8.121)$$

B desorbs partially or fully from the disc. In general, reaction (8.121) can be represented by one of the special cases:

$$A_{ads} \rightarrow B_{ads} + ne^-, \qquad (8.121a)$$

$$A_{ads} \rightarrow B_d + ne^-, \qquad (8.121b)$$

$$A_d \rightarrow B_{ads} + ne^-, \qquad (8.121c)$$

$$A_d \rightarrow B_d + ne^-, \qquad (8.121c)$$

where the subscripts "ads" and "d" designate adsorbed and dissolved state, respectively.

The amount of B_d can be determined from the current at the ring potentiostated at the reduction potential of B:

$$B + ne^- \rightarrow A. \qquad (8.122)$$

The shape of the current (disc or ring) vs disc potential curves, presented schematically in Fig. 8.19, depends on the actual character of reaction (8.121). Curve 1 corresponds to absence of adsorption. Curve 2 has the well-known shape characteristic of the adsorption of the electroactive species A. Curve 3 at the ring ($i_{\mathscr{R}} - \varphi_{\mathscr{D}}$) corresponds either to the simplest case (neither A nor B adsorb) or to equal coverages with both species (in order to differentiate between these two cases it is necessary to analyze the $i_{\mathscr{D}} - \varphi_{\mathscr{D}}$) curves). Curve 4 results from predominant adsorption of A, and curve 5 from the reverse case – predominant adsorption of B.

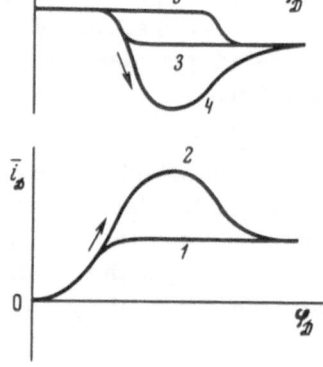

Fig. 8.19. Adsorption studied by the quasi-stationary method I [14]. Schematic representation of $\bar{i}_{\mathscr{D}}$ vs $\varphi_{\mathscr{D}}$ (1, 2) and $\bar{i}_{\mathscr{R}}$ vs $\varphi_{\mathscr{D}}$ (3–5) curves during a linear anodic potential sweep on the disc.

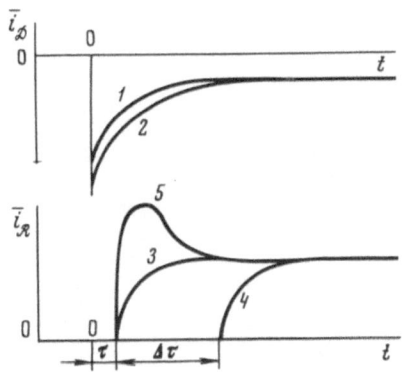

Fig. 8.20. Adsorption studies using the
nonstationary method II [14]. Schematic
representation of disc $\bar{i}_{\mathscr{D}}$ vs t transients
(1, 2) and ring $\bar{i}_{\mathscr{R}}$ vs t transients (3-5)
accompanying a potential step on the disc
from $\varphi_{\mathscr{D}}^{(1)}$ (A does not oxidize) to $\varphi_{\mathscr{D}}^{(2)}$
(A oxidizes to B).

Method II (nonstationary) is based on measurements of transition time, which characterizes the rate of convective diffusion of the species from the disc to the ring. The simplest transient case was discussed in § 8.7; here the discussion is restricted to complications introduced in transient processes by adsorption.

At the initial moment (t = 0) the concentration of A in the bulk is $c_A^* \neq 0$, and B is absent in the solution ($c_B^* = 0$). At the initial disc potential $\varphi_{\mathscr{D}}^{(1)}$ reaction (8.121) does not occur. The ring is potentiostated all the time at a potential corresponding to the reduction of B (reaction 8.122), thus allowing B to be detected when it appears at the ring.

The disc potential is then changed in a stepwise manner from the initial $\varphi_{\mathscr{D}}^{(1)}$ to the final value $\varphi_{\mathscr{D}}^{(2)}$, at which reaction (8.121) occurs. The disc transient corresponds to curve 1 (Fig. 8.20) if A does not adsorb, and 2 if A adsorbs at $\varphi_{\mathscr{D}}^{(1)}$. The ring transient ($i_{\mathscr{R}}$ vs t curve) corresponds in the simplest cases discussed in § 8.7 (neither A nor B adsorbs, or both species adsorb equally) to curve 3. The transition time $\tau_{\mathscr{R}}$ elapsed between potential changeover and the appearance of the ring current is calculated according to Eq. (8.120). However, if A does not adsorb at $\varphi_{\mathscr{D}}^{(1)}$ and B adsorbs at $\varphi_{\mathscr{D}}^{(2)}$ rapidly and with significant coverage, desorption of the latter species from the disc can occur only after saturation has been reached at the electrode surface. This results in an increase of the transition time by a time $\Delta\tau_{\mathscr{R}}$ over the value calculated from Eq. (8.120). The increased transition time indicates adsorption of B. The ring transient corresponds in this case to

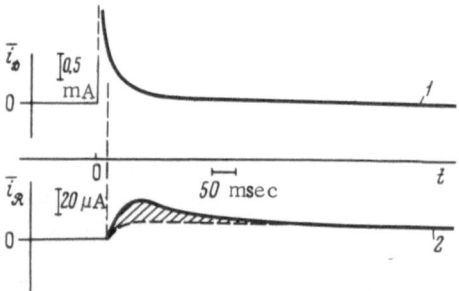

Fig. 8.21. Experimental study of Cu$^+$ adsorption on platinum using method II [14]. Solution: 10^{-3} M CuCl + 0.5 M HCl, potential step from -0.1 to 0.4 V; m = 900 rpm. 1) Disc transient; 2) ring transient; the dashed line is the background curve (without adsorption).

curve 4. The coverage with B can be found either by integration of the disc transient from t = 0 to $t = \Delta\tau_{\mathcal{R}}$, or by integration of the ring transients to obtain the area between curves 3 and 4.[†] Finally, curve 5 corresponds to the case where A adsorbs more strongly at $\varphi_{\mathcal{D}}^{(1)}$ than B at $\varphi_{\mathcal{D}}^{(2)}$. The area between curves 5 and 3 is proportional to the difference of surface coverages with A (at $\varphi_{\mathcal{D}}^{(1)}$) and B (at $\varphi_{\mathcal{D}}^{(2)}$). Bruckenstein and Napp [14] discussed two additional modifications of method II.

An experimental verification of the above methods was carried out using the Cu$^+$/Cu^{2+} couple in 0.5 M HCl solution in which two separate reduction waves are observed. Coverage of platinum with Cu$^+$ determined by method I in 10^{-3} M solution at 0 V was 0.15 μC·cm^{-2}, i.e., about one monolayer.

The same system was also studied using method II. The potential of the platinum disc was changed in a stepwise way from -0.1 V (Cu$^+$ stable at the disc) to 0.4 V (at which Cu$^+$ oxidizes to Cu^{2+}). The ring was potentiostated at -0.1 V, the potential at which Cu^{2+} reduces to Cu$^+$ and can thus be detected.

[†]Curve 3, necessary for this purpose, can be obtained separately in a system in which adsorption is absent, or in the system studied in the presence of high c_A^* values, when the amount adsorbed is such smaller than the amount transferred by convective diffusion.

Experimental disc and ring transients are shown in Fig. 8.21. The transition time was found to be equal to that calculated according to Eq. (8.120). This indicates instantaneous desorption of divalent copper ions formed at the disc.

The shape of the transient (Fig. 8.21, curve 2) is characteristic of the case of the reaction substrate being more strongly adsorbed at the initial potential than is the product at the final potential (cf. Fig. 8.20, curve 5). The crosshatched area in Fig. 8.21 corresponds to the difference in coverages of Pt with Cu^+ at -0.1 V and Cu^{2+} at 0.4 V. The difference was found to be close to 0.15 $\mu C \cdot cm^{-2}$.

The results of adsorption measurements at the ring-disc system are in good agreement with data obtained using radioactive tracers [14]. A similar method has been used to evaluate the rate of hydrogen adsorption on platinum [105].

Study of Inorganic Reactions and Other Applications. The ring-disc system is well suited to investigations of intermediates formed in redox systems involving several oxidation states of reactants. As an example, the results of Bruckenstein et al. [106, 107] obtained for the system $I^- - I_2 - ICl - IO_3^-$ and $Br^- - HOBr - Br_2$ at a platinum electrode can be quoted.

The ring-disc system was used for the determination of supersaturation of the electrode solution layer with the gaseous products (particularly H_2 and O_2) of disc electrode reactions. The surface concentration can be calculated from the magnitudes of the disc currents [108, 109].

Albery and Ulstrup [110] used the ring-disc assembly to control the uniform accessibility of the rotating disc electrode, or, more strictly, to measure deviations from uniform accessibility in dilute solutions with significant ohmic potential drop (cf. § 4.10). The method proposed by the authors supposes that Eq. (8.46) describing the collection efficiency coefficient N is derived satisfactorily assuming a uniform disc current density. If the latter condition is not satisfied, however, the experimental N value differs from the calculated one.

Measurements must be carried out in a system in which the disc reaction product is stable in the bulk solution. For this purpose, the $NaBr - HClO_4$ solution was chosen. Br^- oxidizes at the platinum disc to Br_2 which is transported to the ring and reduced thereon. The electric field does not, of course, affect the motion of electrically neutral bromine which is transported exclusively by convective diffusion. Varying the concentration ratio of NaBr and $HClO_4$, Albery and Bruckenstein found that the experimental and calculated N values agreed within 3% in solutions of concentration at least 0.07 N.

In more dilute solutions, considerable deviations are observed due to the significant ohmic drop in the solution which results in a nonuniform distribution of the current density at the disc.

The electronic apparatus used for polarization of the disc and ring electrodes should be briefly mentioned. Tacussel [111] developed a double potentiostat (BIPAD) which allows constant potentials of the disc and ring to be maintained and varied simultaneously and independently, according to a given program. Schematics of similar arrangements are described in various papers [12, 72]. The trielectrode arrangement (rotating disc with sectioned ring) has been described by Miller [85].

References

1.   A. N. Frumkin and L. N. Nekrasov, Dokl. Akad. Nauk SSSR, 126:115 (1959).
2.   Yu. V. Ivanov and V. G. Levich, Dokl. Akad. Nauk SSSR, 126:1029 (1959); Some Problems of Theoretical Physics, Atomizdat, Moscow (1958), p. 32.
3.   A. N. Frumkin, L. N. Nekrasov, B. G. Levich, and Ju. B. Ivanov, J. Electroanal. Chem., 1:84 (1959-1960).
4.   W. J. Albery, Trans. Faraday Soc., 62:1915 (1966).
5.   W. J. Albery and S. Bruckenstein, Trans. Faraday Soc., 62:1920 (1966).
6.   W. J. Albery, S. Bruckenstein, and D. T. Napp, Trans. Faraday Soc., 62:1932 (1966).
7.   W. J. Albery, S. Bruckenstein, and D. C. Johnson, Trans. Faraday Soc., 62:1938 (1966).
8.   W. J. Albery and S. Bruckenstein, Trans. Faraday Soc., 62:1946 (1966).
9.   W. J. Albery and S. Bruckenstein, Trans. Faraday Soc., 62:2584 (1966).
10.   W. J. Albery and S. Bruckenstein, Trans. Faraday Soc., 62:2596 (1966).
11.   W. J. Albery, Trans. Faraday Soc., 63:1771 (1967).
12.   W. J. Albery, M. L. Hitchman, and J. Ulstrup, Trans. Faraday Soc., 64:2831 (1968).
13.   W. J. Albery, M. L. Hitchman, and J. Ulstrup, Trans. Faraday Soc., 65:1101 (1969).

14.   S. Bruckenstein and D. T. Napp, J. Am. Chem. Soc., 90:6303 (1968).
15.   V. G. Levich, Physicochemical Thermodynamics, Prentice Hall, Inc. (1962).
16.   H. Matsuda, J. Electroanal. Chem., 15:109, 325 (1967); 16:153 (1968).
17.   V. A. Dumkin and A. Prudnikov, Textbook of Operational Computations,
      Vysshaya Shkola, Moscow (1965).
18.   A. Erdelyi, Asymptotic Expansions, Dover (1961).
19.   N. S. Gradshtein and I. M. Ryzhik, Tables of Integrals, Sums, Series, and
      Products, Nauka, Moscow (1971).
20.   S. Bruckenstein, Élektrokhimiya, 2:1085 (1966).
21.   L. N. Nekrasov and I. N. Berezina, Dokl. Akad. Nauk SSSR, 142:855 (1962);
      Theory and Practice of Polarographic Analysis, Shtiintsa, Kishinev (1962),
      p. 289.
22.   L. Müller and L. N. Nekrasov, Dokl. Akad. Nauk SSSR, 154:437 (1964);
      L. N. Nekrasov, Élektrokhimiya, 2:438 (1966).
23.   A. Damjanovic, M. A. Genshaw, and J. O'M. Bockris, J. Phys. Chem., 70:3761
      (1966); J. Chem. Phys., 45:4057 (1967); J. O'M. Bockris and S. Srinivasan,
      J. Electroanal. Chem., 11:350 (1966).
24.   V. S. Bagotskii, M. R. Tarasevich, and V. Yu. Filinovskii, Élektrokhimiya,
      4:1247 (1968).
25.   V. S. Bagotskii, M. R. Tarasevich, and V. Yu. Filinovskii, Élektrokhimiya,
      5:1218 (1969).
26.   M. R. Tarasevich, Élektrokhimiya, 4:210 (1968).
27.   M. R. Tarasevich, Élektrokhimiya, 5:713 (1969).
28.   J. D. McIntyre, J. Phys. Chem., 73:4111 (1969).
29.   V. S. Bagotskii, L. N. Nekrasov, and N. A. Shumilova, Usp. Khim., 34:1697
      (1965); L. N. Nekrasov, Fuel Cells. Kinetics of Electrode Processes, Nauka,
      Moscow (1968), p. 121.
30.   L. N. Nekrasov and L. Miller, Dokl. Akad. Nauk SSSR, 149:1107 (1963);
      Electrochim. Acta, 9:1015 (1964); J. Electroanal. Chem., 9:282 (1965).
31.   M. R. Tarasevich, et al., Élektrokhimiya, 6:372, 376 (1970).
32.   A. Damjanovich, M. A. Genshaw, and J. O'M. Bockris, J. Phys. Chem.,
      70:3761 (1966); J. Electrochem. Soc., 114:466, 1107 (1967).
33.   J. O'M. Bockris and S. Srinivasan, Fuel Cells: Their Electrochemistry,
      McGraw-Hill Book Company (1969).
34.   A. J. Appleby and A. Borucka, J. Electrochem. Soc., 116:1212 (1969).
35.   A. J. Appleby, J. Electrochem. Soc., 117:641 (1970).
36.   K. F. Blurton and E. McMullin, J. Electrochem. Soc., 116:1476 (1969);
      Energy Convers., 9:141 (1969).
37.   L. Müller and L. N. Nekrasov, Dokl. Akad. Nauk SSSR, 157:416 (1964);
      Zh. Fiz. Khim., 38:3028 (1964).
38.   A. N. Frumkin et al., Élektrokhimiya, 1:17 (1965).
39.   E. I. Khrushcheva, N. A. Shumilova, and M. R. Tarasevich, Élektrokhimiya,
      1:730 (1965).
40.   V. I. Tikhomirova, V. I. Luk'yanycheva, and V. S. Bagotskii, Élektrokhimiya,
      3:762 (1967).

41. A. V. Yuzhanina, V. I. Luk'yanycheva, N. A. Shumilova, and V. S. Bagotskii, Élektrokhimiya, 6:1054 (1970).

42. L. Müller and V. V. Sobol', Élektrokhimiya, 1:111 (1965).

43. L. N. Nekrasov and N. I. Dubrovina, Élektrokhimiya, 4:362 (1968).

44. G. W. Tindall, S. H. Caddle, and S. Bruckenstein, J. Amer. Chem. Soc., 91:2119 (1969).

45. V. Sobol', E. Khrushcheva, and I. Dagaeva, Élektrokhimiya, 1:1332 (1965).

46. E. I. Khrushcheva, L. N. Nekrasov, N. A. Shumilova, and M. R. Tarasevich, Élektrokhimiya, 3:831 (1967).

47. L. N. Nekrasov, E. I. Khrushcheva, N. A. Shumilova, and M. R. Tarasevich, Élektrokhimiya, 2:363 (1966).

48. K. A. Radyushkina, M. R. Tarasevich, and R. Kh. Burshtein, Élektrokhimiya, 6:1351 (1970).

49. M. A. Genshaw, A. Damjanovic, and J. O'M. Bockris, J. Phys. Chem., 71:3722 (1967).

50. L. N. Nekrasov and E. I. Khrushcheva, Élektrokhimiya, 3:166 (1967).

51. M. A. Genshaw et al., J. Electroanal. Chem., 15:163, 173 (1967).

52. B. G. Podlibner and L. N. Nekrasov, Élektrokhimiya, 5:340 (1969).

53. M. R. Tarasevich, K. A. Radyushkina, V. Yu. Filinovskii, and R. Kh. Burshtein, Élektrokhimiya, 6:1522 (1970).

54. G. V. Zhumaeva, N. A. Shumilova, and M. R. Tarasevich, Dokl. Akad. Nauk SSSR, 161:151 (1965); Electrochim. Acta, 11:967 (1966).

55. G. V. Zhumaeva, et al., Élektrokhimiya, 4:1253 (1968).

56. Ku Lung-ying, N. Shumilova, and V. Bagotskii, Élektrokhimiya, 3:460 (1967).

57. N. A. Shumilova and V. S. Bagotzky, Electrochim. Acta, 13:285 (1968).

58. G. P. Samoilov, N. A. Shumilova, E. I. Khrushcheva, and V. S. Bagotskii, Élektrokhimiya, 4:1364 (1968).

59. G. P. Samoilov, E. I. Khrushcheva, N. A. Shumilova, and V. S. Bagotskii, Élektrokhimiya, 5:470, 1082 (1969).

60. N. A. Shumilova and G. V. Zhumaeva, Fuel Cells. Kinetics of Electrode Processes, Nauka, Moscow (1968), p. 138.

61. A. N. Doronin and O. L. Kabanova, Élektrokhimiya, 5:1092 (1969).

62. M. R. Tarasevich et al., Élektrokhimiya, 4:432 (1968).

63. F. Z. Sabirov and M. R. Tarasevich, Élektrokhimiya, 5:608 (1969).

64. E. Yeager, Extended Abstracts, 20th CITCE Meeting, Strasbourg (1969), p. 3; E. Yeager, P. Krouse, and K. V. Rao, Electrochim. Acta, 9:1057 (1964).

65. D. C. Johnson, D. T. Napp, and S. Bruckenstein, Anal. Chem., 40:482 (1968).

66. V. Yu. Filinovskii, Élektrokhimiya, 5:991 (1969).

67. L. N. Nekrasov and V. Yu. Filinovsky, Extended Abstracts, 20th CITCE Meeting, Strasbourg (1969), p. 119.

68. L. N. Nekrasov and E. N. Potapova, Élektrokhimiya, 6:806 (1970).

69. L. N. Nekrasov and T. K. Zolotova, Élektrokhimiya, 4:864 (1968).

70. K. B. Prater and A. J. Bard, J. Electrochem. Soc., 117:335 (1970).

71. W. J. Albery and M. L. Hitchman, Extended Abstracts, 20th CITCE Meeting, Strasbourg (1969), p. 124.

72. P. A. Malachesky, K. B. Prater, G. Petrie, and R. N. Adams, J. Electroanal. Chem., 16:41 (1968).

73. K. E. Heusler, Z. Elektrochem., 65:192 (1961).
74. L. Kiss and J. Farkas, Magyar Kem. Folyoirat, 75:11 (1969).
75. S. Bruckenstein and D. C. Johnson, Anal. Chem., 36:2187 (1964).
76. D. T. Napp, D. C. Johnson, and S. Bruckenstein, Anal. Chem., 39:481 (1967).
77. D. C. Johnson and S. Bruckenstein, J. Amer. Chem. Soc., 90:6592 (1968).
78. H. Schurig and K. E. Heusler, Z. Anal. Chem., 224:45 (1967).
79. S. Bruckenstein and G. A. Feldman, J. Electroanal. Chem., 9:395 (1965).
80. K. B. Prater and A. J. Bard, J. Electrochem. Soc., 117:207 (1970).
81. S. W. Feldberg, Electroanalytical Chemistry, Vol. 3, A. J. Bard, ed., Marcel Dekker, New York (1969), p. 199.
82. L. F. Kozin, E. E. Kobrand, and I. A. Sheka, Ukr. Khim. Zh., No. 1, p. 22 (1970).
83. B. Miller and R. E. Visco, J. Electrochem. Soc., 115:251 (1968).
84. L. Kiss, A. Körösi, and J. Farkas, Magyar Kem. Folyoirat, 72:191 (1966).
85. B. Miller, J. Electrochem. Soc., 116:1117 (1969).
86. B. Miller, J. Electrochem. Soc., 116:1675 (1969).
87. B. Miller, J. Electrochem. Soc., 117:491 (1970).
88. R. D. Armstrong and M. Henderson, J. Electroanal. Chem., 26:381 (1970).
89. H. W. Pickering and C. Wagner, J. Electrochem. Soc., 114:698 (1967).
90. H. G. Feller, Z. Metallkunde, 58:875 (1967); Corrosion Sci., 8:259 (1968).
91. D. Jones and N. Nackerman, Corrosion Sci., 8:565 (1968).
92. A. I. Oshe, B. N. Kabanov, et al., Élektrokhimiya, 2:1485 (1966).
93. A. I. Oshe and B. N. Kabanov, Zashchita Metallov, 4:260 (1968).
94. A. I. Oshe, V. A. Lobachev, and B. N. Kabanov, Élektrokhimiya, 5:1383 (1969).
95. K. E. Heusler, Ber. Bunsenges., 72:1197 (1968).
96. B. Cavalier, C. Dezael, and J. Jacq, Bull. Soc. Chim. France, 1966: 3210.
97. J. Jacq, B. Cavalier, and O. Bloch, Electrochim. Acta, 13:1119 (1968).
98. G. W. Tindall and S. Bruckenstein, Anal. Chem., 40:1051, 1637 (1968).
99. D. T. Napp and S. Bruckenstein, Anal. Chem., 40:1036 (1968).
100. G. W. Tindall and S. Bruckenstein, Anal. Chem., 40:1402 (1968).
101. G. W. Tindall and S. Bruckenstein, J. Electroanal. Chem., 22:367 (1969).
102. L. N. Nekrasov et al., Élektrokhimiya, 4:489, 539, 996, 1501 (1968); 5:212, 889 (1969); 6:218, 1155, 1219 (1970).
103. Z. Galus and R. N. Adams, J. Amer. Chem. Soc., 84:2061 (1962).
104. J. T. Maloy, K. B. Prater, and A. J. Bard, J. Phys. Chem., 72:4348 (1968).
105. I. Gonz, Yu. B. Vasil'ev, and V. S. Bagotskii, Élektrokhimiya, 6:325 (1970).
106. P. Beran and S. Bruckenstein, Anal. Chem., 40:1044 (1968).
107. D. C. Johnson and S. Bruckenstein, J. Electrochem. Soc., 117:460 (1970).
108. F. Ludwig and E. Yeager, J. Electrochem. Soc., 113:1109 (1966); Rev. Polarogr., 14:94 (1967).
109. E. A. Khomskaya and A. S. Kolosov, Élektrokhimiya, 6:256 (1970).
110. W. J. Albery and J. Ulstrup, Electrochim. Acta, 13:281 (1968).
111. J. R. Tacussel, Extended Abstracts, 20th CITCE Meeting, Strasbourg (1969), p. 131.

Chapter 9

Design of the Rotating Disc Assembly

Depending on the application required, the design of a rotating assembly must meet different demands. For the majority of analytical work, rotation speeds within 100 to 2000 rpm are sufficient, and the accuracy of their measurements may be within 5%. However, kinetic measurements on electrode reactions often require very intensive stirring (10,000–30,000 rpm), while accurate determinations of diffusion coefficients require the rotation speed to be measured precisely with an error less than 0.5% and with minimum (0.01 mm and less) axial and radial wobble of the electrode. Polarization curves for a number of anodic reactions can be obtained in an open beaker, whereas in other cases hermetic sealing of the cell is necessary, particularly in the case of corrosive media.

Complex assemblies designed for work at high angular velocities (20,000–30,000 rpm) require careful attention to design. Therefore, it is not expedient to aim at a universal tool suitable for all kinds of applications. The design should follow strictly the application requirements of the rotating disc apparatus in relation to the problem under study. The separate elements of the assembly will now be discussed.

§9.1. The Disc Electrode

The electrode material in studies of electropolishing, etching, and oxidation of metals, of the kinetics of electrochemical reactions on the given metals, etc. is obviously determined by the problem under study. In studies of diffusion kinetics, diffusion coefficients, and concentration of the dissolved species, noble metal electrodes (platinum, gold), as well as amalgam and graphite electrodes, are commonly used.

The diameter of the disc is also determined by the character and aims of the study. The theory of the rotating disc is, strictly speaking, developed for an infinite disc. Therefore, very accurate measurements (e.g., of diffusion coefficients) must be carried out at electrodes whose radius r_0 exceeds by at least an order of magnitude the thickness of the boundary layer δ_0. In earlier work, discs 10 cm in diameter were used. However, large electrodes are not very suitable, especially in work requiring high rotation speeds. Therefore the discs presently used are 1-10 mm in diameter, and the rotation speed is chosen so that $\delta_0 \ll r_0$.

Another important factor requiring consideration in relation to the choice of the disc diameter and range of rotation speeds is the necessity for maintaining laminar conditions of liquid flow. The transition to turbulent flow occurs (for a well-centered disc with a smooth surface) at a Reynolds number given by

$$\text{Re} = r_0^2 \omega / \nu \approx 10^5 \text{ (see § 2.9)}.$$

When automatic polarographs are used to record polarization curves, the electrode dimensions are determined by the maximum current for which the polarograph is designed, usually 10^{-4}-10^{-3} A. Therefore microelectrodes (1 mm in diameter) are normally used in polarography.

The rotating electrode consists usually of a metal disc mounted on a shaft in the vertical axis, which serves also as the current lead.

The nonworking parts of the rotating electrode in contact with the solution are insulated or built of an insulating material.

Platinum electrodes are sealed in glass. Microdisc electrodes can be constructed by sealing a platinum wire 1 mm in diameter into glass and grinding off the lower end of the electrode perpendicularly to the rotation axis (Fig. 9.1a).

If the electrode material cannot be sealed in glass, its lower end can be turned in the form of a truncated cone mounted in a ground glass or plastic material (Fig. 9.1b). The electrode can also be cylindrical, pressed into a holder of plastic material (Fig. 9.1c).

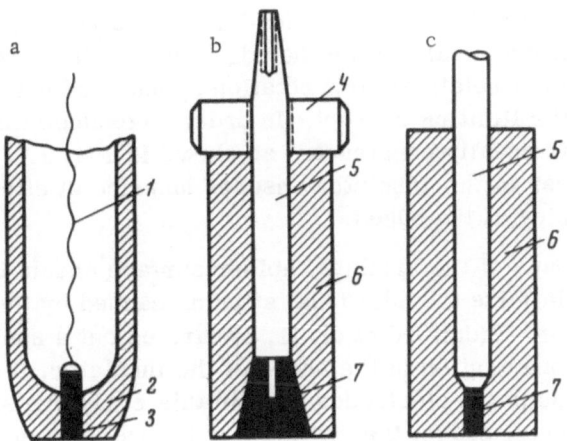

Fig. 9.1. Various designs of rotating disc electrodes. 1) Current lead; 2) glass; 3) platinum; 4) screw; 5) metallic axis; 6) plastic material; 7) investigated metal.

The most convenient insulating materials are Teflon and polyethylene, owing to their inert behavior. Other plastic materials, such as polystyrene, epoxy resin, etc. are also used. It is not recommended to use glass for electrodes rotating at 10,000 and more rpm since balancing and centering of rotating glass systems in work at high angular velocities is difficult.

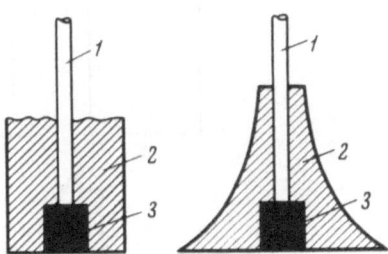

Fig. 9.2. The rotating disc in an insulating cylindrical and conical holder. 1) Axis; 2) plastic material; 3) investigated metal.

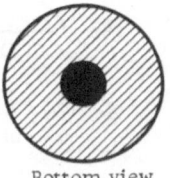

Bottom view

The form of the insulating holder plays a role in highly accurate measurements. Near the edges of the disc turbulence may appear even at relatively low rotation speeds and cause an increase in the limiting current. In order to exclude this error, the electrode is usually constructed as shown in Fig. 9.2. A sufficiently thick layer of the insulator ensures absence of effects of additional stirring at the edges.

The shape of the insulated holder surface should be chosen so as to minimize the additional stirring caused by the rotation of the insulator. Riddiford et al. [1, 2] carried out a special study aimed at optimization of the shape of the insulator. Various types of Teflon-insulated electrodes used in this study are shown in Fig. 9.3. In all cases, the diameter of the working electrode was much smaller (3 mm) than the external diameter of the insulator (15-30 mm). Their experiments consisted in measurements of the limiting current of KI_3 reduction (in excess KI) as a function of the ratio of the upper and lower radii of the insulator. Experimental current values were compared with theoretical ones calculated according to Eq. (2.38). Simultaneously, streamlines were observed in a colored electrolyte.

It was shown that limiting currents closest to the theoretical values were obtained at 50 and 240 rpm at electrodes of type B (1% deviation). A somewhat larger error, especially at high stir-

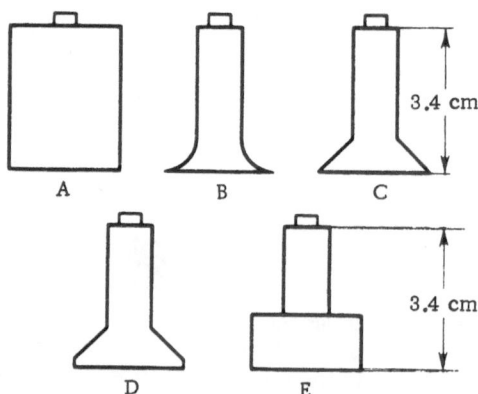

Fig. 9.3. Various types of the insulating corpus of the rotating electrode.

ring rates, arises when electrodes of types C and A are used (2-3%). Electrodes of types D and E are unsatisfactory.

Evidently, the criterion for absence of errors due to unsuitable hydrodynamic conditions is the "mutual insulation" of the liquid volumes below and above the plane of the disc. The flow in a real cell corresponds most closely to the conditions assumed in calculations where streamlines do not cross this plane. This has been observed in the case of electrode B with a correctly chosen ratio of its upper and lower radii (Fig. 9.4).

Nevertheless, the simplest [3] construction involving a cylindrical holder A can be successfully used in the study of a great majority of problems.

Some methods of surface preparation and construction of disc electrodes will now be considered.

Preliminary vacuum treatment of platinum electrodes immediately preceding measurements requires sealing of electrodes in molybdenum glass surrounded by an additional jacket with a thin-walled bead around the disc itself. The bead is then sealed to the vacuum system, and the electrode surface is reduced by heating in a hydrogen atmosphere followed by degassing. The electrode, separated from the vacuum system, is then placed in its holder in the electrochemical cell, and the glass bead is broken under the solution. In this way the reduced platinum surface is brought into contact with deoxygenated solution without being exposed to air [4].

Studies of oxidation reactions often require use of graphite paste electrodes. Graphite or amorphous carbon powder is mixed with an organic liquid (e.g., bromoform or bromonaphthalene), and the paste is placed in a small Teflon dish mounted on a glass rod with a sealed-in current lead [5, 6]. Landsberg et al. [7] used anodes prepared from a mixture of azobenzene with graphite. The resistance of electrodes containing the optimum ratio of components (3:2) does not exceed a few ohms. The mixture is heated to the melting point of azobenzene and poured into a plastic mold. The electrodes are polished before the measurements are commenced. The electrode can be used to obtain well-defined oxidation waves of a variety of ions, and it is not poisoned during several consecutive measurements. The overpotential for oxygen evolution increases with increasing azobenzene content in the electrode.

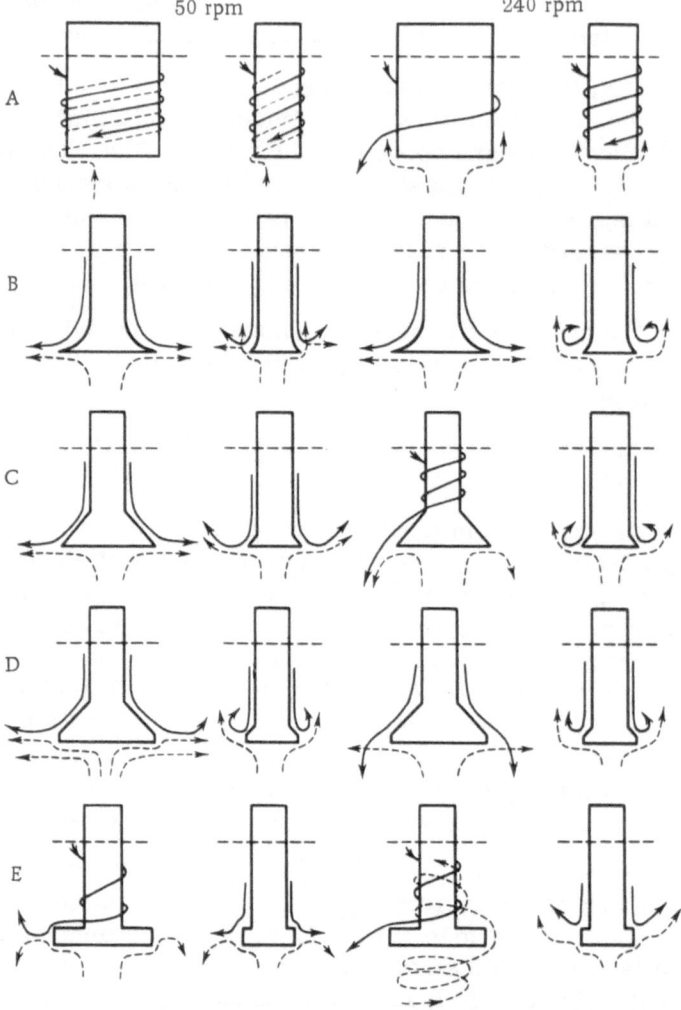

Fig. 9.4. Streamlines near the rotating electrode [2]. Letters indicate
the type of the insulating holder (according to Fig. 9.3). The horizontal dashed line indicates the solution level in the cell.

The cathodic polarization curves are accompanied by the reduction
of azobenzene to hydroazobenzene; therefore, when used as a
cathode, the electrodes described must be periodically polarized
anodically to regenerate azobenzene.

In studies of dissolution of nonmetallic samples under conditions of convective diffusion, pressed pellets are usually used.

The pellets are glued to the lower end of a shaft made of epoxy resin [8], or they are pressed in a metallic mold mounted on the shaft [9]. Krichevskii and Tsekhanskaya used pellets of terephthalic acid 10 mm in diameter and 10 mm high, compressed under a pressure of a few thousand kilograms per square centimeter into a steel dish mounted by an adjustable screw on the shaft.

The electrode itself can be mounted on the shaft by means of a screw; it is then difficult, however, to avoid considerable radial wobbles. Better results are obtained with cylindrical or conical joints. The upper end of the electrode shaft is turned in the form of a truncated cone (cf. Fig. 9.1b) which fits accurately into a corresponding bore in the lower end of the drive shaft. The final mounting of the electrode is accomplished by using a setscrew or clamping device [10].

Sometimes the electrode is insulated from the shaft by plastic sleeves, and the current lead is placed in a vertical channel inside the shaft (cf. Fig. 9.9). Sometimes the whole shaft is insulated from other parts of the asembly in order to minimize electric noise. The sliding contact is made by a brush or mercury contact at the upper part of the shaft. In the latter case, a drop of mercury is placed in a hole in the upper face of the shaft, and a stationary current lead is placed in the drop [11].

For work not requiring very high rotation speeds, the current lead can be taken out through the wall of the shaft to a copper ring, insulated from the shaft by an ebonite sleeve. Sliding contact is made between the rotating ring and a stationary copper−graphite brush pressed in contact with the ring by a spring. Sliding contacts of this type ensure sufficiently low (for the usual types of studies) levels of noise, viz. 0.1 mV at 40,000 rpm [12].

A ring–disc assembly, used for studies of intermediates, consists of two concentric electrodes, disc and ring, located in the same plane and separated by a narrow insulator. The assembly is shown in Fig. 9.5a [13]. The disc (made of gold, platinum, or another metal) is attached to a metal rod. The latter is placed inside a metal tube with the ring attached to its lower end. The rod with the disc is separated from the tube by a thin (0.1-0.2 mm) Teflon sleeve. The external surface of the tube is covered with Teflon. The shaft on which the electrode is mounted consists also of two concentric mutually insulated parts with separate current leads (cf. Fig. 9.5b).

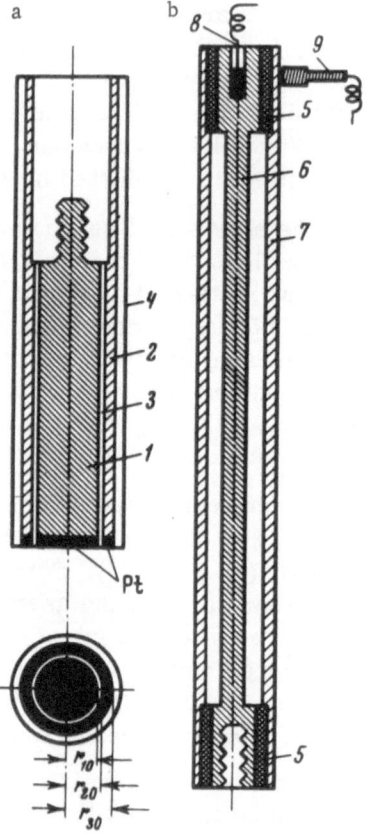

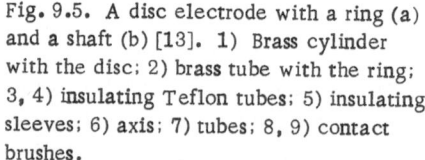

Fig. 9.5. A disc electrode with a ring (a) and a shaft (b) [13]. 1) Brass cylinder with the disc; 2) brass tube with the ring; 3, 4) insulating Teflon tubes; 5) insulating sleeves; 6) axis; 7) tubes; 8, 9) contact brushes.

It must be stressed that quantitative measurements using the ring-disc assembly require extemely careful removal of eccentricity of the electrodes and of wobbles during rotation.

Some experiments require separate preliminary treatment of the disc and ring. For this purpose, a detachable system can be constructed which can easily be taken apart so that the disc and ring can then be separately treated and mounted together immediately before measurements [14].

New variants of a ring-disc assembly have been recently introduced: a split-ring rotating electrode [15] and a disc electrode with a sectioned ring [16]. The technical details for assembling these devices can be found in the original papers.

Finally, a ring-disc assembly has been constructed which allows [17] the products of electrode reactions to be recorded optically without polarizing the ring. The ring "electrode" made of optically transparent material is cut across a light guide which passes through the length of the shaft rotating under a holder with a photoelectric cell. The appearance in the solution near the ring of a product of the disc electrode reaction which absorbs light of the given frequency is recorded by the photocell as a decrease of light intensity.

§ 9.2. Drive Mechanism; Measurements of Rotation Speed

The electrode is usually rotated by means of a 20-200 W electric motor. The electrode is mounted on the motor shaft, and the rotation speed is varied by changing the voltage applied to the motor. The dc motors are most convenient and allow speed variations between hundreds to thousands of revolutions per minute. The voltage of the motor must be stabilized in order to maintain the rotation speed constant.

Synchronous singlephase or threephase motors can be used with stable speeds. The electrode is joined to the motor by a system of pulleys using a transmission made, for example, of a spun polycaprolactam fiber, thin belt, or a steel spring.

The most convenient system consists of an asynchronous motor with the electrode mounted on its shaft and of an electronic power generator [18]. The rotation speed is varied by means of frequency changes of the ac current. Very high rotation speeds (20,000-80,000 rpm) may be obtained using an air turbine [12, 19].

The rotation speed can usually be measured by means of mechanical tachometers with an accuracy of 5-10%. The tachometer shaft is attached for a short time to the upper end of the electrode shaft. It should be remembered that the tachometer uses a certain part of the motor's power. Therefore, when motors of small power are used, the measured rotation speed may be lower than its normal value. The shaft can also be joined by means of a reducer to a mechanical revolution counter [10]; this method, however, is unsuitable for controlling short-time measurements of the rotation speed.

A very convenient and accurate method consists in attaching the upper end of the shaft to a magnetoelectric tachometer or a mirror, reflecting a beam from a light source, and a photocell. The frequency of impulses is measured by a frequency counter.

One of the most accurate procedures for velocity measurement is the stroboscopic method based on the following principle: The rotating shaft is illuminated by short light flashes from the impulse lamp of a stroboscope. The frequency of flashes is made equal to that of the shaft rotation. The shaft then makes exactly one revolution between two flashes, and each flash finds the shaft always in the same position. The inertia of the eye makes it register the separate flashes as a constant illumination, and the shaft seems to be stationary. For convenience, the upper end of the shaft is provided with a marked indicator disc. The error is within 1%; it can be reduced to 0.1% by careful measurements of the frequency of the ac current fed to the stroboscope.

For automatic stabilization of the rotation speed, electronic systems are used [20]. A differential amplifier is included in the feedback circuit between the tachometer and the electronic device controlling the speed of the motor. The signal from the tachometer, determined by the rotation speed, is compared with the signal from the calibrated voltage source. When the velocity deviates from the desired value an off-balance signal appears in the system which, by means of the controlling device, results in a compensating change of the rotation speed. The accuracy of stabilization is within 0.5% for 30–20,000 rpm and 0.1% for 1000–10,000 rpm.

Speed variations can be programmed by means of a controller which, apart from the feedback signal, is supplied from a special power source by varying voltage according to a desired function. In this way, "current vs rotation speed" curves at a constant potential can be automatically obtained ("rotoamperometry" [21]).

Automatic recording of "current vs square root of rotation speed" curves is most commonly desired since tedious graphical processing of the data can then be avoided. For this purpose, the tachometer voltage, proportional to the rotation speed, is transmitted to an analog function generator (which converts the rotation speed into its square root) and further to the input X of an X–Y recorder. The voltage, proportional to the cell current, is transmitted to the Y input [22]. The rate of change of the rotation speed ω should not be too high in this case, otherwise a steady-state is not established at the disc and the quantitative theory developed for stationary conditions becomes invalid (cf. Chap. 5).

§ 9.3. Electrochemical Cell

Although the theory of the rotating disc is, strictly speaking, developed for an infinite volume of solution, the cell dimensions are usually of little importance. The hydrodynamic conditions near the surface of the disc remain virtually unchanged by placing a stationary solid body at a distance of a few millimeters from the disc surface. Nevertheless, for very precise measurements, cells of not too small dimensions are required. Moreover, a small volume of solution may result in errors connected with the decrease of reactant concentration during electrolysis.

The design of the reference electrode and the position of the Luggin capillary should be chosen so as to minimize the ohmic drop in the solution [29]. Usually, currents at the rotating electrode are relatively high, resulting in high ohmic drops in dilute solutions. Zholudev and Stender [23] suggested that the counter and reference electrodes be placed on two opposite sides of the disc. The same principle was used for electrodes designed by Belyaeva, Gusev, and Rakcheev [24]. The reference electrode placed in a vertical glass tube is mounted inside the rotating disc and contacts the solution through a narrow vertical channel in the electrode coinciding with the rotation axis.

All these improvements, although seemingly useful, do not solve the problem of potential measurements. The difficulties consist in the nonequipotentiality of the disc surface in the presence of high ohmic drop in the solutions (cf. § 4.10). Therefore, a potential measurement at the given point of the disc does not reflect the existing potential distribution.

On the other hand, experimental evidence shows that such complex designs are not necessary for experiments which do not demand high accuracy. It is enough to place the tip of the Luggin capillary at a distance of 0.2-0.5 mm, without disturbing the hydrodynamic conditions. This disturbance is minimized when the Luggin capillary is placed vertically along the rotation axis.

The various methods of sealing the cell from the ambient atmosphere can be divided into three groups: rotating joint, hydraulic seal, and gasket seal or flat joint.

A rotating glass joint [25] can be used for rotation speeds in the range up to 3000 rpm. It ensures a good seal and excludes contamination of the solution, although it somewhat complicates the work. The upper part of the cell then consists of a conical joint; the other part (female) is attached to the shaft and rotates around the immobile cone. Sulfuric acid (1 N) is placed between the rubbing parts instead of grease.

Hydraulic seals filled with water or mercury have the same advantages as the rotating joint. A system described by Nekrasov [26] is shown in Fig. 9.6. The seal, made of Plexiglas, is filled with mercury. The latter is slowed down in its rotation by vertical ribs on the internal wall of the seal jar. The seal is attached tightly to the cell by a Plexiglas connector. This system can be used for rotation speeds up to 15,000 rpm. At higher speeds

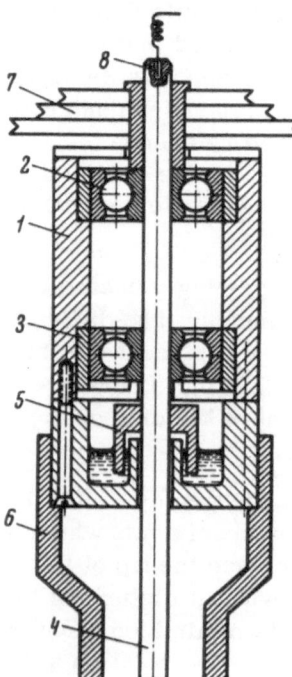

Fig. 9.6. Arrangement with a hydraulic seal [26]. 1) Frame; 2) bearing; 3) insulating sleeve; 4) shaft; 5) seal; 6) part fitting the connector; 7) pulley; 8) current lead.

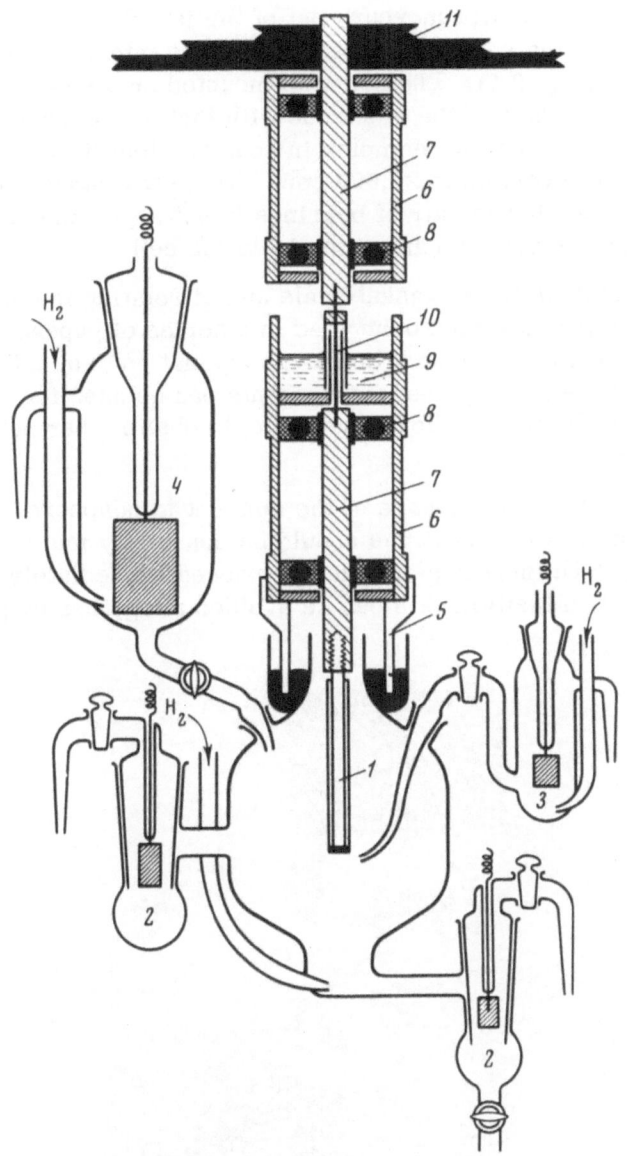

Fig. 9.7. Arrangement with a hydraulic seal [27]. 1) Rotating electrode; 2) counterelectrodes; 3) reference electrode; 4) cell for preliminary purification; 5) connector; 6) corpus; 7) shaft; 8) bearing; 9) seal; 10) needle; 11) pulley.

centrifugal force spills mercury out of the jar, disturbing the
hermeticity of the seal. Aikazyan [27] used a microseal ~0.5 mm
in diameter (Fig. 9.7). The bell was mounted on a steel needle
connecting the shaft of the electrode with that of the apparatus.
The system is somewhat complex in construction; it ensures a
good seal, however, up to 20,000 rpm. Its disadvantages lie in
the fact that the lower pair of bearings is not separated from the
cell, and grease vapors can contaminate the cell.

The construction of gasket seals and of rotating metallic joints
of various types has been described in a series of papers [18, 28-
30]. They require special precautions against contamination with
particles of the solid grease. Flat joints can be used for high
rotation speeds; it is difficult, however, to obtain a hermetic seal
in this way.

In order to avoid leakage of the ambient atmosphere, an excess pressure of the inert gas should be constantly maintained
in the cell. The inert gas can also be passed immediately outside
the seal [29]. Finally, seals can be avoided altogether by providing

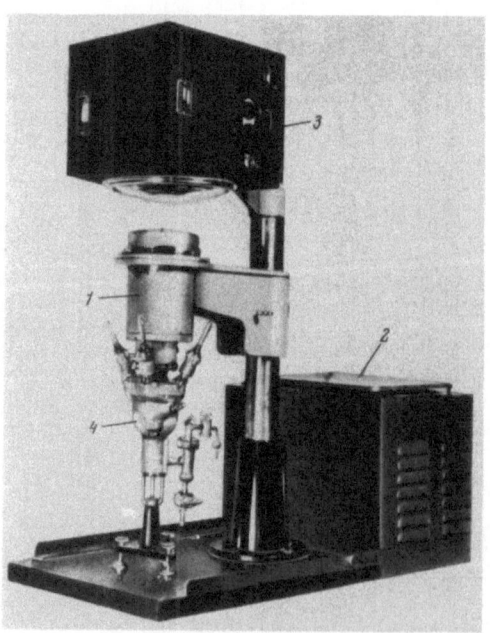

Fig. 9.8. Apparatus with a rotating disc electrode designed for speeds up to 33,000 rpm [18].

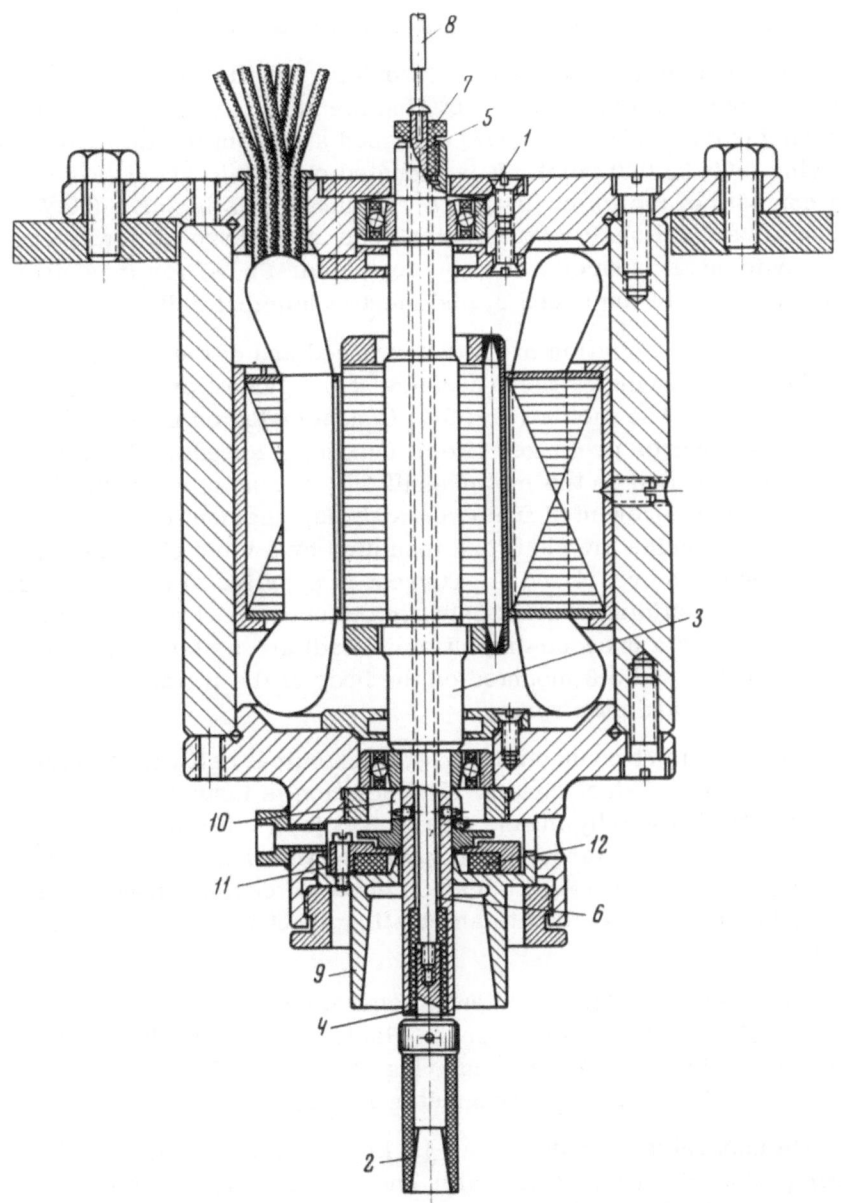

Fig. 9.9. Electric motor for the rotating disc electrode (to 33,000 rpm) [18].

a narrow (0.1 mm) opening between the stationary and moving part of the cell, which serves as the outlet of the inert gas [12].

We conclude this section by describing briefly the design and main characteristics of some rotating electrode systems. Pleskov, Belyanchikov, and Pominov [18] designed a system for intensive stirring (900-33,000 rpm). It is mounted on a table stand (Fig. 9.8) and consists of the following parts: asynchronous motor 1 (50 W) having a vertical shaft with the electrode attached to its lower end, electronic power source 2, strobotachometer PST-1 for measurements of the rotation speed 3, and electrochemical cell 4.

Continuous variation of the rotor speed and of the electrode is achieved by changing the frequency of the generator in the 15-550 Hz range. The motor (Fig. 9.9) is assembled on radial bearings 1 and ensures prolonged life at maximum speeds. The electrode 2 is mounted on the motor shaft 3 by means of an insulating sleeve 4 and is prevented from vertical displacement by a brass rod 5 placed inside the shaft and insulated by a vinyl chloride tube 6 and ebonite sleeves 4 and 7. Current is supplied to the electrode by means of a sliding contact between the upper part of the rod 5 (which contains a pressed-in platinum head) and a stationary copper–graphite needle 8 mounted on the body of the motor and pressed downward by a spring.

A ground joint connects the electrochemical cell with the internal surface of the steel cone 9. The solution is insulated from the ambient atmosphere by means of a gasket joint consisting of a brass sleeve with a flange [10] mounted on the shaft and a stationary copper–graphite ring 1 located above the cone 9. Ring 11 is pressed on the sleeve 10 with an elastic gasket 12 made of spongy resin.

This apparatus is convenient to operate and ensures high stability (1%) of the rotation speed. Radial wobbles of the electrode do not exceed 0.01-0.02 mm, ensuring strictly laminar flow conditions up to 22,000 rpm (disc diameter 10 mm).

The apparatus of Tobias et al. [31] consists of two parts: the upper part is fixed (electric motor and tachometer), and the lower part is detachable (shaft with bearings, electrode, and a glass cell in a steel holder). The cell is assembled before experiments in a box containing an inert dehydrated gas. It is sealed (for this purpose the shaft is raised in the bearings and fastened with a Teflon ring), taken out of the box, and connected to the upper part. The shaft is then released by lowering it back, and the apparatus is ready to use.

Zambonin [32] and Wojtowicz and Conway [29] constructed systems suitable for work at higher temperatures, particularly in melts. The cell can be made of borosilicate glass, stable up to 380°C. Zambonin's electrode is rotated employing the principle of a magnetic stirrer. The shaft glass tube with the electrode is completely enclosed in the cell and rotates in a brass and Teflon sliding bearing. The upper part of the shaft contains a magnet: the second magnet is placed outside, above the first one, on the rotor shaft. Rotation of one magnet brings the second one into rotation. In this way, special seals can be avoided and hermeticity ensured.

Commercial rotating disc assemblies (Beckman, [33]) as well as double potentiostats for disc-ring assemblies (Taccusel, [34]) are available.

All the designs described above are constructed basically for isothermal conditions, i.e., for the same temperature of the disc and solution. When a temperature gradient between the disc and solution is desired (measurements of thermoelectric force, effect of heat transfer on mass transfer, etc.) a special heater is placed inside the rotating electrode [35].

Sometimes, a system consisting of a stationary disc in a rotating liquid (cf. also Appendix 1) is used. The bottom of a cylindrical cell serves then as the electrode, and the solution is rotated by means of a stirring disc [36, 37] placed parallel to and above the electrode. This modification of the method is used to obtain the desired stirring conditions in those cases where the sample studied (e.g., a liquid) cannot be rotated.

References

1.   S. Azim and A. C. Riddiford, Anal. Chem., 34:1023 (1962).
2.   K. F. Blurton and A. C. Riddiford, J. Electroanal. Chem., 10:457 (1965).
3.   K. B. Prater and R. N. Adams, Anal. Chem., 38:153 (1966).

4.   V. I. Tikhomirova, V. I. Luk'yanycheva, and V. S. Bagotskii, Élektrokhimiya, 3:762 (1967).

5.   R. Galus, C. Olson, H. Y. Lee, and R. N. Adams, Anal. Chem., 34:164 (1962); R. N. Adams, Rev. Polarogr., 11:71 (1963).

6.   H. S. Swofford and R. L. Carman, Anal. Chem., 38:966 (1966).

7.   R. Landsberg and W. Geissler, Mitteilungsblatt Chem. Ges. DDR, Sonderheft 1960, Anal. Chem., S. 301; R. Landsberg and S. Müller, Wiss. Z. Techn. Hochschule Chemie, Leuna−Merseburg, 3:319 (1960-1961).

8.   R. Landsberg, H. Fürtig, and L. Müller, Z. Phys. Chem., 216:199 (1961).

9.   I. R. Krichevskii and Yu. V. Tsekhanskaya, Zh. Fiz. Khim., 33:2331 (1959).

10.   M. I. Bobrova and A. N. Kudasheva, Tr. Leningr. Inzh-Ekon. Inst., 36:104 (1961).

11.   G. F. Dezider'ev, S. I. Berezina, and G. A. Gorbachuk, Izv. Kazansk. Fil. Akad. Nauk SSSR, Ser. Khim. Nauk, No. 5, p. 99 (1959).

12.   H. E. Hintermann and E. Suter, Rev. Sci. Instr., 36:1610 (1965).

13.   A. N. Frumkin and L. N. Nekrasov, Dokl. Akad. Nauk SSSR, 126:115 (1959).

14.   G. V. Zhumaeva and N. A. Shumilova, Élektrokhimiya, 2:606 (1966); A. N. Doronin, Élektrokhimiya, 4:1193 (1968).

15.   G. Trimborn, A. Heindrichs, and W. Vielstich, Messtechnik, 75:224 (1968); K. E. Heusler and H. Schurig, Z. Phys. Chem., N. F., 47:117 (1965).

16.   B. Miller, J. Electrochem. Soc., 116:1117 (1969); R. H. Sonner, B. Miller, and R. E. Visco, Anal. Chem., 41:1498 (1969).

17.   J. E. McClure, Anal. Chem., 42:551 (1970).

18.   M. P. Belyanchikov, Yu. V. Pleskov, and V. G. Pominov, Zh. Fiz. Khim., 34:1368 (1960).

19.   M. Daguenet, I. Epelboin, and M. Froment, Compt. Rend., 258:3694 (1960).

20.   J. D. E. McIntyre and W. F. Peck, Anal. Chem., 41:1713 (1969).

21.   K. Prater, J. Electrochem. Soc., 115:27C (1968).

22.   S. G. Creason and R. F. Nelson, J. Electroanal. Chem., 21:549 (1969).

23.   M. D. Zholudev and V. V. Stender, Ukr. Khim. Zh., 23:200 (1957).

24.   V. A. Belyaeva, Zh. Fiz. Khim., 36:1385 (1962); I. I. Gusev and P. V. Rakchev, Zh. Fiz. Khim., 42:1172 (1968); Tr. MKhTI im. D. I. Mendeleeva, No. 54, p. 176 (1967).

25.   K. I. Rozental' and V. I. Veselovskii, Zh. Fiz. Khim., 27:1163 (1953); A. A. Rakov and K. I. Rozental', Zav. Lab., 19:495 (1953).

26.   L. N. Nekrasov, Dissertation, Moscow State University (1961).

27.   A. N. Frumkin and E. A. Aikazyan, Izv. Akad. Nauk SSSR, Otd. Khim. Nauk, 1959:202.

28.   G. Faita, G. Fiori, and J. W. Augustinsky, Electrochim. Metallorum., 2:437 (1967).

29.   J. Wojtowicz and B. E. Conway, J. Electroanal. Chem., 13:333 (1967).

30.   J. Ulstrup, Electrochim. Acta, 13:535 (1968); H. Nord, E. E. Foverskov, and G. Bech Nielsen, Acta Chem. Scand., 18:681 (1964).

31.   F. G. Baucke, D. Landolt, and C. W. Tobias, Rev. Sci. Instr., 39:1753 (1968).

32.   P. G. Zambonin, Anal. Chem., 41:868 (1969).

33.   K. Cammann, Beckman Rep., No. 2, p. 5 (1969).

34.   J. R. Tacussel, Extended Abstracts, 20th CITCE Meeting, Strasbourg (1969), p. 131.
35.   P. I. Zarubin, L. A. Poluboyartseva, and V. M. Novakovskii, Zashchita Metallov, 1:297 (1965); B. R. Sundheim and W. Sauerwein, Rev. Sci. Instr., 38:229 (1967).
36.   S. Okada, S. Yoshizawa, F. Hine, and K. Asada, J. Electrochem. Soc. Japan, 27:E69 (1959).
37.   J. Suzuki, Rev. Polarogr., 15:21 (1968).

Articles on the Rotating Disc Electrode†

1. Hydrodynamics

G. Schmieden, Z. Angew. Math. Mech., 8:460 (1928).

D. Ryabushinskii, Byull. Aérodinam. Inst. v Kuchine, 5:5 (1934); J. Roy. Aero. Soc., 39:340 (1935); Compt. Rend., 233:899 (1951).

S. Goldstein, Proc. Camb. Philos. Soc., 31:232 (1935).

N. Gregory, J. T. Stuart, and W. S. Walker, Philos. Trans. Roy. Soc. (London), Math. Phys. Sci., Ser. A, 248:155 (1956).

C. S. Wu, Appl. Sci. Res., Sec. A, 8:140 (1959).

M. H. Rogers and G. N. Lange, J. Fluid Mech., 7:617 (1960).

M. H. Rogers and G. N. Lange, Quart. J. Mech. Appl. Math., 17:319 (1964).

E. W. Schwiderski and H. J. Lugt, Phys. Fluids, 7:867 (1964).

H. Espig and K. Hoyle, J. Fluid Mech., 22:671 (1965).

A. A. Hayday, Appl. Sci. Res., Sec. A, 14:405 (1965).

D. D. Joseph, Quart. J. Mech. Appl. Math., 18:325 (1965).

P. Mitschka and J. Ulbrecht, Coll. Czech. Chem. Commun., 30:2511 (1965).

T. L. Kaminskii and P. Sverner, Trans. ASAE, 9:875 (1966).

D. K. Edwards, AIAA J., 5:333 (1967).

H. J. Lugt and E. W. Schwiderski, J. Appl. Mech. Trans. ASME, Ser. E., 34:829 (1967).

V. A. Yurchenko, A. A. Kopteev, and Yu. P. Popov, in: Mass Transfer Processes in Chemical Engineering, No. 3, Khimiya, Moscow (1968), p. 12.

D. G. Barbee and Tso-Shin Shih, Teploperedacha, No. 3, p. 85 (1968).

K. G. Meithal, Ganita, 9:95 (1968).

V. A. Yurchenko, A. A. Kopteev, A. I. Zaitseev, A. K. Zhebrovskii, and Ya. S. Yanev, Teor. Osnov. Khim. Tekhnol., 3:412 (1969).

---

†Many experimental and theoretical papers involve nonelectrochemical studies of the hydrodynamics of the rotating disc, as well as of mass and heat transfer, mainly in the gas phase. They are often useful for electrochemists working on new developments and applications of the rotating disc electrode. A bibliography of such papers is given below.

T. S. Cham and M. R. Head, J. Fluid Mech., 37:129 (1969).
L. B. Zung, Phys. Fluids, 12:18 (1969).

2. Mass Transfer

E. M. Sparrow and J. L. Gregg, Trans. ASME, Ser. C, J. Heat Transfer, 82:294 (1960).
P. Arva, Veszprémi Vegyipari Egyet. Közl., 8(3):171 (1964).
D. E. Rosner, AIAA J., 2:593 (1964).
Shian-Shaw Koong and P. L. Blackshear, Teploperedacha, No. 3, p. 116 (1965).
E. R. Benton, J. Fluid Mech., 24:781 (1966).
G. S. Hamford and M. Litt, Chem. Eng. Sci., 23:849 (1968).
M. S. Krishna, C. V. Rama Raju, G. J. V. Jagannagha Raju, and C. Venkato Rao,
       Ind. J. Technol., 6:50 (1968).
D. W. Zeh and W. N. Gill, AIChE J., 14:715 (1968).
G. E. Verevochkin and Yu. P. Konakov, Fiz. Khim. Osnov. Materialoved., No. 6,
       p. 130 (1969).
Khr. Iv. Noninski and H. Am. Nankov, Khim. i Indust. (NRB), 41:362 (1969).
G. Gognet and M. Daguenet, Compt. Rend., Ser. C, 270:142 (1970).
T. Misek and J. Marek, Brit. Chem. Eng., 15:202 (1970).

3. Heat Transfer

C. Wagner, J. Appl. Phys., 19:837 (1948).
K. Millsaps and K. Pohlhausen, J. Aero. Sci., 19:120 (1952).
E. C. Cobb and O. A. Saunders, Proc. Roy. Soc., A236:343 (1956).
R. L. Young, Trans. ASME, 78:1163 (1956).
L. A. Dorfman, Dokl. Akad. Nauk SSSR, 119:1012 (1958).
D. R. Davies, Quart. J. Mech. Appl. Math., 12:14 (1959).
D. R. Davies, Quart. J. Mech. Appl. Math., 12:211 (1959).
F. Kreith, J. H. Taylor, and J. P. Chong, J. Heat Transfer, Trans. ASME, Ser. C,
       81:95 (1959).
E. M. Sparrow and J. L. Gregg, J. Heat Transfer, Trans. ASME, Ser. C, 81:249 (1959).
D. R. Olander, J. Heat Transfer, Trans. ASME, Ser. C, 84:185 (1962).
N. Riley, Quart. J. Mech. Appl. Math., 17:319, 331 (1964).
D. A. Dorfman and A. Z. Sepazemdinov, Intern. J. Heat Mass Transfer, 8:317 (1965).
V. M. Kapinos, Izv. Vuzov, Aviats. Tekhnika, No. 2, p. 76 (1965).
A. Iguchi and T. Maki, Mem. Fac. Eng. Nagoya Univ., 19:144 (1967).
P. I. Zarubin and L. A. Poluboyartseva, Inzh. Fiz. Zh., 14:558 (1968).

4. Transient Processes

H. K. Thiriot, Z. Angew. Math. Mech., 20:1 (1940).
E. M. Sparrow and J. L. Gregg, J. Aero Sci., 27:252 (1960).
D. R. Olander, Intern. J. Heat Mass Transfer, 5:825 (1962).
D. J. Benney, J. Fluid Mech., 18:385 (1964).

R. D. Cess, Appl. Sci. Res., Sec. A, 13:233 (1964).

R. J. Hanold and J. R. Moszynski, Teploperedacha, No. 2, p. 69 (1967).

G. M. Homsey and J. L. Hudson, Appl. Sci. Res., 18:384 (1968).

R. D. Andrews and N. Riley, Quart. J. Mech. Appl. Math., 22:19 (1969).

J. P. Macey and E. J. Wellman, Phys. Fluids, 11:20 (1969).

G. M. Homsey and J. L. Hudson, Teploperedacha, No. 1, p. 133 (1969).

A. F. Jones, J. Fluid Mech., 39:257, 269 (1969).

A. S. Popel', Izv. Akad. Nauk SSSR, Mekhan. Zhidk. i Gaza, 1970:181.

5. Heat and Mass Transfer from a Nonisothermic Disc

J. P. Hartnett, J. Appl. Mech., Trans. ASME, Ser. E, 81:672 (1959).

J. P. Hartnett and E. C. Deland, J. Heat Transfer, Trans. ASME, Ser. C., 83:95 (1961).

V. M. Kapinos, Inzh. Fiz. Zh., 6:12 (1963).

N. M. Schnurr, J. Heat Transfer, Trans. ASME, Ser. C., 86:467 (1964).

C. L. Tien and J. Tsuji, J. Heat Mass Transfer, 7:247 (1964).

J. P. Hartnett. Shing-Hwa Tsai, and H. N. Jantscher, Teploperedacha, No. 3, p. 47 (1965).

A. A. Hayday, Teploperedacha, No. 4, p. 19 (1965).

H. J. Lugt and E. W. Schwiderski, Q. Appl. Math., 23:133 (1965).

C. L. Tien, Intern. J. Heat Mass Transfer, 8:411 (1965).

M. Daguenet and M. Garreau, Compt. Rend., Ser. A, 263:421 (1966).

A. Cezairliyan, Intern. J. Heat Mass Transfer, 10:97 (1967).

E. L. Koschmieder, J. Fluid Mech., 33:515 (1968).

J. Mabuchi, T. Tanaka, and M. Kumada, Bull. JSME, 11:885 (1968).

J. Mabuchi, T. Tanaka, M. Kumada, and Y. Sakakibara, Trans. Japan. Soc. Mech. Eng., 34:336 (1968).

E. G. Hauptmann and W. Ramsey, Appl. Sci. Res., 20:436 (1969).

6. The Rotating Disc Immersed in a Flowing Liquid or Gas

H. Schlichting and E. Truckenbrodt, J. Aero Sci., 18:639 (1951).

H. Schlichting and E. Truckenbrodt, Z. Angew. Math. Mech., 32:97 (1952).

A. N. Tifford and S. T. Chu, J. Aero. Sci., 19:284 (1952).

J. T. Stuart, Quart. J. Mech. Appl. Math., 7:446 (1954).

E. Truckenbrodt, Z. Angew. Math. Mech., 34:150 (1954).

H. H. Sogin, Trans. ASME, 80:61, 593 (1958).

J. Yamaga, Proc. 10th Japan National Congress Appl. Mech., 1960, p. 279.

B. S. H. Rarity, Quart. J. Mech. Appl. Math., 18:455 (1965).

C. L. Tien and J. Tsuji, J. Heat Transfer, Trans. ASME, Ser. C, 87(2) (1965).

I. Mabuchi, T. Tanaka, and Y. Sakakibara, Bull. JSME, 10:104 (1967).

C. G. Richards and W. P. Graebel, J. Basic Eng., Trans. ASME, Ser. D, 89:807 (1967).

Wen-Jei Yang and Hsu-Chien Yeh, Raket. Tekhnika i Kosmonavtika, No. 5, p. 205 (1967).

I. Mabuchi, T. Tanaka, M. Kumada, and Y. Sakakibara, Trans. Japan Soc. Mech. Engrs., 34:125 (1968).

J. Mabuchi, T. Tanaka, M. Kumada, and Y. Sakakibara, Bull. JSME, 11:875 (1968).

7. The Rotating Disc with a Screen

F. Schultz-Grunow, Z. Angew. Math. Mech., 15:191 (1935).

H. Föttinger, Z. Angew. Math. Mech., 17:356 (1937).

K. Pantell, Forschg. Ing.-Wes., 16:97 (1950).

S. L. Soo, Trans. ASME, 80:287 (1958).

S. L. Soo, R. W. Besant, and Z. N. Sarata, Z. Angew. Math. Phys., 13:297 (1962).

F. Kreith, E. Doughman, and H. Kozlowski, J. Heat Transfer, Trans. ASME, Ser. C, 85:153 (1963).

F. J. Bayley and L. Conway, J. Mech. Eng. Sci., 6:164 (1964).

J. W. Mitchell and D. E. Metzger, Teploperedacha, No. 4, p. 68 (1965).

D. K. Petree, W. L. Dukley, and J. M. Smith, Ind. Eng. Chem. Fundamentals, 4:171 (1965).

S. K. Sharma and H. G. Sharma, Appl. Sci. Res., Sec. A, 15(4/5) (1965).

D. E. Metzger and J. W. Mitchell, Teploperedacha, No. 1, p. 157 (1966).

S. K. Sharma and R. S. Agarwal, Appl. Sci. Res., 16:204 (1966).

L. A. Dorfman, Inzh. Fiz. Zh., 12:788 (1967).

R. A. Conover, J. Basic Engng., Trans. ASME, Ser. D, 90:325 (1968).

G. L. Mellor, P. J. Chapple, and V. K. Stokes, J. Fluid Mech., 31:95 (1968).

F. Rieger, Chem. Eng. Sci., 24:1017 (1969).

8. Two Parallel Rotating Discs; Immobile Disc in a Rotating Liquid

G. K. Batchelor, Quart. J. Mech. Appl. Math., 4:29 (1951).

H. P. Greenspan, and L. N. Howard, J. Fluid Mech., 17:385 (1963).

L. A. Dorfman, Inzh. Fiz. Zh., 12:788 (1967).

M. H. Rogers and G. N. Lange, Quart. J. Mech. Appl. Math., 17:319 (1964).

J. C. J. Nihoul, Appl. Sci. Res. Sec. B, 11:412 (1964/65).

V. M. Kapinos, Izv. Akad. Nauk SSSR, Énergetika i Transport, No. 3, p. 111 (1965).

V. M. Kapinos, Intern. J. Chem. Eng., 5:461 (1965).

F. Kreith and J. L. Peube, Compt. Rend., 260:5184 (1965).

C. E. Pearson, J. Fluid Mech., 21:623 (1965).

V. M. Kapinos, Izv. Vuzov, Aviats. Tekhnika, No. 1, p. 123 (1966).

B. Duncan, J. Fluid Mech., 24:417 (1966).

F. Kreith, Intern. J. Heat Mass Transfer, 9:265 (1966).

P. V. Shubba Raju, Appl. Sci. Res., 16:395 (1966).

V. M. Buzhnik, G. A. Artemov, V. N. Bandura, and A. M. Fedorovskii, Izv. Vuzov, Énergetika, No. 8, p. 113 (1967).

R. W. Bartlett, W. E. Nelson, and F. A. Halden, J. Electrochem. Soc., 114:1149 (1967).

F. Kreith and H. Viviand, J. Appl. Mech., Trans. ASME, Ser. E, 34:541 (1967).

L. P. Kholpanov, Zh. Fiz. Khim., 41:2034 (1967).

L. A. Dorfman, Izv. Akad. Nauk SSSR, Mekhan. Zhidk. i Gaza, No. 1, p. 40 (1968).

J. L. Hudson, Intern. J. Heat Mass Transfer, 11:407 (1968).

J. L. Hudson, Chem. Eng. Sci., 23:1007 (1968).

M. A. A. Khan, J. Mécanique, 7:575 (1968).

L. Matsch and W. Rice, J. Appl. Mech., Trans. ASME, Ser. E, 35:155 (1968).
C. Boyd and W. Rice, Prikl. Mekhan., No. 2, p. 22 (1968); No. 2, p. 238 (1969).
M. B. Bardin, A. N. Dikusar, and M. Kh. Kishinevskii, Élektrokhimiya, 6:212 (1970).

9. Magnetohydrodynamics

V. V. Sychev, Zh. Prikl. Matem. Mekhan., 24:1360 (1961).
E. M. Sparrow and R. D. Cess, J. Appl. Mech., E29:181 (1962).
A. C. Srivistava and S. K. Sharma, J. Phys. Soc. Japan, 19:1390 (1964).
I. Pon, Inzh. Fiz. Zh., 12:793 (1967).
A. S. Popel', Izv. Akad. Nauk SSSR, Mekhan. Zhidk. i Gaza, No. 6, p. 71 (1967).
E. W. Schwiderski and H. J. Lugt, J. Appl. Mech. Trans. ASME, Ser. E, 34:563 (1967).
P. A. Gilman and E. R. Benton, Phys. Fluids, 11:2389 (1968).
E. R. Dsa, Proc. Camb. Phil. Soc., 65:807 (1969).
C. J. Stephenson, J. Fluid Mech., 38:335 (1969).

10. Miscellaneous Applications

F. Kelemen, Studia Univ. Bades-Bolyai. Math.-Phys., No. 1, p. 211 (1961).
F. Kelemen, Studia Univ. Bades-Bolyai. Math.-Phys., No. 1, p. 219 (1961).
L. A. Bromley, R. F. Humphreys, and W. Murray, Teploperedacha, No. 1, p. 87 (1966).
J. K. Marshall, M. Hull, and J. A. Kitchener, J. Coll. Interface Sci., 22:342 (1966);
      Trans. Faraday Soc., 65:3093 (1969).

Supplementary References

Chapter 2

M. B. Bardin and A. I. Dikusar, "The effect of the eccentricity of the rotating disc electrode on the diffusion limiting current," Élektrokhimiya, 6:1147 (1970).

M. B. Bardin, Yu. S. Lalikov, and V. S. Temyanko, "Polarographic determination of some noble metals using platinum electrodes," in: Analysis of Noble Metals, Izd. AN SSSR, Moscow (1959), p. 80.

G. I. Barinov, "Kinetics of metal dissolution in liquid ammonia," Technology of Materials Used in Electronics, Krasnoyarsk (1970), p. 82.

G. I. Barinov and P. M. Shurygin, "Diffusion in liquid tellurium," Technology of Materials Used in Electronics, Krasnoyarsk (1970), p. 87.

R. Yu. Bek et al., "Thiourea and thiocyanate electrolytes," Izv. Sibirsk. Otd. Akad. Nauk SSSR, Ser. Khim. Nauk, No. 3, pp. 42, 47, 50 (1970).

V. N. Boronenkov, O. A. Esin, and O. M. Shurygin, "Kinetics of electrochemical processes at a disc electrode in oxide melts," Physical Chemistry of Molten Slags, Naukova Dumka, Kiev (1970), p. 129.

V. N. Velichko, "Study of mass transfer using the rotating disc electrode," Zh. Fiz. Khim., 45:2364 (1971).

G. P. Girina and L. G. Feokitstov, "Polarization curves of haloproprionitriles and dibromoethylenes on a lead rotating electrode. Comparison with data obtained at mercury electrodes," Élektrokhimiya, 9:462 (1973).

G. P. Girina, L. G. Feoktistov, and V. Yu. Filinovskii, "Criteria of comparison for half-wave potentials at rotating disc and dropping electrodes. Reduction of β-halopropionitriles at mercury," Élektrokhimiya, 9:177 (1973).

A. M. Golovin, N. M. Rubinina, and V. M. Khokhrin, "Convective diffusion from the surface of a rotating disc," Teor. Osnov. Khim. Tekhnol., 5:651 (1971).

M. R. Dausheva and O. A. Songina, "Electrochemical behavior of difficultly soluble suspensions," Usp. Khim., 42:323 (1973).

M. R. Dausheva, M. R. Tarasevich, and S. I. Zhdanov, "Electrochemistry of activated carbon suspensions. III. Electrochemical behavior of activated carbon suspensions at an amalgam disc electrode in the presence of molecular oxygen," Élektrokhimiya, 7:165 (1971); "Electrochemistry of activated carbon suspen-

sions. IV. Reduction of hydrogen peroxide at activated carbon suspensions,"
   Élektrokhimiya, 7:284 (1971).

Yu. K. Delimarskii and I. I. Penkalo, "Effect of vibrations on the diffusion flux to the
   rotating disc," Ukr. Khim. Zh., 36:1279 (1970).

T. K. Zolotova, I. V. Shelepin, and Yu. B. Vasil'ev, "Kinetics and mechanism of
   electroreduction of anthraquinone-2,6-disulfonic acid," Élektrokhimiya,
   9:1211 (1973).

L. M. Izrailev and P. M. Shurygin, "A study of anodic dissolution of iron in NaCl−KCl
   melts. Cathodic processes in electrolysis of iron chloride melts," Technology
   of Materials Used in Electronics, Krasnoyarsk (1970), pp. 93, 98.

V. A. Kokorekina, L. G. Feokitstov, S. A. Shevelev, and A. A. Fainzil'berg, "Electrooxidation of aliphatic polynitro compounds. 1. Voltammetric curves of
   polynitroalkanes at a rotating disc electrode," Élektrokhimiya, 6:1849 (1970).

G. S. Kornienko and M. Kh. Kishinevskii, "Diffusional flux to the ring electrode on
   a rotating disc," Élektrokhimiya, 8:1759 (1972).

V. S. Krylov and V. N. Malinenko, "Hydrodynamics and ionic mass transfer during
   fast electrodissolution of the rotating disc," Élektrokhimiya, 9:3 (1973).

L. M. Pis'men, "Macrokinetics of electrochemical reactions at a suspended electrode. II. Electrode with a rotating disc current collector," Élektrokhimiya,
   9:1328 (1973).

N. P. Skvortsov, Yu. V. Vodzinskii, and I. A. Korshunov, "Determination of the number
   of electrons transferred in oxidation of abietic acid at a graphite electrode,"
   Élektrokhimiya, 8:1078 (1972).

N. M. Trepak, L. K. Il'ina, A. L. L'vov, and V. N. Rodnikova, "Concerning the
   cathodic reduction of nitrates on cadmium," Élektrokhimiya, 8:939 (1972).

V. I. Chernenko and K. I. Litovchenko, "Solution of the equation of convective
   diffusion to the surface of a rotating disc electrode," Ukr. Khim. Zh., 38:868
   (1972).

N. G. Chovnik and M. V. Myshalov, "Polarograms of the reduction of trivalent to
   divalent iron ions in molten $AlCl_3$−NaCl system," Élektrokhimiya, 6:1659
   (1970).

P. M. Shurygin, "Study of the kinetics of heterogeneous reactions at the uniformly
   accessible surface of a rotating sample," Physical Chemistry of Molten
   Slags, Naukova Dumka, Kiev (1970), p. 147.

F. Aiment, M. Daguenet, F. Kermiche, and M. Mekati, "The theory and application
   of microelectrodes," Electrochim. Acta, 18:87 (1973).

J. C. Bazan and J. A. Schmidt, "A solid porous silver iodide electrode as silver
   ion indicator in molten nitrates," Electrochim. Acta, 18:459 (1973).

F. Beck, "Cathodic behavior of concentrated sulfuric acid," Electrochim. Acta,
   17:2317 (1972).

J. W. Breitenbach, O. P. Olaj, and F. Sommer, "Anodic behavior of acetate ions in
   a nonaqueous solution at a rotating platinum disc electrode," Monatsh.
   Chem., 101:1435 (1970).

M. Brezina, J. Koryta, and M. Musilova, "Electrode reactions of oxygen and hydrogen peroxide on silver in alkaline solution," Coll. Czech. Chem. Commun.,
   33:3397 (1968).

A. J. Calandra, C. Tamayo, J. Herrera, and A. J. Arvia, "Kinetics and mechanism of the electrochemical reduction of NO_2^+ in concentrated sulfuric acid," Electrochim. Acta, 17:2035 (1972).

Der Tau Chin, "Concentration recovery in the downstream of a rotating disc electrode," J. Electrochem. Soc., 120:628 (1973).

I. Cornet, W. N. Lewis, and R. Kappesser, "The effect of surface roughness on mass transfer to a rotating disc," Trans. Inst. Chem. Eng., 47:T222 (1969).

M. Daguenet, M. H. Meklati, and G. Cognet, "A study of the diffusion-limiting flux to a rough surface under turbulent conditions: A rough rotating disc with a working microelectrode," Compt. Rend., Ser. C, 272:1355 (1971).

B. R. Eggins, "One-electron intermediate in the anodic oxidation of hydroquinone in acetonitrile," J. Chem. Soc., Chem. Commun. 1972(7):427.

B. T. Ellison and I. Cornet, "Mass transfer to a rotating disc," J. Electrochem. Soc., 118:68 (1971).

C. T. Garcia, A. J. Calandra, and A. J. Arvia, "The electrochemical reduction of nitrosonium ion in concentrated sulfuric acid: The NO^+/NO redox couple," Electrochim. Acta, 17:2181 (1972).

B. Gostiša-Mihelčić, W. Vielstich, and A. Heindrichs, "Precise measurements of the diffusion coefficient of protons using the rotating disc electrode," Ber. Bunsenges. Phys. Chem., 76:19 (1972).

R. Greif, R. Kappesser, and I. Cornet, "Mass transfer in non-Newtonian flow," J. Electrochem. Soc., 119:717 (1972).

J. A. Harrison and D. A. Philippart, "The reduction of Ru complexes to metal," J. Electroanal. Chem., 40:357 (1972).

I. V. Kadija and V. M. Nakic, "Ring electrode on rotating disc as a tool for investigation of gas-evolving electrochemical reactions," Electroanal. Chem., 34:15 (1972).

H. Kolny and Z. Zembura, "Limiting current, diffusion coefficient, and activation energy of diffusion of hydrogen ions in acidified 0.1 M NaCl solution at 5-75°C," Roczn. Chem., 45:1593 (1972).

K. Korinek, J. Koryta, and M. Musilova, "Electrooxidation of hydrazine on mercury, silver, and gold electrodes in alkaline solution," J. Electroanal. Chem., 21:319 (1969).

J. Koryta, M. Brezina, and N. Kriz, "Electrode reactions of hydrogen peroxide on platinum in alkaline and acid solutions," Rev. Roum. Chim., 17:171 (1972).

U. Künkel, R. Landsberg, and S. Müller, "Electrooxidation of 4-methoxystyrene," Z. Chem., 10:303 (1970).

J. Kuta and E. Yeager, "The study of charge transfer in the $U(VI)-U(V)$ couple on a rotating gold disc electrode," J. Electroanal. Chem., 31:119 (1971).

H. Löwe and I. Barry, "Kinetics of germanium dissolution in alkaline ferricyanide solutions," Z. Phys. Chem. (Leipzig), 249:73 (1972).

C. Martinez, A. J. Arvia, and J. A. Wargon, "Kinetics of the cathodic discharge of ammonium ion dissolved in dimethyl sulfoxide on platinum," Electrochim. Acta, 18:485 (1973).

Y. Miyoshi and W. J. Lorenz, "Proton transfer in sulfuric acid solutions," Ber. Bunsenges. Phys. Chem., 74:412 (1970).

D. Möller and K.-H. Heckner, "Kinetics of irreversible electrode processes on solid electrodes; analysis of complex processes in the anodic oxidation of hydroxylamine," Z. Chem., 11:32 (1971).

D. Möller and K.-H. Heckner, "Oxidation of hydroxylamine on a rotating platinum disc electrode," Z. Phys. Chem. (Leipzig), 251:80 (1972).

V. D. Parker and L. Eberson, "Anodic oxidation of hydroquinone in acetonitrile," Chem. Commun., 1970:1289.

R. Pereiro, A. J. Arvia, and A. J. Calandra, "Kinetics of the $SCN^-/(SCN)_2$ couple on platinum in acetonitrile," Electrochim. Acta, 17:1723 (1972).

J. Postlethwaite, K. L. Oug, and D. J. Pickett, "Determination of diffusion coefficients of p-nitrobenzenesulfonic acid with a rotating disc electrode," Can. J. Chem. Eng., 50:245 (1972).

Z. Samec and I. Nemec, "The voltammetric study of some phenanthroline and ferrocene-type complexes with a rotating platinum disc electrode in acetonitrile," J. Electroanal. Chem., 31:161 (1971).

W. H. Smyrl and J. Newman, "Limiting current on a rotating disc with radial diffusion," J. Electrochem. Soc., 118:1079 (1971).

F. Sommer, J. W. Breitenbach, and O. F. Olah, "The cathodic behavior of acrylonitrile in the presence of quaternary ammonium salts and phosphonium," Monatsh. Chem., 99:2422 (1968).

K. Szabo and J. Mica, "Electrochemical investigations on a solid rotating gallium electrode," Mag. Kem. Folyoirat., 78:381 (1972).

W. E. Triaca, H. A. Videla, and A. J. Arvia, "Electrochemical oxidation of iodide dissolved in sodium nitrate — potassium nitrate eutectic melt on a platinum rotating disc electrode," Electrochim. Acta, 16:1671 (1971).

F. Vydra et al., "Voltammetry with disc electrodes and its analytical applications," J. Electroanal. Chem., 24:379 (1970); 31:175 (1971); 33:161 (1971); 38:349 (1972); 39:229 (1972).

F. Vydra and M. Stulikova, "Application of a rotating disc electrode to study of sorption processes. Ionic exchange of iron(III) on Dowex 50 W X8 in acidic solution," Coll. Czech. Chem. Commun., 37:123 (1972).

J. A. Wargon and A. J. Arvia, "The electrochemical oxidation of nitrite dissolved in dimethyl sulfoxide on platinum electrodes," Electrochim. Acta, 16:1619(1971).

P. G. Zambonin, "On the voltammetric behavior of the carbon dioxide—oxygen—carbonate system in molten alkali nitrates," Anal. Chem., 44:763 (1972).

Z. Zembura and H. Kolny, "Limiting currents of the cathodic hydrogen evolution from acidified 0.1 M Na_2SO_4 at 9-75°C," Roczn. Chem., 46:301 (1972).

Chapter 3

I. E. Barbasheva, Yu. M. Povarov, and P. D. Lukovtsev, "Electrochemical behavior of the iodine—iodide system in water—dimethylformamide," Élektrokhimiya, 8:1275 (1972).

V. A. Belyaeva, "Cathodic reduction of the ferro-ferri sulfates from ethylenediaminetetraacetate on a rotating disc electrode," Élektrokhimiya, 7:90 (1971).

S. V. Gorbachev and E. I. Martynycheva, "Analysis of the method of determination
    of limiting chemical currents," Physical Chemistry and Electrochèmistry,
    Tr. MKhTI im. Mendeleeva. No. 67, p. 269 (1970).

S. V. Gorbachev and I. E. Tishchenko, "The effect of the rotation speed of the
    electrode on the value of the half-wave potential in the mixed I_2/KI +
    $K_3Fe(CN)_6$ system," Zh. Fiz. Khim., 46:2055 (1972).

L. Müller and H. Prühmke, "Determination of the kinetic parameters for reaction from
    the ideal adsorption layer by means of the rotating disc electrode," Élektro-
    khimiya, 8:1377 (1972).

Yu. M. Povarov, L. V. Eroshkina, and P. D. Lukovtsev, "The effect of adsorbed cat-
    ions on cathodic and anodic processes in the iodine–iodide system," Élek-
    trokhimiya, 6:1450 (1970).

Yu. M. Povarov, A. M. Trukhan, and P. D. Lukovtsev, "On limiting diffusion cur-
    rents at electrodes with inhomogeneous surface," Élektrokhimiya, 7:1704 (1971).

Yu. M. Povarov and P. D. Lukovtsev, "Temperature coefficient of the limiting cur-
    rent on electrodes with inhomogeneous surface," Élektrokhimiya, 7:1715 (1971).

O. P. Osipov, M. A. Novitskii, Yu. M. Povarov, and P. D. Lukovtsev, "Kinetics of
    electrode reactions in the bromine–bromide system on platinum electrode,"
    Élektrokhimiya, 8:327 (1972).

A. M. Trukhan, Yu. M. Povarov, and P. D. Lukovtsev, "Special features of iodine
    reduction at a palladium electrode," Élektrokhimiya, 7:147 (1971).

A. V. Chikov, Yu. M. Povarov, P. D. Lukovtsev, and S. D. Pirozhkov, "Temperature
    dependence of limiting currents on Pt and Ir in the iodine–iodide system
    in acetonitrile," Élektrokhimiya, 8:1089 (1972).

D. H. Angell and T. Dickinson, "The kinetics of the ferrous/ferric and ferro/ferri-
    cyanide reactions at platinum and gold electrodes," J. Electroanal. Chem.,
    35:55 (1972).

R. D. Amstrong, R. Edmondson, and R. E. Firman, "The anodic dissolution of tungsten
    in alkaline solutions," J. Electroanal. Chem., 40:19 (1972).

A. J. Arvia, R. C. V. Piatti, and J. J. Podesta, "Kinetics and mechanism of H_2 evolu-
    tion at Ni electrodes from acid and alkaline solutions," An. Asoc. Quim.
    Argent., 57:1 (1969).

R. G. Barradas, N. C. Giordano, and W. H. Sheffield, "Electrochemical oxidation of
    piperidine at mercury and platinum," Electrochim. Acta, 16:1235 (1971).

H. Bartelt and H. Skilandat, "Kinetic parameters of the Co(ethylenediamine)$_3^{2+}$/Co-
    (ethylenediamine)$_3^{3+}$ redox system," J. Electroanal. Chem., 23:407 (1969).

H. Bartelt, "Effect of hydrolysis and ionic pair formation on charge-transfer reac-
    tions in the Co(NH$_3$)$_6^{2+}$/Co(NH$_3$)$_6^{3+}$ redox system," Electrochim. Acta, 16:307 (1971).

H. Bartelt, "Determination of the exchange current density of the Co(Phen)$_3^{3+}$/Co(Phen)$_3^{2+}$
    redox system on platinum electrode," Electrochim. Acta, 16:629 (1971).

H. Bartelt, R. Landsberg, and M. Prügel, "Effect of surface coverage on electron-
    transfer reactions in the Co(diethylenetriamine)$_3^{2+}$/Co(ethylenediamine)$_3^{3+}$
    system on platinum, "Coll. Czech. Chem. Commun., 36:1898 (1971).

H. Bartelt and M. Prügel, "Determination of the surface coverage of platinum electrode
    in the Co(diethylenetriamine)$_2^{2+}$/Co(diethylenetriamine)$_2^{3+}$ redox system,"
    J. Electroanal Chem., 32:309 (1971).

M. Breitenbach and K.-H. Heckner, "Ring-disc study of the kinetics of the anodic oxidation of aniline in acetonitrile on platinum electrode," J. Electroanal. Chem., 29:309 (1971).

M. Breitenbach and K.-H. Heckner, "Electrochemical study of the formation and properties of polyaniline films on platinum and carbon electrodes," J. Electroanal. Chem., 43:267 (1973).

M. Brezina, J. Koryta, and Pham-Thi Lang Phuong, "Decomposition of hydrogen peroxide at a rotating silver disc in alkaline medium," J. Electroanal. Chem., 40:107 (1972).

Der Tau Chin and M. Litt, "Mass transfer to point electrodes on the surface of a rotating disc," J. Electrochem. Soc., 119:1338 (1972).

L. Dunsch, E. Keller, and G. Henze, "On the anodic oxidation of aniline in saturated NaCl solution: Kinetic study with a rotating disc electrode," Z. Chem., 13:118 (1973).

W. R. Fawcett, P. A. Forte, R. O. Loutfy, and J. M. Prokipcak, "Electroreduction of substituted benzofurazans," Can. J. Chem., 50:263 (1972).

M. Fleischmann, D. Pletcher, and A. Rafinski, "Kinetics of the silver(I)/silver (II) system at a platinum electrode in perchloric and nitric acids," J. Appl. Electrochem., 1:1 (1971).

H. Fuchs and R. Landsberg, "Reduction of BrO^-, BrO_2^-, and BrO_3^- on a smooth platinum electrode," Z. Phys. Chem. (Leipzig), 247:132 (1971).

J. A. Harrison and Z. A. Khan, "The oxidation of hydrogen," J. Electroanal. Chem., 30:327 (1971).

J. A. Harrison and J. Thompson, "The reduction of gold cyanide complexes," J. Electroanal. Chem., 40:113 (1972).

K.-H. Heckner and D. Möller, "The kinetics of irreversible reactions on solid electrodes; derivation of the basic equations," Z. Chem., 10:477 (1970).

G. L. Holleck, "The reduction of chlorine on carbon in $AlCl_3-KCl-NaCl$ melts," J. Electrochem. Soc., 119:1158 (1972).

F. Kermiche-Aouanouk and M. Daguenet, "The theory of microelectrodes," J. Chim. Phys., 69:1705 (1972).

M. Kulanek and P. Stutz, "On the dissolution of difficultly soluble materials during complex formation. The rate of dissolution of silver halides in ammonium thiousulfate solutions," Z. Phys. Chem., 244:340 (1970).

R. H. Landsberg and F. Sheller, "Problem of nonlinear diffusion at heterogeneous electrodes," Anal. Chem., 45:420 (1973).

H. Löwe and J. Hoyer, "Kinetics of germanium dissolution in hydrofluoric and nitric acid mixtures," Z. Phys. Chem. (Leipzig), 248:235 (1971).

C. Martinez, A. J. Calandra, and A. J. Arvia, "The anodic oxidation of thiocyanate ion dissolved as KSCN in dimethyl sulfoxide," Electrochim. Acta, 17:2153 (1972).

I. Morcos and E. Yeager, "Kinetic studies of the oxygen−peroxide couple on pyrolytic graphite," Electrochim. Acta, 15:953 (1970).

L. Müller, "The order of electrochemical reactions; its determination with a rotating disc electrode and its meaning," Z. Chem., 11:275 (1971).

L. Müller, "Determination of the rate of an electrochemical reaction at the interface using a rotating disc electrode; the case of reactions in a real adsorbed layer (Temkin isotherm)," J. Electroanal. Chem., 34:451 (1972).

L. Müller and P. Janietz, "The mechanism of the catalytic decomposition of H_2O_2 on silver electrodes in alkaline solutions," Coll. Czech. Chem. Commun., 36:906 (1971).

L. Müller and R. Wetzel, "Determination of the activating reaction in the adsorption of hypochlorite ions on an oxygen-covered platinum electrode," Z. Phys. Chem. (Leipzig), 245:436 (1970).

C. A. Nunez and S. Müller, "Dependence of the dissolution currents of a passive zinc disc electrode in alkaline solution on the rotation speed of the disc," Rev. CENIC, Cienc. Fis., 2:1, 19 (1970).

Yu. M. Povarov and P. D. Lukovtsev, "Surface inhomogeneity of platinum-metal electrodes in redox reactions," Electrochim. Acta, 18:13 (1973).

F. Scheller and R. Landsberg, "On investigation of electrode kinetics at partially blocked electrodes," Z. Phys. Chem. (Leipzig), 244:273 (1970).

L. Sereno, V. A. Macagno, and M. C. Giordano, "Electrochemical behavior of the chloride/chlorine system at platinum electrodes in acetonitrile solutions," Electrochim. Acta, 17:561 (1972).

J. A. Wargon and A. J. Arvia, "Kinetics and mechanism of the electrochemical oxidation of nitrite ion dissolved as sodium nitrite in dimethyl sulfoxide solutions on platinum electrodes," Electrochim. Acta, 17:649 (1972).

H. Wolf and R. Landsberg, "The dependence of blocking parameters on the graphite electrode surface on conditions of the cathodic reduction of hypochlorite," J. Electroanal. Chem., 28:295 (1970).

H. Wolf and R. Landsberg, "On the cathodic reduction of hypochlorite on graphite electrodes covered with paraffin," Electrochim. Acta, 16:1627 (1971).

Z. Zembura and W. Ziolkowska, "Kinetics of oxygen reduction on the surface of iron corroding in acid solutions," Roszn. Chem., 45:1053 (1971).

Chapter 4

Ya. V. Durdin and E. S. Tsvetasheva, "Study of limiting currents in chloroacetic acid solutions with a rotating disc electrode. 1. Solutions with excess current," Vestn. LGU, No. 1, Ser. 4, p. 64 (1971).

V. T. Ivanov, "Integral equations of electric fields in electrolytes," Élektrokhimiya, 8:883 (1972).

V. T. Ivanov and A. I. Shafeev, "Numerical calculations of the field of a ring-disc electrode under galvanostatic conditions," Élektrokhimiya, 9:1191 (1973).

W. J. Albery and M. L. Hitchman, "Current distribution at a rotating disc electrode," Trans. Faraday Soc., 67:2408 (1971).

M. Bierowski, "Limiting current on the rotating disc electrode in solutions containing several kinds of ions," Rocz. Nauk.-Dydakt. WSP Rzeszowie. Nauki Mat.-Przyrod. i Techn., 1967(3):113.

S. Bruckenstein and B. Miller, "An experimental study of nonuniform current distribution at rotating disc electrodes," J. Electrochem. Soc., 117:1044 (1970).

I. Epelboin, C. Gabrielli, M. Keddam, J.-C. Lestrade, and H. Takenouti, "Passivation of iron in sulfuric acid medium," J. Electrochem. Soc., 119:1632 (1972).

C. Gabrielli, M. Keddam, J.-C. Lestrade, and H. Takenouti, "Nonuniform dissolu-
    tion of an iron anode in sulfuric acid solutions in the active—passive transi-
    tion region," Compt. Rend., Ser. C., 274:123 (1972).
L. Nanis and W. Kesselman, "Engineering applications of current and potential dis-
    tributions in disc electrode systems, J. Electrochem. Soc., 118:454 (1971).
W. H. Tiedemann, J. Newman, and D. N. Bennion, "The error in measurements of
    electrode kinetics caused by nonuniform ohmic-potential drop to a disc
    electrode," J. Electrochem. Soc., 120:256 (1973).

Chapter 5

Kh. Z. Brainina, Inversion Voltammetry of Solid Phases, Khimiya, Moscow (1972).
M. S. Makharov, V. A. Antip'eva, and V. I. Bakanov, "Amalgam chronopotentio-
    metry with accumulation on the mercury rotating disc electrode," Tr. Tyu-
    mensk. Industr. Inst., Khim. i Khim. Tekhnol., p. 3 (1972).
V. S. Krylov and V. N. Babak, "Nonstationary diffusion to the surface of the rotating
    disc," Élektrokhimiya, 7:649 (1971).
V. A. Lopatin, B. M. Grafov, and V. G. Levich, "Mass transfer to a rotating disc
    electrode in the case of time-dependent bulk concentration of the re-
    actant," Élektrokhimiya, 7:123 (1971).
S. G. Ogryz'ko-Zhukovskaya, N. A. Fedotov, and V. P. Belokopytov, "Determination
    of diffusion coefficient and concentration of the electroactive species in the
    solution with a rotating disc electrode," Élektrokhimiya, 8:1191 (1972).
V. I. Chernenko, "Chronopotentiometry at a rotating disc electrode; discharge com-
    plicated by a preceding chemical reaction," Élektrokhimiya, 7:820 (1971).
V. I. Chernenko, K. I. Litovchenko, and V. F. Kislenko, "Transient processes at
    a rotating disc electrode polarized with ac," Élektrokhimiya, 7:1100 (1971).
V. I. Chernenko and K. I. Litovchenko, "Transient phenomena at a rotating disc
    electrode accompanying a reversible electrode reaction with a catalytic
    reaction in the solution bulk," Élektrokhimiya, 7:1470 (1971).
V. I. Chernenko, K. I. Litovchenko, and Yu. E. Udovenko, "Cyclic chronopoten-
    tiometry at a rotating disc electrode," Élektrokhimiya, 7:1476 (1971).
K. I. Yuodkazis, E. B. Davidovichyus, and R. N. Vishomirskis, "Determination of
    the transition time at a rotating disc electrode," Tr. Akad. Nauk Lit. SSR,
    B, No. 5(72), p. 23 (1972).
C. Deslouis, I. Epelboin, M. Keddam, and J.-C. Lestrade, "Diffusion impedance
    of a rotating disc under laminar flow conditions. Experimental study and
    comparison with the Nernst equation," J. Electroanal. Chem., 28:57 (1970).
Der Tau Chin, "Anodic mechanism of electrochemical machining. Study of current
    transient in a rotating disc electrode," J. Electrochem. Soc., 118:174 (1971).
I. Epelboin and M. Keddam, "Faradaic impedances: Diffusion impedance and re-
    action impedance," J. Electrochem. Soc., 117:1052 (1970).
M. L. Hitchman and W. J. Albery, "Diffusion coefficient measurements and rotating
    disc electrodes," Electrochim. Acta, 17:787 (1972).
M. Keddam, "Concerning the study of anodic iron dissolution by analysis of the
    faradaic impedance," Trait. Surf., 11(95):39 (1970).

E. Levart and D. Schuhmann, "A general method of determination of the behavior of a rotating disc electrode under transient conditions; perturbations caused by electrical signal of low amplitude," J. Electroanal. Chem., 28:45 (1970).

E. Levart and D. Schuhmann, "Diffusion impedance of a rotating disc electrode in the region of very low frequencies," Coll. Czech. Chem. Commun., 36:866 (1971).

B. Miller, M. I. Bellavance, and S. Bruckenstein, "Application of isosurface concentration of voltammetry at a rotating disc electrode to simple electron-transfer kinetics," J. Electrochem. Soc., 118:1082 (1971).

L. Nanis and I. Klein, "Transient mass transfer at the rotating disc electrode," J. Electrochem. Soc., 119:1683 (1972).

V. J. Puglisi and A. J. Bard, "Controlled potential coulometry employing a rotating disc electrode," Anal. Lett., 3:443 (1970).

Chapter 6

V. A. Vekslina, V. V. Vashchenko, and V. F. Torpova, "Determination of the rate constant of a coupled chemical reaction using the catalytic wave of chlorate ions on a rotating graphite disc electrode," Élektrokhimiya, 8:1484 (1972).

S. A. Kabakchi and V. Yu. Filinovskii, "Modeling of convective diffusion processes near a rotating disc electrode using AVM MN-7," Élektrokhimiya, 8:1428 (1972).

L. F. Kozin and D. V. Sokol'skii, "Ions of lower valence and the kinetics of electrode reactions," Electrode Processes, Izd. Nauka, Kazakhskoi SSR, Alma-Ata (1971), p. 3.

V. A. Kokorekina, L. G. Feoktistov, V. Yu. Filinovskii, and S. A. Shevelov, "Determination of the rate constant of dissociation and recombination of trinitromethane in acetonitrile with a rotating disc electrode," Élektrokhimiya, 7:1196 (1971).

V. I. Chernenko, "Electron transfer with subsequent chemical reaction under conditions of convective diffusion," Ukr. Khim. Zh., 38:749 (1972).

V. I. Chernenko and K. I. Litovchenko, "Determination of kinetic parameters of multistep reactions from nonstationary polarization curves on a rotating disc electrode," Problems of Chemistry and Chemical Technology. Republican Interdepartmental Thematic Scientific-Technical Collection, No. 26 (1972), p. 60.

C. P. Andrieux, L. Nadjo, and J. M. Saveant, "Electrodimerization," J. Electroanal. Chem., 42:223 (1973); 44:327 (1973).

R. D. Armstrong and J. A. Harrison, "Dissolution precipitation at a rotating disc electrode," J. Electroanal. Chem., 36:79 (1972).

M. Daguenet and F. Aouanouk, "The theory of a rotating disc microelectrode. I. The electrokinetic equation," J. Chim. Phys. Phys. Chim. Biol., 67:1956 (1970); M. Daguenet and F. Aouanouk, "The theory of a rotating disc microelectrode. II. Study of reaction kinetics," J. Chim. Phys. Chim. Biol., 67:1959 (1970).

K. Holub, "Regeneration of a substance undergoing an electrode reaction by disproportionation at a rotating disc electrode," J. Electroanal. Chem., 30:71 (1971).

J. Jordan, "Remarkable redox equilibria in fused salts," J. Electroanal. Chem., 29:127 (1971).

F. Kermiche-Aouanouk and M. Daguenet, "The theory of a rotating disc micro-electrode. Study of catalytic reaction and diffusion under transient conditions," Electrochim. Acta, 17:723 (1972).

H. Matsuda, "The theory of stationary current—voltage curves of redox electrode reactions in hydrodynamic voltammetry. VI. The ring electrode," J. Electroanal. Chem., 35:77 (1972).

H. Mishima, T. Iwasita, V. A. Macagno, and M. C. Giordano, "The electrochemical oxidation of nitrate ion on platinum from silver nitrate in acetonitrile solutions. 1. Kinetic analysis," Electrochim. Acta, 18:287 (1973).

D. Möller and K.-H. Heckner, "Electrode kinetics of irreversible reactions on solid electrolytes," J. Electroanal. Chem., 46:277 (1972); 38:337 (1972).

F. Opekar and P. Beran, "The use of the rotating disc electrode for measuring the rate of Fe(II) reaction with tert-butylhydroperoxide and hydrogen peroxide in hydrochloric acid and acrylonitrile medium," J. Electroanal. Chem., 32:49 (1971).

R. H. Philp, "Half-wave potential shifts at rotating disc electrode oxidation of N,N,N',N'-tetramethyl-p-phenyldiamine," J. Electroanal. Chem., 27:369 (1970).

G. Schmid, M. A. Lobeck, and H. Keiser, "Behavior of nitrous and nitric acids at a rotating disc electrode," Ber. Bunsenges., 73:189 (1969); 74:1035 (1970); 76:151 (1972).

P. G. Zambonin and J. Jordan, "Chemistry of electron transfer and oxygen transfer in fused salts," J. Amer. Chem. Soc., 89:6365 (1967).

Chapter 7

V. A. Volkov and S. S. Kruglikov, "Reactions of disemicarbazide at a nickel electrode," Élektrokhimiya, 9:1298 (1973).

A. D. Davydov, V. D. Kashcheev, and V. P. Kriven'kii, "Anodic dissolution of molybdenum at high current densities," Élektron. Obrabotka Materialov, 1973(1):5.

B. M. Dikova and N. T. Kudryavtsev, "On the nature of the cathodic polarization in some (noncyanide) complex copper plating baths," Khim. i Indust. (NRB), 42:255 (1970).

A. N. Doronin, "Stationary potentials of a thallium electrode in hydrogen peroxide solutions," Élektrokhimiya, 7:1694 (1971).

A. N. Doronin, "Stationary potentials of a thallium electrode in hydrochloric and perchloric acids," Zashchita Metallov, 8:469 (1972).

I. A. Kakovskii and N. I. Sorokina, "Some peculiarities of silver telluride dissolution in cyanide solutions," Dokl. Akad. Nauk SSSR, 206:387 (1972).

V. P. Karshin and V. A. Grigoryan, "Kinetics of gypsum dissolution in water," Zh. Fiz. Khim., 44:1356 (1970).

Ya. M. Kolotyrkin, Yu. A. Popov, and Yu. V. Alekseev, "On the mechanism of iron electrode passivation in phosphate solutions," Élektrokhimiya, 8:1725 (1972).

V. I. Kravtsov, E. G. Tsventamyi, and N. Yu. Myzlov, "Study of the kinetics of
    complex rhodium chloride reduction on a rhodium electrode," Élektrokhimiya,
    8:941 (1972).
S. S. Kruglikov and V. A. Volkov, "On the behavior of p-toluenesulfamide during
    deposit of nickel from sulfate electrolyte," Élektrokhimiya, 6:1033 (1970).
S. S. Kruglikov, G. I. Yakub, and L. M. Antipova, "Application of a rotating disc
    electrode to the evaluation of the levelling properties of some additives to
    a sulfuric acid copper-plating bath," in: Physical Chemistry and Electro-
    chemistry, Tr. MKhTI im. Mendeleeva, No. 67, p. 236 (1970).
V. A. Kuznetsov and Yu. G. Kotlov, "Equations of diffusion at stationary and rotating
    disc electrodes applied to the study of the kinetics of heterogeneous chemi-
    cal reactions and in analysis of nonconducting solutions," Élektrokhimiya,
    9:670 (1973).
V. I. Mel'nikov and L. I. Antropov, "Corrosion of a disc iron electrode in sulfuric
    acid solutions," Zashchita Metallov, 7:583 (1971).
I. P. Nelaev and V. G. Sorokin, "On corrosion inhibition of a disc steel electrode in
    neutral solutions," Zashchita Metallov, 9:598 (1973).
B. A. Purin and O. B. Kiselev, "A study of the cathodic polarization of cobalt in
    pyrophosphate electrolyte," Izv. Akad. Nauk Latv. SSR, Ser. Khim., 1971(6):681.
A. A. Filippov and G. V. Volkova, "A study of the kinetics of gold dissolution in a
    chloride solution of tetravalent iridium," in: Synthesis, Purification, and
    Analysis of Inorganic Materials, Nauka, Sibirsk. Otd., Novosibirsk (1971), p. 163.
P. C. A. Bailey and G. A. Wright, "Kinetics of the anodic dissolution of thallium in
    perchlorate solutions," Electrochim. Acta, 16:865 (1971).
A. F. M. Barton and N. M. Wilde, "Dissolution rates of polycrystalline samples of
    gypsum and orthorhombic form of calcium sulfate measured with a rotating
    disc electrode," Trans. Faraday Soc., 67:3590 (1971).
A. Caprani and I. Epelboin, 'Electrochemical behavior of titanium in sulfuric and
    hydrofluoric acid solutions containing oxygen," J. Electroanal. Chem.,
    29:335 (1971).
J. A. Harrison, R. P. J. Hill, and J. Thompson, "Deposition of lead on a rotating
    silver disc," J. Electroanal. Chem., 44:445 (1973).
F. Hine and M. Yasuda, "Studies on cathodic reaction in the diaphragm-type
    chlorine cell," J. Electrochem. Soc., 118:170 (1971).
G. L. Holleck and J. Giner, "The aluminum electrode in $AlCl_3$—alkali halide melts,"
    J. Electrochem. Soc., 119:1161 (1972).
P. Javet and H. E. Hintermann, "Study of mass transfer and effect of thiourea on
    electrodeposition," Microtechnic, 24:230 (1970).
D. C. Johnson and S. Bruckenstein, "Effect of passivation of platinum electrodes
    on oxygen reduction in sulfuric acid," Anal. Chem., 43:1313 (1971).
L. I. Krishtalik, N. M. Alpatova, and M. G. Fomicheva, "Electrochemical regenera-
    tion of solvated electrons and hydrogen evolution in hexamethylphosphor-
    triamide," Croat. Chem. Acta, 44:1 (1972).
D. D. Macdonald and G. A. Wright, "Dissolution of bismuth by triiodide ion in
    acidified potassium iodide solution," Can. J. Chem., 48:2847 (1970).

L. S. Marcoux, A. Lomax, and A. J. Bard, "Electrochemistry and electron spin reso-
nance spectroscopy of 9,10-di(α-naphthyl)anthracene," J. Amer. Chem. Soc.,
92:243 (1970).

D. Möller and K.-H. Heckner, "Electrode kinetics of irreversible processes on solid
electrodes. IV. Effects of blocking and adsorption during voltammetric
study of hydroxylamine," Z. Chem., 11:356 (1971).

L. Müller and P. Janietz, "Effect of the physicochemical surface state of silver
electrode on the mechanism and rate of electroreduction of perbenzoic acid,"
J. Electroanal. Chem., 31:287 (1971).

L. Müller and R. Wetzel, "Separation of several parallel electrochemical processes
using a rotating disc electrode," Z. Phys. Chem. (Leipzig), 247:41 (1971).

Chapter 8

L. I. Antropov, G. G. Vrzhosek, M. P. Tarasevich, and M. A. Marinich, "Effect of
tetrabutylammonium cation on cathodic reduction of oxygen in platinum
in acid solutions," Élektrokhimiya, 8:149 (1972).

L. I. Antropov, M. R. Tarasevich, G. G. Vrzhosek, M. A. Marinich, and L. L. Knots,
"Effect of amines on oxygen electroreduction on platinum in acid solutions,"
Élektrokhimiya, 8:1846 (1972).

V. S. Bagotskii, M. R. Tarasevich, and V. Yu. Filinovskii, "Calculations of kinetic
parameters of oxygen and hydrogen peroxide reactions taking into account
the adsorption step," Élektrokhimiya, 8:84 (1972).

V. Glodzinska and L. N. Nekrasov, "Effect of zinc, cadmium, and thallium cation
adsorption on oxygen reduction and hydrogen-evolution overpotential on
rhodium in sulfuric acid solutions," Élektrokhimiya, 7:905 (1971).

V. A. Gromykov, Yu. V. Vasil'ev, and V. S. Bagotskii, "A ring-disc study of the
adsorption of molecular hydrogen and oxygen on platinum and palladium,"
Élektrokhimiya, 8:914 (1972).

V. A. Gromyko, Yu. V. Vasil'ev, and V. S. Bagotskii, "Separation of electrochemical
and catalytic processes on oxidation of organic species on palladium,"
Élektrokhimiya, 8:1238 (1972).

N. I. Dubrovina and L. N. Nekrasov, "Ring-disc study of oxygen electroreduction
in alkaline solutions containing surface-active substances," Élektrokhimiya,
8:1503 (1972).

G. V. Zhutaeva, V. S. Bagotskii, and N. A. Shumilova, "Double ring-disc rotating
electrode and its application to studies of oxygen reduction," Élektrokhimiya,
7:1707 (1971).

N. V. Iodelene, V. Yu. Skominskas, and Yu. Yu. Matulis, "Mechanism of Cr^{3+} ion
electroreduction," Tr. Akad. Nauk Lit. SSR, B, No. 4(63), p. 77, 89 (1971);
B, No. 2(65), p. 43 (1971).

L. Kish and J. Farkas, "Ring-disc study of metal ionization," Zashchita Metallov,
9:433 (1973).

A. D. Korsun, V. M. Kutyurin, and I. Yu. Artamkina, "Mechanism of electrooxidation of chlorophyl "a" in water—acetone mixtures," Khim. Prirodn.
Soedin., No. 4, p. 563 (1972).

O. V. Kuz'michev, V. N. Kabanov, T. I. Popova, and N. A. Simonova, "Concerning the existence of univalent magnesium," Élektrokhimiya, 8:947 (1972).

S. Sh. Leites, V. I. Luk'yanycheva, V. S. Bagotskii, A. V. Yuzhanina, and V. F. Konanykina, "The effect of pH on the cathodic reduction of oxygen on a smooth platinum electrode," Élektrokhimiya, 9:620 (1973).

S. Sh. Leites, V. S. Bagotskii, V. I. Luk'yanycheva, and V. F. Konanykina, "The effect of solution pH on the electrochemical behavior of hydrogen peroxide on platinum," Élektrokhimiya, 9:1166 (1973).

V. I. Luk'yanycheva, A. V. Yuzhanina, B. I. Lentsner, L. L. Knots, N. A. Shumilova, and V. S. Bagotskii, "The state of adsorbed oxygen and its effect on the mechanism of reduction of molecular oxygen on platinum electrodes in alkaline solution," Élektrokhimiya, 7:1287 (1971).

N. D. Merkulova, G. V. Zhutaeva, N. A. Shumilova, and V. S. Bagotskii, "Behavior of hydrogen peroxide on silver electrode in alkaline solution," Élektrokhimiya, 8:727 (1972).

N. D. Merkulova, G. V. Zhutaeva, N. A. Shumilova, and V. S. Bagotskii, "Behavior of oxygen and hydrogen peroxide on silver amalgam in alkaline solution," Élektrokhimiya, 8:1337 (1972).

L. N. Nekrasov and N. I. Dubrovina, "Detection of the products of one-electron reduction of oxygen in aqueous alkaline solutions containing surface-active substances," Élektrokhimiya, 8:946 (1972).

L. N. Nekrasov, L. A. Dukhanova, N. I. Dubrovina, and L. N. Vykhodtseva, "Ring-disc study of the cathodic reduction of oxygen in dimethylformamide solutions," Élektrokhimiya, 6:388 (1970).

L. N. Nekrasov and A. D. Korsun, "Formation of intermediates in the cathodic reduction of benzophenone," Élektrokhimiya, 6:1753 (1970).

L. N. Nekrasov and B. G. Podlibner, "A study of disproportionation of ferrocenyl-hydroxylamine, the initial product of the cathodic reduction of nitroferrocene," Élektrokhimiya, 7:379 (1971).

L. N. Nekrasov, I. P. Rybkina, and B. G. Podlibner, "A ring-disc study of the cathodic reduction of p-nitroaniline in acid solutions," Élektrokhimiya, 8:1404 (1972).

L. N. Nekrasov, D. N. Soshchin, and V. N. Gramenitskaya, "Electroreduction of some alicyclic and aromatic carbonyl compounds," Élektrokhimiya, 6:1577 (1970).

A. I. Oshe, Z. B. Rozhdestvenskaya, and S. A. Levitskaya, "Ring-disc study of the anodic oxidation of zinc in 10 N KOH solutions," Élektrokhimiya, 7:742 (1971).

B. G. Podlibner and L. N. Nekrasov, "Cathodic reduction of p-nitroaniline in alkaline media," Élektrokhimiya, 7:1191 (1971).

G. P. Samoilov, E. I. Khrushcheva, N. A. Shumilova, and V. S. Bagotskii, "Electrochemical behavior of O_2 and H_2O_2 on thermally treated nickel in alkaline solution," Élektrokhimiya, 6:1347 (1970).

M. R. Tarasevich and K. A. Radyushkina, "A study of the parallel-consecutive steps of oxygen and hydrogen peroxide reactions. II," Élektrokhimiya, 7:248 (1971).

M. R. Tarasevich and V. S. Vilinskaya, "A study of the parallel-consecutive steps of oxygen and hydrogen peroxide reactions. III, IV, V," Élektrokhimiya, 8:1489 (1972); 9:98, 1187 (1973).

M. R. Tarasevich, G. I. Zakharina, and R. M. Smimova, "A study of oxygen and hydrogen peroxide reactions using O^{18}. III. Decomposition of hydrogen peroxide on platinum in the presence of various cations and anions," Élektrokhimiya, 9:645 (1973).

E. A. Khomskaya, A. S. Kolosov, and V. V. Polishuk, "On the change of supersaturation in the vicinity of the anode during electrolytic oxygen evolution from KOH solutions," Élektrokhimiya, 7:1074 (1971).

A. V. Yuzhanina, V. I. Luk'yanycheva, B. I. Lentsner, L. L. Knots, N. A. Shumilova, and V. S. Bagotskii, "The effect of HSO_4^- anions on the kinetics of oxygen reduction at a smooth platinum electrode in 1 N H_2SO_4," Élektrokhimiya, 8:877 (1972).

W. J. Albery and M. L. Hitchman, The Ring-Disc Electrode, Clarendon Press, Oxford (1971).

W. Albery, "The ring-disc electrode," Trans. Faraday Soc., 67:153 (1971).

W. Albery, J. S. Drury, and M. L. Hitchman, "The ring-disc electrode. 12. Application to ring current transients," Trans. Faraday Soc., 67:161 (1971).

W. Albery, J. S. Drury, and M. L. Hitchman, "The ring-disc electrode. 13. Laplace transformation of transient currents," Trans. Faraday Soc., 67:166 (1971).

W. J. Albery, J. S. Drury, and M. L. Hitchman, "The ring-disc electrode. 14. Kinetic and Transient Parameters," Trans. Faraday Soc., 67:2162 (1971).

W. J. Albery, J. S. Drury, and A. P. Hutchinson, "The ring-disc electrode. 15. Alternating current measurements," Trans. Faraday Soc., 67:2414 (1971).

W. Albery and J. S. Drury, "The ring-disc electrode. 16. Comparison of analytical and numerical solutions," J. Chem. Soc. Faraday Trans. I, 1972:456.

G. Archdale and J. A. Harrison, "Anodic dissolution of lead in sulfuric acid," J. Electroanal. Chem., 39:357 (1972).

R. D. Armstrong, "Diagnostic criteria for distinguishing between the dissolution-precipitation and the solid-state mechanism of passivation," Corrosion Sci., 11:693 (1971).

R. D. Armstrong and I. Baurhoo, "Solution-soluble species in the operation of the iron electrode in alkaline solution," J. Electroanal. Chem., 34:41 (1972).

R. D. Armstrong and I. Baurhoo, "The dissolution of iron in concentrated alkali," J. Electroanal. Chem., 40:325 (1972).

R. D. Armstrong, J. A. Harrison, H. R. Thirsk, and R. Whitfield, "The anodic dissolution of titanium in sulfuric acid," J. Electrochem. Soc., 117:1003 (1970).

R. D. Armstrong and M. Henderson, "The transpassive dissolution of chromium," J. Electroanal. Chem., 32:1 (1971).

R. D. Armstrong and G. D. West, "The anodic behavior of cadmium in alakline solutions," J. Electroanal. Chem., 30:385 (1971).

V. S. Bagotzky, N. A. Shumilova, G. P. Samoilov, and E. I. Khrushcheva, "Electrochemical oxygen reduction on nickel electrodes in alkaline solutions," Electrochim. Acta, 17:1625 (1972).

M. Breitenbach and K.-H. Heckner, "Ring-disc study of the anodic oxidation kinetics of aniline in acetonitrile and water," J. Electroanal. Chem., 33:45 (1971).

S. Bruckenstein and M. Z. Hassan, "Rotating ring-disc study of the reduction of oxidized platinum by mercurous mercury and its adsorption on reduced platinum," Anal. Chem., 43:928 (1971).

S. H. Cadle, "A ring-disc study of the reduction of bismuth on platinum and on gold electrodes," J. Electrochem. Soc., 118:39C (1971).

S. H. Cadle and S. Bruckenstein, "Ring-disc electrode study of the underpotential deposition of copper on platinum in 0.5 M hydrochloric acid," Anal. Chem., 43:932 (1971).

S. H. Cadle and S. Bruckenstein, "Inhibition of hydrogen adsorption by submonolayer deposition of metals," Anal. Chem., 43:1858 (1971).

S. H. Cadle and S. Bruckenstein, "Ring-disc electrode study of bismuth reduction on platinum," Anal. Chem., 44:1993 (1972).

S. H. Cadle and S. Bruckenstein, "Ring-disc electrode study of reduction and oxidation of bismuth on gold," J. Electrochem. Soc., 119:1166 (1972).

R. D. Cowling and H. E. Hintermann, "The anodic oxidation of titanium carbide," J. Electrochem. Soc., 118:1912 (1971).

Der Tau Chin, "A rotating ring-hemispherical electrode for electroanalytical applications," J. Electrochem. Soc., 120:631 (1973).

A. B. Djordjevic, B. Z. Nikolic, I. V. Kadija, and A. R. Despic, "Kinetics and mechanism of electrochemical oxidation of hypochlorite ions," Electrochim. Acta, 18:465 (1973).

G. Durand and B. Tremillon, "Electrochemical study of iodine in anhydrous acetic acid," Anal. Chim. Acta, 49:135 (1970).

J. Eckert and W. Forker, "Ring-disc electrode study of the anodic dissolution of beryllium in aqueous chloride ion solutions," Z. Phys. Chem. (Leipzig), 253:153 (1973).

A. Fujishima and K. Honda, "Ring-disc electrode study of photoelectrochemical reactions on a semiconducting TiO_2 electrode," Seisan Kenkyu, Mon. J. Inst. Ind. Sci. Univ. Tokyo, 22:478 (1970).

A. Fujishima and K. Honda, "Ring-disc electrode study of the photoelectrochemical behavior of a monocrystalline zinc oxide semiconductor electrode," Seisan Kenkyu, Mon. J. Inst. Ind. Sci. Univ. Tokyo, 22:524 (1970).

A. Fujishima and K. Honda, "Mechanism of the anodic dissolution of monocrystalline ZnO electrode under illumination," Denki Kagaku, 40:33 (1972).

A. Fujishima, H. Iketani, and K. Honda, "Ring-disc electrode study of the $Cu-Cu^+-Cu^{2+}$ system," Seisan Kenkyu, Mon. J. Inst. Ind. Sci. Univ. Tokyo, 22:243 (1970).

A. Fujishima, H. Iketani, and K. Honda, "Relationship between the dimension of the rotating disc electrode and the rotation speed," Bull Chem. Soc. Japan, 43:3949 (1970).

A. Fujishima, E. Sugiyama, and K. Honda, "Photosensitized electrooxidation of iodine ions on monocrystalline cadmium sulfide electrode," Bull. Chem. Soc. Japan, 44:304 (1971).

C. Gabrielli, M. Keddam, and H. Takenouti, "A study of potential distribution of the ring-disc electrode," J. Chim. Phys. Phys. Chim. Biol., 67:737 (1972).

C. H. Hamann and W. Vielstich, "Detection of formate traces in the presence of carbon monoxide in alakline solution," J. Electroanal. Chem., 32:459 (1971).

C. H. Hermann and W. Vielstich, "Oxidation of carbon monoxide on platinum in alkaline solutions," Ber. Bunsenges., 75:918 (1971).

M. Z. Hassan, D. F. Untereker, and S. Bruckenstein, "Ring-disc study of thin mercury films on platinum," J. Electroanal. Chem., 42:161 (1973).

H. E. Hintermann, A. C. Riddiford, R. D. Cowling, and J. Malyszko, "Anodic behavior of titanium carbide in sulfuric acid solution," Electrodepos. Surf. Treat., 1:59 (1972).

M. W. Hull, "On the anodic dissolution of molybdenum in acidic and alkaline electrolytes," J. Electroanal. Chem., 38:143 (1972).

M. W. Hull, "Concerning the production of unipositive magnesium in aqueous electrolytes," J. Electroanal. Chem., 38:App. 1 (1972).

M. W. Hull and H. I. James, "Production of soluble intermediates at platinum electrodes in sulfate electrolytes," J. Electroanal. Chem., 45:11 (1973).

D. C. Johnson, "A study of the adsorption and desorption of iodine and iodide at platinum electrodes in 1.0 M sulfuric acid," J. Electrochem. Soc., 119:331 (1972).

D. C. Johnson, D. T. Napp, and S. Bruckenstein, "A ring-disc electrode study of the current−potential behavior of platinum in 1.0 M sulfuric acid and 0.1 M perchloric acid," Electrochim. Acta, 15:1493 (1970).

D. C. Johnson and E. W. Resnick, "Rotating photoelectrode for electrochemical study of the products of photochemical reactions," Anal. Chem., 44:637 (1972).

I. V. Kadija and V. M. Nakic, "Rotating two-ring electrodes as a tool for electrochemical investigations of gas-evolving reactions," J. Electroanal. Chem., 35:177 (1972).

T. Kihara, K. Sasaki, and H. Shiba, "Anodic oxidation of anthracene in acetonitrile at a rotating ring-disc electrode," Bull. Chem. Soc. Japan, 44:3457 (1971).

L. Kiss and J. Farkas, "Ionization of metals and reduction of metal ions studied using a ring-disc electrode," Acta Chim. Acad. Sci. Hung., 64:241 (1970); 65:7, 141 (1970); 66:33, 395 (1970); 67:179 (1971); 68:359 (1971); 69:167 (1971); 74:123 (1972).

J. T. Maloy, K. B. Prater, and A. J. Bard, "Electrogeneration of chemiluminescence. V. Rotating ring-disc electrode. Digital simulation and experimental evaluation," J. Amer. Chem. Soc., 93:5959 (1971).

J. T. Maloy and A. J. Bard, "Electrogeneration of chemiluminescence. VI. Efficiency and mechanism of 9,10-diphenylanthracene, rubrene, and pyrene systems at a rotating ring disc electrode," J. Amer. Chem. Soc., 93:5968 (1971).

J. Margarit, G. Dabosi, and M. Levy, "Ring-disc electrode in electrochemical kinetics; a study of the geometric factor and of two redox systems in aqueous solutions," Bull. Soc. Chim. France, 1972(5):2096.

H. Matsuda, "Theory of stationary current−voltage curves of redox electrode reactions in hydrodynamic voltammetry. Stationary disc and ring electrodes in a uniformly rotating solution," J. Electroanal. Chem., 38:159 (1972).

K. Tokuda and H. Matsuda, "Theory of stationary current−voltage curves of redox electrode reactions in hydrodynamic voltammetry. VIII. Ring-disc electrode," J. Electroanal. Chem., 44:199 (1973).

N. D. Merkulova, G. V. Zhutaeva, N. A. Shumilova, and V. S. Bagotzky, "Reactions of hydrogen peroxide on a silver electrode in alkaline solutions," Electrochim. Acta, 18:169 (1973).

B. Miller, M. I. Bellavance, and S. Bruckenstein, "Feasibility and applications of programmed speed control at rotating disc electrodes," Anal. Chem., 44:1983 (1972).

B. Miller and M. I. Bellavance, "Rotating ring-disc electrode studies of corrison rates and partial currents: Cu and Cu_3OZn in oxygenated chloride solutions," J. Electrochem. Soc., 119:1510 (1972).

B. Miller and M. I. Bellavance, "Measurements of current and potential distribution at rotating disc electrodes," J. Electrochem. Soc., 120:42 (1973).

S. Miller, R. Landsberg, and U. Künkel, "A voltammetric study of 4-methoxy-styrene," Z. Chem., 11:358 (1971).

G. Neubert, E. Gorman, R. van Reet, and K. B. Prater, "Current distribution at the rotating ring-disc electrode," J. Electrochem. Soc., 119:677 (1972).

Y. Okinaka, "On the oxidation-reduction mechanism of the cadmium metal—cadmium hydroxide electrode," J. Electrochem. Soc., 117:289 (1970).

K. B. Prater, "Digital simulation," Chem. Instr., 3:259 (1972).

K. B. Prater and A. J. Bard, "Rotating ring-disc electrodes. III. Catalytic and ECE reactions," J. Electrochem. Soc., 117:1517 (1970).

V. J. Puglisi and A. J. Bard, "Rotating ring-disc electrodes. IV. Dimerization and second-order ECE reactions," J. Electrochem. Soc., 119:833 (1972).

V. J. Puglisi and A. J. Bard, "Electrohydrodimerization reactions. II. Rotating ring-disc electrodes, Voltammetric and coulometric studies of dimethyl-fumarate, cinnamonitrile, and fumaronitrile," J. Electrochem. Soc., 119:829 (1972).

V. J. Puglisi and A. J. Bard, "Electrohydrodimerization reactions. III. Rotating ring-disc electrodes. Voltametric and coulometric studies of mixed reductive coupling of dimethylfumarate in the presence of cinnamonitrile and acrylo-nitrile in dimethylformamide solutions," J. Electrochem. Soc., 120:748 (1973).

W. H. Smyrl and J. Newman, "Detection of nonuniform current distribution on a disc electrode," J. Electrochem. Soc., 119:208 (1972).

W. H. Smyrl and J. Newman, "Ring-disc and sectioned disc electrodes," J. Electrochem. Soc., 119:212 (1972).

G. W. Tindall and S. Bruckenstein, "Voltammetric rotating ring-disc studies of silver deposition on platinum at underpotential," Electrochim. Acta, 16:245 (1971).

V. A. Vicente and S. Bruckenstein, "Rotating gold ring-disc study of tin(II) in 4.0 M hydrochloric acid," Anal. Chem., 44:297 (1972).

Chapter 9

R. Kh. Brushtein, V. S. Vilinskaya, L. L. Knots, V. V. Kushnev, B. I. Lentsner, and M. R. Tarasevich, "Automatic potential control of the disc and ring electrodes," Élektrokhimiya, 8:1183 (1972).

A. P. Dzhyuve, O. K. Gal'dikene, and Yu. Yu. Matulis, "Concerning the method of investigation of the crystal structure of electrodeposits obtained on a rotating electrode," Tr. Akad. Nauk Lit. SSR, B, No. 1(68), p. 27 (1972).

Yu. V. Efremov, I. E. Krasnova, and A. V. Abrosimov, "Contact between the electrolytic bridge and a rotating electrode," Zh. Fiz. Khim., 46:2432 (1972).

P. Beran and F. Opekar, "Rotating disc electrode," Chem. Listy, 65:855 (1971).

S. Bruckenstein and B. Miller, "Circuit for transient-free current–potential control conversion," J. Electrochem. Soc., 117:1040 (1970).

S. C. Creason and R. F. Nelson, "Rapid data acquisition in rotating disc voltammetry,' J. Electroanal. Chem., 27:189 (1970).

G. W. Harrington, H. A. Laitinen, and V. Trendafilov, "Demountable ring-disc electrode," Anal. Chem., 45:433 (1973).

B. Miller and S. Bruckenstein, "Hydrodynamic potentiometry and amperometry at ring-disc electrodes," J. Electrochem. Soc., 117:1032 (1970).